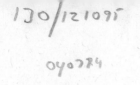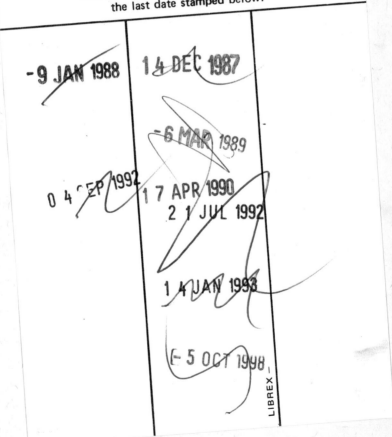

J. Chinal

Design Methods for Digital Systems

Translated from French
by A. Preston and A. Sumner

Springer-Verlag Berlin Heidelberg New York 1973

Jean Chinal
Directeur des Etudes,
Ecole Nationale Supérieure
de l'Aéronautique et de l'Espace,
Toulouse, France.

Alan Preston
Head of English Department,
E. N. S. A. E.

Arthur Sumner
Engineer,
S. N. I. A. S.

Translation of the French edition:
Techniques Booléennes et Calculateurs Arithmétiques
© by Dunod, Paris 1967

An edition of this book is published by the Akademie-Verlag, Berlin, for distribution in socialist countries

With 310 Figures and Tables

AMS Subject Classifications (1970) 49A20, 02F10, 02J05.

ISBN 3-540-05871-0 Springer-Verlag Berlin Heidelberg New York
ISBN 0-387-05871-0 Springer-Verlag New York Heidelberg Berlin

This work is subjekt to copyright. All rights are reserved, whether the whole or part of the material is concerned, specifically those of translation, reprinting, re-use of illustrations, broadcasting, reproduction by photocopying machine or similar means, and storage in data banks. Under § 54 of the German Copyright Law where copies are made for other than private use, a fee is payable to the publisher, the amount of the fee to be determined by agreement with the publisher.
© by Springer-Verlag Berlin · Heidelberg 1973. Library of Congress Catalog Card Number 72-86277
Printed in Germany

Foreword

This book constitutes an introduction to the theory of binary switching networks (binary logic circuits) such as are encountered in industrial automatic systems, in communications networks and, more particularly, in digital computers.

These logic circuits, with or without memory, (sequential circuits, combinational circuits) play an increasing part in many sectors of industry. They are, naturally, to be found in digital computers where, by means of an assembly (often complex) of elementary circuits, the functions of computation and decision which are basic to the treatment of information, are performed. In their turn these computers form the heart of an increasing number of digital systems to which they are coupled by interface units which, themselves, fulfil complex functions of information processing. Thus the digital techniques penetrate ever more deeply into industrial and scientific activities in the form of systems with varying degrees of specialization, from the wired-in device with fixed structure to those systems centered on a general-purpose programmable computer.

In addition, the present possibility of mass producing microminiaturised logic circuits (integrated circuits, etc.) gives a foretaste of the introduction of these techniques into the more familiar aspects of everyday life.

The present work is devoted to an exposition of the algebraic techniques nesessary for the study and synthesis of such logic networks. No previous knowledge of this field of activity is necessary: any technician or engineer possessing an elementary knowledge of mathematics and electronics can undertake its reading. Mathematicians will discover, apart from recent algebraic theories, a large number of applications, mostly borrowed from the field of computers, and thus an exposition of the practical motivations of these theories, which constitute an ever more important aspect of applied mathematics.

Also to be found are a large number of recent methods taken from technical literature. The bibliography and footnotes indicate the sources of the information used. An effort has been made to give to the presentation and to the proofs of the corresponding theorems, as original an aspect as possible and, at the same time, to integrate the results of diverse origins and of unequal theoretical levels into a single work. In many instances the original proofs have been modified or replaced by new proofs better adapted to the overall structure of the work. On the whole, no systematic effort has been made to give a strict mathematical turn to the proofs: this, in particular, is the case for certain proofs presenting little difficulty and which would otherwise have necessitated a relatively complicated formalism. In every case, however, the reader will find a reasonable justification of the results stated.

Apart from the bibliography, there are other sources to which I am particularly indebted. In particular, I must mention the instruction received at the University of Michigan (USA) some years ago with Dr N. Scott on computers and Dr M. Harrison on logic networks and the theory of finite automata. Later, during a study visit in 1963 to the Institute of Mathematics at Bucharest (Rumania), it was with great interest that I attended the lectures conducted by Academician Gr. Moisil and the interviews we had with the other specialists there proved very fruitful.

My thanks are due to Professor J.-C. Gille, for the help and encouragement given to me throughout the period of preparation of this work.

I am also grateful to M.l'Ingénieur Général de l'Air Vialatte, Director of STAé, for having accorded me the possibility of taking part in the work of the Centre d'Etudes et de Recherches en Automatisme in the field of sequential systems. My thanks go also to Professor L. Rauch of the University of Michigan (USA) who provided me with certain references, and to Mr G. Cohen, Ingénieur Principal de l'Air, who was kind enough to read the various chapters during their preparation and whose comments and suggestions enabled me to incorporate various improvements: furthermore, thanks to the understanding and help afforded me by M. Charron, and for which I am extremely grateful, a large part of the typing and preparation of the diagrams was undertaken by CERA and by the Ecole Nationale Supérieure de l'Aéronautique. Finally, I wish to acknowledge the help given to me by Mrs C. G. Chinal in the preparation of the final manuscript, and to my wife Gretchen, for her understanding and patience during the period of preparation of this work.

J. Chinal

Contents

CHAPTER 1 General Concepts

1.1 Digital Systems. 1
1.2 Sets 3
    1.2.1 The Concept of a Set 3
    1.2.2 Equality of Two Sets 4
    1.2.3 Subsets. Inclusion Relation 5
    1.2.4 Complement of a Set 5
    1.2.5 Union and Intersection of Sets 6
        1.2.5.1 Union 6
        1.2.5.2 Intersection 6
    1.2.6 Empty Set. Universal Set 7
    1.2.7 Set of Subsets 7
    1.2.8 Cover of a Set 7
    1.2.9 Partition of a Set 7
    1.2.10 Cartesian Product of Sets 8
    1.2.11 Application. Counting the Elements of a Set 10
1.3 Relations 11
    1.3.1 Relations between the Elements of a Number of Sets . . . 11
    1.3.2 Notation 11
    1.3.3 Graphical Representation of a Binary Relation 12
    1.3.4 Some Properties of Binary Relations 13
        1.3.4.1 Reflexivity 13
        1.3.4.2 Symmetry 13
        1.3.4.3 Transitivity 13
    1.3.5 Equivalence Relations 13
    1.3.6 Equivalence Classes 14
1.4 Univocal Relations or Functions 17
1.5 Concept of an Algebraic Structure 18

CHAPTER 2 Numeration Systems. Binary Numeration

2.1 Numbers and Numeration Systems 21
2.2 Decimal System of Numeration 21
2.3 Numeration System on an Integer Positive Base 22
2.4 Binary Numeration System 24
    2.4.1 Binary Numeration 24
    2.4.2 Binary Addition and Subtraction 25
        2.4.2.1 Addition 25
        2.4.2.2 Subtraction 26

Contents

- 2.4.3 Binary Multiplication and Division 28
  - 2.4.3.1 Multiplication 28
  - 2.4.3.2 Division. 28
- 2.4.4 Systems to the Base 2^p ($p \geq 1$). 29
- 2.5 Passage from One Numeration System to Another. 30
  - 2.5.1 First Method 30
  - 2.5.2 Second Method 33
- 2.6 Decimal-Binary and Binary-Decimal Conversions 34
  - 2.6.1 Conversion of a Decimal Number to a Binary Number. 34
  - 2.6.2 Conversion of a Binary Number to a Decimal Number. 35
  - 2.6.3 Other Conversion Methods 35
    - 2.6.3.1 Successive Subtraction Method 35
    - 2.6.3.2 Mixed Method (Binary-Octal-Decimal) 37
    - 2.6.3.3 Mixed Method (Binary-Hexadecimal) 39

CHAPTER 3 Codes

- 3.1 Codes 44
- 3.2 The Coding of Numbers 44
  - 3.2.1 Pure Binary Coding 44
  - 3.2.2 Binary Coded Decimal Codes (BCD Codes) 44
    - 3.2.2.1 Principle 44
    - 3.2.2.2 The Number of 4-Positions BCD Codes 45
    - 3.2.2.3 4-Bit Weighted BCD Codes 46
    - 3.2.2.4 Non-Weighted 4-Bit BCD Codes 47
    - 3.2.2.5 BCD Codes Having More than 4 Positions 48
- 3.3 Unit Distance Codes. 50
  - 3.3.1 Definition (Binary Case) 50
  - 3.3.2 Reflected Code (Gray Code, Cyclic Code). 50
    - 3.3.2.1 Construction of a Reflected Code with n Binary Digits . . 50
    - 3.3.2.2 Conversion Formulae 51
  - 3.3.3 The Usefulness of Unit Distance Codes. 54
- 3.4 Residue checks 56
  - 3.4.1 Residue Checks 56
  - 3.4.2 Residue Checks Computation 57
    - 3.4.2.1 Modulo m Residues of the Powers of an Integer 57
    - 3.4.2.2 Examples 58
  - 3.4.3 Examples of Congruence 60
    - 3.4.3.1 Problem 60
    - 3.4.3.2 Counters 61
    - 3.4.3.3 True Complement 64
- 3.5 The Choice of a Code 65
  - 3.5.1 Adaptation to Input-Output Operations 65
  - 3.5.2 Adaptation to Digital Operations 65
  - 3.5.3 Aptitude for the Automatic Detection or Correction of Errors. . . 66

CHAPTER 4 Algebra of Contacts

- 4.1 Electromagnetic Contact Relays. 67
- 4.2 Binary Variables Associated with a Relay 69
- 4.3 Complement of a Variable 70

4.4. Transmission Function of a Dipole Contact Network	72
4.4.1 Transmission of a Contact	73
4.4.2 Transmission of a Dipole Comprising 2 Contacts in Series	73
4.4.3 Transmission of a Dipole Consisting 2 Contacts in Parallel	74
4.5 Operations on Dipoles	75
4.5.1 Complementation	76
4.5.2 Connections in Series	76
4.5.3 Connections in Parallel	76
4.5.4 Properties of Operations $-$, $\cdot$ and $+$	76
4.5.5 2 Variable Functions	81
4.5.6 Examples of Application	83

CHAPTER 5   Algebra of Classes. Algebra of Logic

5.1 Algebra of Classes	87
5.2 Algebra of Logic (Calculus of Propositions)	91
5.2.1 Propositions. Logic Values of a Proposition	91
5.2.2 Operations on Propositions	92
5.2.2.1 Operation AND (Logical Product)	92
5.2.2.2 Operation OR (Or-Inclusive)	94
5.2.2.3 Negation of a Proposition	95
5.2.3 Compound Propositions and Functions of Logic	95
5.2.4 Numeric Notation of Logic Values	96
5.2.5 Other Logic Functions	97
5.2.5.1 2 Variable Functions	97
5.2.5.2 Function "Exclusive-OR" $(P \oplus Q)$	97
5.2.5.3 Equivalence $(P \leftrightarrow Q)$	97
5.2.5.4 Implication $(P \to Q)$	99
5.2.5.5 Sheffer Functions (P/Q) ("Function NAND")	100
5.2.5.6 Peirce Function $(P \downarrow Q)$ ("Function NOR")	101
5.2.6 Tautologies. Contradictions	101
5.2.7 Some Properties of the Functions AND, OR and Negation	103
5.2.8 Examples and Applications	104
5.2.9 Algebra of Logic (Propositional Calculus)	105
5.2.10 Functional Logic	106
5.2.10.1 Propositional Functions	106
5.2.10.2 Operations on Propositional Functions	106
5.2.10.3 Quantifiers	107
5.2.11 Binary Relations and Associated Propositions	107
5.2.12 Classes Associated with a Property or with a Relationship	110
5.2.13 Operations on Propositions and on Classes	111
5.3 Algebra of Contacts and Algebra of Logic	111
5.4 Algebra of Classes and Algebra of Contacts	112
5.5 Concept of Boolean Algebra	114

CHAPTER 6   Boolean Algebra

6.1 General. Axiomatic Definitions	116
6.2 Boolean Algebra	117
6.2.1 Set. Operations	117
6.2.2 Axioms for a Boolean Algebra	117

- 6.3 Fundamental Relations in Boolean Algebra 118
- 6.4 Dual Expressions. Principle of Duality. 122
  - 6.4.1 Dual Expressions 122
  - 6.4.2 Principle of Duality 123
  - 6.4.3 Notation 124
- 6.5 Examples of Boolean Algebra. 124
  - 6.5.1 2-Element Algebra 124
  - 6.5.2 Algebra of Classes 125
  - 6.5.3 Algebra of Propositions 126
  - 6.5.4 Algebra of n-Dimensional Binary Vectors 126
- 6.6 Boolean Variables and Expressions 129
  - 6.6.1 Boolean Variables. 129
  - 6.6.2 Boolean Expressions. 129
  - 6.6.3 Values for a Boolean Expression. Function Generated by an Expression 130
  - 6.6.4 Equality of 2 Boolean Expressions. 131
  - 6.6.5 Canonical Forms Equal to an Algebraic Expression 131
    - 6.6.5.1 Expressions Having 1 Variable 131
    - 6.6.5.2 Expressions Having n Variables. 133
    - 6.6.5.3 Dual Expressions 134
  - 6.6.6 A Practical Method for the Determination of a Complement ... 135
  - 6.6.7 Reduction to one Level of the Complementations Contained in an Expression. 136
  - 6.6.8 Relationship between a Dual Expression and its Complement . 136

CHAPTER 7 Boolean Functions

- 7.1 Binary Variables 138
- 7.2 Boolean Functions 138
  - 7.2.1 Definition 138
  - 7.2.2 Truth Table for a Boolean Function of n-Variables 138
  - 7.2.3 Other Means of Defining a Boolean Function 140
    - 7.2.3.1 Characteristic Vector 140
    - 7.2.3.2 Decimal Representations 141
  - 7.2.4 Number of Boolean Functions Having n Variables 141
- 7.3 Operations on Boolean Functions 142
  - 7.3.1 Sum of two Functions $f + g$ 142
  - 7.3.2 Product of two Functions $f \cdot g$ 143
  - 7.3.3 Complement of a Function $\bar{f}$ 143
  - 7.3.4 Identically Null Function (0) 143
  - 7.3.5 Function Identically Equal to 1 (1) 143
- 7.4 The Algebra of Boolean Functions of n Variables 144
- 7.5 Boolean Expressions and Boolean Functions 144
  - 7.5.1 Boolean Expressions. 144
  - 7.5.2 Function Genenrrated by a Boolean Expression 144
  - 7.5.3 Equality of 2 Expressions 146
  - 7.5.4 Algebraic Expressions for a Boolean Function 147
    - 7.5.4.1 Fundamental Product 147
    - 7.5.4.2 Disjunctive Canonical Form of a Function with n Variables 151
    - 7.5.4.3 Fundamental Sums 151

		7.5.4.4 Conjunctive Canonical Form of a Function with n Variables	153
		7.5.4.5 Some Abridged Notations	154
		7.5.4.6 Partial Expansions of a Boolean Function	155

7.6 Dual Functions 156
   7.6.1 Dual of a Dual Function 156
   7.6.2 Dual of a Sum 156
   7.6.3 Dual of a Product 156
   7.6.4 Dual of a Complement 157
   7.6.5 Dual of a Constant Function 157

7.7 Some Distinguished Functions 157
   7.7.1 n-Variables AND Function 157
   7.7.2 n-Variables OR Function 158
   7.7.3 Peirce Function of n-Variables (OR-INV-NOR) 158
   7.7.4 Sheffer Function in n-Variables (AND-INV-NAND) 159
   7.7.5 Some Properties of the Operations $\downarrow$ and $/$. Sheffer Algebra . . . 160
      7.7.5.1 Expressions by Means of Operations $(+)$, $(\cdot)$, $(-)$ 160
      7.7.5.2 Properties of Duality for $\downarrow$ and $/$ 160
      7.7.5.3 Sheffer Algebrae 161
      7.7.5.4 Expressions for the Operations $\cdot$, $+$, and $-$ with the aid of the Operations $\downarrow$ and $/$ (n-Variables) 162
      7.7.5.5 Canonical Forms 163
      7.7.5.6 Some General Properties of the Sheffer and Peirce Operators 165
   7.7.6 Operation $\oplus$ 167

7.8 Threshold Functions 172
   7.8.1 Definition 172
   7.8.2 Geometrical Interpretation 172
   7.8.3 Particular Case: Majority Function 173

7.9 Functionally Complete Set of Operators 174

7.10 Determination of Canonical Forms 177

7.11 Some Operations on the Canonical Forms 180
   7.11.1 Passage from one Canonical Form to Another 180
   7.11.2 Complement of a Function f Given in Canonical Form 182

7.12 Incompletely Specified Functions 183

7.13 Characteristic Function for a Set 184

7.14 Characteristic Set for a Function 184

CHAPTER 8 Geometric Representations of Boolean Functions

8.1 The n-Dimensional Cube 185
   8.1.1 Example: — 2-Variable Functions 185
   8.1.2 3-Variables Functions 186
   8.1.3 Functions of n-Variables (any n) 186
      8.1.3.1 Sets (E_1) and E_2) 186
      8.1.3.2 Relation of Adjacency between 2 Vertices 187
      8.1.3.3 Geometric Representation 187
   8.1.4 Other Representations and Definitions 188
      8.1.4.1 The n-Dimensional Cube 188
      8.1.4.2 The k-Dimensional Sub-Cube of an n-Dimensional Cube $(0 \leq k \leq n)$ 188
      8.1.4.3 Distance of 2 Vertices of an n-Dimensional Cube . . . 190

8.2	Venn Diagrams		192
8.3	The Karnaugh Diagram		195
	8.3.1	The Case of 4 Variables (or less)	195
	8.3.2	Karnaugh Diagrams in 5 and 6 Variables	197
		8.3.2.1 The 5-Variable Case	197
		8.3.2.2 The Case of 6-Variables	198
	8.3.3	Decimal Notation	199
	8.3.4	The Karnaugh Diagram and the Representation of Functions in Practice	199
		8.3.4.1 Notations	199
		8.3.4.2 Derivation of a Karnaugh Diagram	199
8.4	The Simplification of Algebraic Expressions by the Karnaugh Diagram Method		202
	8.4.1	Karnaugh Diagram Interpretations for the Sum, Product and Complementation Operations	202
		8.4.1.1 Sum	202
		8.4.1.2 Product	202
		8.4.1.3 Complement	202
	8.4.2	Representation of Certain Remarkable Functions	202
		8.4.2.1 Variables	202
		8.4.2.2 Product of k Variables (k > 1)	202
		8.4.2.3 Fundamental Sums	204
	8.4.3	The Use of the Karnaugh Diagram for the Proof of Algebraic Identities	205

CHAPTER 9   Applications and Examples

9.1	Switching Networks. Switching Elements		213
9.2	Electronic Logic Circuits. Gates		213
	9.2.1	Gates	213
	9.2.2	Logics	214
	9.2.3	Logic Conventions	215
	9.2.4	Diode Logic	216
	9.2.5	Direct Coupled Transistor Logic (DCTL)	217
	9.2.6	Diode-Transistor Logic (DTL)	218
9.3	Combinational Networks. Function and Performance. Expressions and Structure		219
9.4	Examples		222
	9.4.1	An Identity	222
	9.4.2	Binary Adder and Subtractor	224
		9.4.2.1 Addition	224
		9.4.2.2 Subtraction	228
	9.4.3	Restricted Complement of a Number (Complement to '1')	228
	9.4.4	True Complement of a Number	229
	9.4.5	Even Parity Check	231
	9.4.6	Addition of 2 Numbers Having Even Parity Digits	232
	9.4.7	Comparison of 2 Binary Numbers	233
	9.4.8	V-Scan Encoder Errors	235
		9.4.8.1 Error Pattern	235
		9.4.8.2 Principle of the V-Scan Encoder	236

Contents

    9.4.8.3 Relation between the Value Available at a Brush and the Digits of the Number x to be Read Out 237
    9.4.8.4 Error Caused by the Failure of a Single Brush 238
    9.4.8.5 The length of an Error Burst 239
    9.4.8.6 Numerical Value of the Error $e = y - x$ 240
    9.4.8.7 Examples 241
  9.4.9 Errors of a Reflected Binary Code Encoder 242
  9.4.10 Detector of Illicit Combinations in the Excess-3 Code 243

CHAPTER 10 The Simplification of Combinational Networks

10.1 General 245

10.2 Simplification Criterion. Cost Function 246

10.3 General Methods for Simplification 249
  10.3.1 Algebraic Method 249
  10.3.2 Karnaugh Diagram Method 250
  10.3.3 Quine-McCluskey Algorithm 250
  10.3.4 Functional Decomposition Method (Ashenhurst, Curtis, Povarov) 250

10.4 The Quine-McCluskey Algorithm 251
  10.4.1 Single-Output Networks 251
    10.4.1.1 Statement of the Problem 251
    10.4.1.2 Geometrical Interpretation of the Minimisation Problem 251
    10.4.1.3 n-Dimensional Cube and Sub-Cubes 251
    10.4.1.4 Maximal k-Cubes 253
    10.4.1.5 Statement of the Problem of Minimisation. 253
    10.4.1.6 Properties of the Minimal Cover 253
    10.4.1.7 Essential Maximal k-Cubes 254
    10.4.1.8 Quine-McCluskey Algorithm (Geometrical Interpretation) 254
    10.4.1.9 Incompletely Specified Functions 269
    10.4.1.10 Product of Sums 269
  10.4.2 Multi-Output Circuits 269

10.5 Functional Decompositions 271
  10.5.1 Simple Disjoint Decompositions 272
    10.5.1.1 Definition 272
    10.5.1.2 Number of Partitions (YZ) of Variables 274
    10.5.1.3 Necessary and Sufficient Condition for the Existence of a Simple Disjoint Decomposition 274
    10.5.1.4 Other form of Criterion of Decomposability. Column Criteria 279
    10.5.1.5 Theorem 281
    10.5.1.6 Theorem 283
    10.5.1.7 Theorem 286
    10.5.1.8 Theorem 287
    10.5.1.9 Theorem 290
  10.5.2 Simple Disjoint Decompositions and Maximal Cubes 291
  10.5.3 Simple Non-Disjoint Decompositions 300

CHAPTER 11 Concept of the Sequential Network

11.1 Elementary Example. The Ferrite Core 303
  11.1.1 Read-Out 305
  11.1.2 Writing 307

Contents

11.2 Eccles-Jordan Flip-Flop. 309
    11.2.1 Type S-R Flip-Flop (Elementary Register) 309
    11.2.2 Type T Flip-Flop (Elementary Counter, Symmetric Flip-Flop) 311

11.3 Dynamic Type Flip-Flop 311
    11.3.1 Type S-R Flip-Flop 311
    11.3.2 Type T Flip-Flop 313

11.4 Some Elementary Counters 313
    11.4.1 Dynamic Type Flip-Flop Pure Binary Counter. 313
    11.4.2 Gray Code Counter with Eccles-Jordan Type T Flip-Flops. . . 314
    11.4.3 Gray Code Continuous Level Counter. 314
    11.4.4 Random Impulse Counter. 317

CHAPTER 12   Sequential Networks. Definitions and Representations

12.1 Quantization of Physical Parameters and Time in Sequential Logic Networks. 318
    12.1.1 Discrete Time 319
    12.1.2 Operating Phases. Timing Signals 319
        12.1.2.1 Synchronous Systems 319
        12.1.2.2 Asynchronous Systems 320

12.2 Binary Sequential Networks. General Model 321
    12.2.1 Inputs 321
    12.2.2 Outputs. 321
    12.2.3 Internal States 321
    12.2.4 Operation. 321
    12.2.5 Present and Future States. Total State. Transitions 322
    12.2.6 Concept of a Finite Automaton 323
    12.2.7 Synchronous and Asynchronous Sequential Networks. 323
        12.2.7.1 Synchronous Sequential Networks 324
        12.2.7.2 Asynchronous Sequential Networks. 335

12.3 Sequential Networks and Associated Representations 340
    12.3.1 Logic Diagrams 340
    12.3.2 Equations and Truth Tables. 340
    12.3.3 Structural Representations 341
        12.3.3.1 Alphabets 341
        12.3.3.2 Words, Sequences, Sequence Lengths 341
        12.3.3.3 Transitions. Transition and Output Functions . . . 341
        12.3.3.4 Successors. Direct Successors Indirect Successors . . 342
        12.3.3.5 Transition Table. Table of Outputs 343
        12.3.3.6 Graphical Representation of a Sequential Network . . 345
        12.3.3.7 Connection Matrix 348
    12.3.4 Representations as a Function of Time 349
        12.3.4.1 Literal Representation 349
        12.3.4.2 Sequence Diagram 349

12.4 Study of the Operation of Sequential Networks 350
    12.4.1 Synchronous Circuits 352
    12.4.2 Asynchronous Networks 352
        12.4.2.1 Sequential Network Diagram 353
        12.4.2.2 Equations 353
        12.4.2.3 Truth Tables. 353
        12.4.2.4 Differential Truth Table. 354

12.4.2.5 Stable States	354
12.4.2.6 Table of Transitions and Outputs. Graph	354
12.4.2.7 Cycles with Constant Input	354
12.4.2.8 Hazards	355
12.4.2.9 Races	357
12.4.2.10 Phase Diagram (Huffmann)	367
12.5 Adaptation of the Theoretical Model to the Physical Circuit	368
12.6 Some Supplementary Physical Considerations	370
12.6.1 Synchronous Networks	370
12.6.2 Asynchronous Networks	371
12.6.3 Physical Duration of States	371
12.6.4 Delay Elements	371
12.6.5 Synchronisation	371
12.6.6 State Stability	373
12.6.7 Memory	373
12.6.8 Synchronous and Asynchronous Viewpoints	374
12.7 Incompletely Specified Circuits	375
12.8 Sequential Networks and Combinatorial Networks	376

CHAPTER 13 Regular Expressions and Regular Events

13.1 Events	378
13.1.1 Definition	378
13.1.2 Empty Sequence λ	379
13.1.3 Empty Set of Sequences $\varnothing$	379
13.2 Regular Expressions. Regular Events	379
13.2.1 Regular Operations	379
13.2.1.1 Union of A and B (A$\cup$B)	379
13.2.1.2 Product (Concatenation) of A and B (A·B)	380
13.2.1.3 Iteration of an Event A (A*)	381
13.2.2 Function $\delta(A)$	381
13.2.3 Properties of Regular Operations	382
13.2.4 Generalised Regular Operations	383
13.2.4.1 Intersection of 2 Events A and B (A$\cap$B)	384
13.2.4.2 Complement of an Event A $(\overline{A})$	384
13.2.4.3 Boolean Expression $E(A, B)$	384
13.2.4.4 Event $E_1 = (t)_{t \leq p} E$	384
13.2.4.5 Event $E_2 = (Et)_{t \leq p} E$	384
13.2.5 Regular Expressions	385
13.2.6 Regular Events	386
13.2.7 Set Derived from a Regular Event. Derivation of a Regular Expression	389
13.2.7.1 Derivatives with Respect to a Sequence of Length 1	390
13.2.7.2 Derivatives with Respect to a Sequence of Length >1	392
13.2.8 Properties of Derivatives	392
13.2.8.1 Examples of the Determination of Derivatives	394
13.3 Regular Expressions Associated with a States Diagram	399
13.3.1 Moore Diagram	399
13.3.2 Mealy Diagram	400
13.3.3 Example: Series Adder	402

CHAPTER 14 The Simplification of Sequential Networks and Minimisation of Transition Tables

14.1 Introduction. 405
    14.1.1 Performance. 405
        14.1.1.1 Completely Specified Circuits 405
        14.1.1.2 Incompletely Specified Circuits. 406
    14.1.2 Cost Function 406
14.2 Minimisation of the Number of States for a Completely Specified Table 407
    14.2.1 Some Preliminary Considerations. 407
    14.2.2 Equivalent States 410
        14.2.2.1 States in the Same Table T 410
        14.2.2.2 State Appertaining to 2 Tables T and T' 410
    14.2.3 Covering of Table T by T'. 410
    14.2.4 Successor to a Set of States 410
    14.2.5 Reduction of a Table. Successive Partitions P^k. 411
    14.2.6 Algorithm for the Minimum Number of Internal States and its Practical Application 412
    14.2.7 Example 413
14.3 Minimisation of the Number of Internal States for an Incompletely Specified Table. 416
    14.3.1 Interpretation of Unspecified Variables 416
    14.3.2 Applicable Input Sequences 416
    14.3.3 Compatibility of 2 Partially Specified Output Sequences 417
    14.3.4 Compatibility of 2 States q_i and q_j. 417
    14.3.5 Compatibility of a Set of n States ($n > 2$). 418
    14.3.6 Compatibility of States and Simplification of Tables 419
        14.3.6.1 Covering of a Sequence Z by a Sequence Z' 419
        14.3.6.2 Covering of a State q_i of a Table T by a State q_i' of a Table T' 420
        14.3.6.3 Covering of a Table T by a Table T' 420
    14.3.7 Properties of Compatibles 420
    14.3.8 Determination of Compatibles 421
        14.3.8.1 Compatible Pairs 421
        14.3.8.2 Compatibles of the Order ≥ 2. Maximal Compatibles . . 423
    14.3.9 Determination of a Table T" Covering T, Based on a Closed Set of Compatibles 427
    14.3.10 Determination of a Minimal Table T. 428
    14.3.11 Another Example of Minimisation. 430

CHAPTER 15 The Synthesis of Synchronous Sequential Networks

15.1 General. 434
    15.1.1 The Direct Method 434
    15.1.2 The State Diagram Method 435
15.2 Direct Method. Examples 435
15.3 State Diagram Method 441
15.4 Diagrams Associated with a Regular Expression. 456
    15.4.1 Development of a Regular Expression from a Base 456
    15.4.2 Interpretation in Terms of Derivatives. 463
    15.4.3 Algorithm for the Synthesis of a Moore Machine. 464

15.4.4		Algorithm for the Construction of a Mealy Machine	465
15.4.5		Examples of Synthesis	466
15.4.6		Networks Having Several Binary Outputs	477

15.5 Coding of Initial States. Memory Control Circuits. 478

CHAPTER 16 counters

16.1 Introduction. 483

16.2 Pure Binary Counters. 484
    16.2.1 Up-Counting and Down-Counting in Pure Binary Code. 484
        16.2.1.1 Up-Counting 484
        16.2.1.2 Down-Counting 485
        16.2.1.3 Counting in Pure Binary: Other Formulae 486

16.3 Decimal Counters 489

16.4 Reflected Binary Code Counters 490
    16.4.1 Recurrence Relations in Gray Code Counting 490
    16.4.2 Down-Counting. 493
    16.4.3 Examples of Counters 494

Bibliography 497

Index 504

Chapter 1

General Concepts

1.1 Digital Systems

Modern digital computers, communications networks, (such as telephone exchanges), automatic control systems for complex industrial processes (e. g. nuclear reactors) are either totally or partially *digital systems*: their essential purpose being the processing of discrete data, i. e. the input signals applied to the system or the output signals provided by the system contain only a finite number of distinct values, between which there is no continuity. That the signals are *discrete* may stem from the system conception: the abacus, pebbles or the fingers used in counting are examples of these. More usually, however, it is the effect of a special convention whereby physical values are *quantised* which, on a macroscopic scale, would normally appear to be *continuous*. Indeed the most common examples — voltages, currents, positions and displacements of mechanical devices only vary in a continuous way by reason of the various *inertias* (mechanical, electrical) inherent within themselves. Computers provide an example of this quantification procedure. They can be divided into two categories:

1. *Analog* computers: in which each mathematical (or physical) magnitude is represented by a proportional physical magnitude (hence the name "analog"), where the proportionality factor is known as the scale factor. This procedure is well adapted to the representation of continuous values and for the resolution of problems involving the intervention of such values (differential equations, partial differential equations). Apart from the difficulty of working on discontinuous phenomena, the procedure reveals an inherent disadvantage: the accuracy of the calculations is limited to that of the physical elements of which the computer is comprised. The only way of improving the degree of accuracy is to improve the quality of the elements themselves. Beyond a certain level of accuracy (1% for example) the operation is confronted with a "law of diminishing returns" and becomes increasingly difficult.

2. *Digital* computers: in which the *direct analogy* as used in the analog computer is replaced by an *indirect analogy*. By means of a *digital* representation of the magnitudes, the coding operation (i. e. the analogy referred to earlier) is applied to the digits which make up the numbers instead of being applied to the magnitudes represented by the latter.

Thus, in order to represent magnitudes to an accuracy of 1/1000 in terms of analog systems, it is necessary to divide the overall scale into 1000 small, contiguous zones; these zones must always be identifiable, no matter what the distance between them, and particularly when they are adjacent. With a digital coding using binary numbers, the problem is, on the face of it, exactly the same, but with only two zones. It should be noted that the use of a digital representation involves the presence of a large number of circuits and that the indirect effect of this complexity is to reduce the apparent latitude available. Experience in the *design of electronic circuits* has shown the necessity for avoiding the 'piling-up' of tolerances in individual circuits, since this could be such that the zones (2 in the binary case) assigned to the digit values might no longer be identifiable. However, experience together with the number of digital computers currently in use confirms that, with care, the desired standard of accuracy can be achieved. Moreover, the use of *digital codings* is not limited merely to ordinary calculation procedures, but can be used in numerous other systems to provide similar advantages of accuracy and versatility in use. Finally, in a great many cases the signals are, in the nature of things, discontinuous: signals indicating whether or not an event is taking place, command signals for the execution or non-execution of a particular task, switching operations, etc.

In fact, it will later be shown that the discrete signals of digital systems, whether or not they correspond with figures, can, more generally, be considered as the analog counterparts of magnitudes, always discrete, of another kind. These magnitudes are the two values, falsity or truth, of statements such as:

— "the figure $x_k = 1$",
— "the operation $f(A, B)$ is proceeding",
— "the transmission of the binary sequence S is complete", etc., by which the operation of the system can be described.

In much the same way that the physics of traditional mathematics gave rise to numerous *analytical* works whose object has been the formulation and study of phenomena which are essentially continuous (e. g. integration, differential equations, partial differential equations) the requirements of the designers of digital systems were responsible for new developments and theoretical methods of study and synthesis. However, because of the discontinuous character of the signals used, it is to *algebra*, in the most general sense of the term, that these theories are linked. This present work is devoted to the switching networks made up from binary elements i. e. those for which the characteristic parameters (inputs, outputs, internal parameters) take only two values (or "states"). Boolean algebrae constitute one of the most important tools for the study and synthesis of these networks.

It should be noted that this present theory is not just confined to the binary case: a variety of works, some of them of great importance, have been devoted to the so-called "n value" networks ($n > 2$) (in particular those of Moisil, Ghazalé). To give an account of these works would, unfortunately, exceed the scope of this present book. The above theories are interesting for the following reasons:

1. They pave the way for the generalised use (still uncertain at the present time) of 'n' value elements.

2. They enable (e. g. in the case of Lukasiewicz's 3-value algebrae) the treatment in mathematical form, but limited merely to 'discrete' aspects, of transient phenomena: in the binary case for example, the passage from one state to another is carried out by passing through an intermediary state, whose existence cannot always be ignored, to which may be attributed diverse qualifications ("transient regime", "waiting phase", "doubtful state" etc.,) or, better still, a mathematical symbol of state. This is how the so-called Lukasiewicz's algebrae, i. e. which make up the algebraic structures associated with the 'n'-value logics of the Polish mathematician whose name they bear, have been defined and studied by *Gr. Moisil* who has demonstrated their application to the analysis and synthesis of sequential networks.

3. Finally, from a theoretical standpoint, it is often advantageous to rephrase the problem within a general framework, and under this heading, the 'n' value algebraic networks, as they are called, can often throw new light on questions apparently pertaining only to binary algebrae.

The reader will be able to consult references [54,107] on this subject. As the use of these algebrae is but one possibility among others, this work will be confined to the binary structures. In the remainder of this chapter will be found some reminders of the concepts of set, of relation, function and algebraic structure which will prove useful in due course.

1.2 Sets

1.2.1 The Concept of a Set

The concept of a set is a first notion which can only be demonstrated by examples: thus one speaks of the set of Frenchmen, the set of whole numbers, the set of polynomials, the set of binary sequences of length n, etc. The items which form part of the set are known as *elements*, *members* or *points*. That an element x belongs to a set A is expressed as:

$$x \in A \qquad (1.1)$$

(read as: "x belongs to A") and, more generally, for n elements:

$$x_1, x_2, ..., x_n \in A \tag{1.2}$$

It can also be said that A contains x (or $x_1, x_2, ..., x_n$). Conversely, given that an element x is not included in A it can be expressed as:

$$x \notin A \tag{1.3}$$

(read as: "x is not included in A")

For example, if Z denotes the set of integers, it may be written that:

$$3 \in Z \tag{1.4}$$

$$2/5 \notin Z \tag{1.5}$$

As has been shown, a set is made up of points which possess a common identity, as for example: to be French, to be a whole number, a polynomial, a binary sequence of length n, etc., In addition, a set can be represented by brackets within which can be found either the elements themselves or the symbol for the general element with an indication of the property common to all the elements forming the set. Thus, if R is the set of the roots of the equation:

$$x^3 - 6x^2 + 11x - 6 = 0 \tag{1.6}$$

it can be expressed as:

$$R = \{1, 2, 3\} \tag{1.7}$$

or:

$$R = \{x \mid x^3 - 6x^2 + 11x - 6 = 0\} \tag{1.8}$$

this second notation indicating that R is the set of elements x such that it may be said that:

$$x^3 - 6x^2 + 11x - 6 = 0 \tag{1.9}$$

1.2.2 Equality of Two Sets

Two sets A and B are said to be *equal* (or identical) if they are made up of the same elements, and can be written as:

$$A = B \tag{1.10}$$

In other words, any element x which belongs to A also belongs to B and vice versa.

Examples:

a) the sets:
$$A = \{1, 2, 3\}$$
$$B = \{x \mid x^3 - 6x^2 + 11x - 6 = 0\} \quad (1.11)$$

are equal: indeed it has already been shown that they are formed of the same elements 1, 2, 3.

b) the sets:
$$A = \{x \mid x \text{ is an even prime number}\} \quad (1.12)$$
$$B = \{2\}$$

($\{2\}$ is the set which comprises only one element: 2) are equal.

1.2.3 Subsets. Inclusion Relation

A set A is said to be a *subset* of a set B if every element of A belongs to B. It can further be said that A is included in B, or that A is a part of B or, finally, that B contains A and can be expressed as:

$$A \subset B \quad (1.13)$$

or
$$B \supset A \quad (1.14)$$

Examples:
— the set of even integers is a subset of the set of integers.
— the set of squares, in geometry, is a subset of the set of rectangles.
— the set of ellipses is a subset of the set of conics (second order curves).
— the set (01, 10) is a subset of {00, 01, 10, 11}, the set of binary words of length 2.

Note: Neither the definition given above nor the notations (1.13) and (1.14) exclude the case where $A = B$. Where there are elements which belong to B but not to A, it is sometimes said that A is a *true subset* of B.

1.2.4 Complement of a Set

Given that set A is a subset of a set U, the *complement* of A with respect to U, is called the set of those points in U which do not belong to A. This is expressed as:

$$U - A \quad \text{or} \quad C_U(A)$$

Using the notation already introduced it can be stated that:

$$U - A = \{x \mid x \in U \text{ and } x \notin A\}. \tag{1.15}$$

Example: if U is the set of real numbers and A that of rational numbers, $U - A$ is the set of irrational numbers.

The set $U - A$ is a relative complement. When there is no ambiguity in the consideration of the set U it can be defined as the set of points which do not belong to A. This is frequently written as $\bar{A}$.

$$\bar{A} = \{x \mid x \notin A\}. \tag{1.16}$$

It should be noted that, starting from A, the *complementary* operation whereby the term $\bar{A}$ is derived is a one-term operation (A) or *unary*. It s now necessary to define two-term operations (*binary* operations).

1.2.5 Union and Intersection of Sets

1.2.5.1 Union

The *union* of two sets A and B is the set of points which belong to A or B or to both at the same time. The union of A and B is written $A \cup B$ (and reads: "A union B").

$$A \cup B = \{x \mid x \in A \text{ or } x \in B \text{ or } x \in A \text{ and } x \in B\} \tag{1.17}$$

Thus the union of the set of points of the segment $0 \leq x \leq 1$ and of the segment $0{,}5 \leq x \leq 1{,}5$ is the set of points of the segment $0 \leq x \leq 1{,}5$.

1.2.5.2 Intersection

The *intersection* of the two sets A and B is the set of the points which belong to both A and B. It is written $A \cap B$ (and reads "A intersection B").
This can be expressed as:

$$A \cap B = \{x \mid x \in A \text{ and } x \in B\}. \tag{1.18}$$

For example, the intersection of the set of prime numbers and the set of even numbers is the set $\{2\}$.

1.2.6 Empty Set. Universal Set

As will later be seen, it is convenient to introduce a set which possesses no elements, known as the *empty set*, and written as $\varnothing$. In these conditions, the equality
$$A \cap B = \varnothing \tag{1.19}$$
signifies that the sets A and B have nothing in common. It can also be said that they are *mutually exclusive*.

Similarly, it is convenient to introduce a set containing every possible element, and called a *universal set*, or *universe* (which will be written as U, I etc.).

1.2.7 Set of Subsets

For a given set E, the subsets can be considered as being the elements of a set which is denoted by $\mathscr{P}(E)$ or 2^E (the set of parts of E).

Example:
$$E = \{1, 2, 3\} \tag{1.20}$$
$$\mathscr{P}(E) = \{\varnothing, \{1\}, \{2\}, \{3\}, \{1,2\}, \{2,3\}, \{1,3\}, \{1, 2, 3\}\} \tag{1.21}$$

$\mathscr{P}(E)$ is therefore a set of sets. In the present example it has $2^3 = 8$ elements, if the empty set $\varnothing$ is included. Some of its sub-sets are specially interesting: these are the *covers* of E and the *partitions* of E.

1.2.8 Cover of a Set

Given a set E (belonging to a universal set U), a cover of E is called a set of parts of E so that E is equal to their union.

Example: the set of whole numbers, rational numbers, irrational numbers and transcendant numbers form a cover of the set of real numbers. The sets $\{00, 01\}$, $\{00, 10\}$, $\{01, 11\}$ and $\{11\}$ cover $\{00, 01, 10, 11\}$, the set of binary words of length 2.

1.2.9 Partition of a Set E

This is a cover of E in which any 2 sets are mutually exclusive (disjoint). Any point of E is then found in one set of the partition and in one only.

Thus, the set of rational and irrational numbers constitutes a partition of the set of real numbers.

Note: These different definitions can be interpreted graphically by means of what are known as Venn diagrams. Here the universal set is represented by a rectangle and the sets A, B etc., described earlier are represented by circles. Figure 1.1 interprets the different operations defined above (in each figure the shaded portion corresponds to the operation indicated).

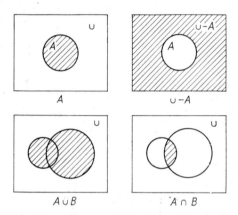

Fig. 1.1. Venn diagram of 3 fundamental operations.

1.2.10 Cartesian Product of Sets

For two given sets A and B the *cartesian product* or, more simply the *product* of the sets A and B is the set denoted by $A \times B$ of *ordered pairs* (x, y), where $x \in A$ and $y \in B$.

Example: if A and B are the set R of real numbers, the set $R \times R$ is the set of pairs (x, y). It can be considered as a set of points in the plane (x, y), since each of these points is characterised by a certain pair (x, y) of coordinates.

The cartesian product of n sets $E_1, E_2, \ldots, E_n$ is similarly defined: it is the set P of ordered combinations

$$(x_{i_1}, x_{i_2}, \ldots, x_{i_k}, \ldots, x_{i_n}),$$

where $x_{i_k} \in E_k$ for every k. This is expressed as:

$$P = E_1 \times E_2 \times \cdots \times E_n.$$

1.2 Sets

In the case where all the E_k are equal to the same set A, it can then be said that:
$$A^n = A \times A \times \cdots \times A$$

Graphic representation of the product of 2 sets.

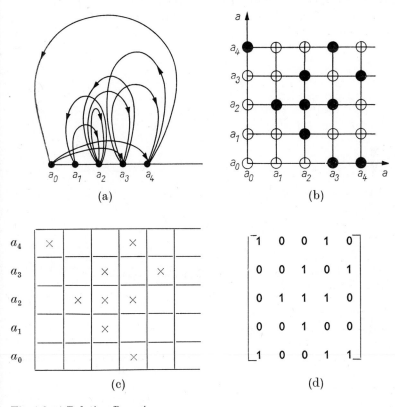

Fig. 1.2. a) Relation R on A.
b, c, d) Product $A \times A$ a and sub-set R (relation R).

In the case of a finite set A, the elements can be represented on a straight line: for example, with the points of the complete abscissa $0, 1, 2, 3, \ldots, k, \ldots$ are associated the elements $a_0, a_1, \ldots a_k, \ldots$ of A.

If it is desired to represent the product $A \times B$, it is necessary to consider the regular grid made up from the two families of straight lines parallel with the axes, of integer coordinates, in a plane xOy (straight line equation $x = i$, $y = j$, with $i, j = 0, 1, 2, \ldots$). The straight line $x = i$ and the straight line $y = j$ intersect at a point (i, j) with which the pair (a_i, b_j) are associated. The circles (plain or shaded) of Fig. 1.2 b are in this way associated with the elements of A^2, with

$$A = \{a_0, a_1, a_2, a_3, a_4\}$$

1.2.11 Application. Counting the Elements of a Set

If $n(A)$ is the number of elements (or "power") of a set A, given two mutually exclusive (disjoint) sets A and B, the following relation can be expressed as:

$$n(A \cup B) = n(A) + n(B) \qquad (A \cap B = \varnothing). \qquad (1.22)$$

Next consider any 2 sets A and B (not necessarily mutually exclusive sets):

Here:
$$A \cup B = (A \cap B) \cup (\bar{A} \cap B) \cup (A \cap \bar{B}) \qquad (1.23)$$

(the 3 sets in line are disjoint).

Moreover
$$A = (A \cap B) \cup (A \cap \bar{B}),$$
whence
$$n(A) = n(A \cap B) + n(A \cap \bar{B})$$
and similarly:
$$n(B) = n(A \cap B) + n(\bar{A} \cap B)$$

hence, by substitution in (1.23)

$$n(A \cup B) = n(A \cap B) + n(B) - n(A \cap B) + n(A) - n(A \cap B)$$
$$= n(A) + n(B) - n(A \cap B). \qquad (1.24)$$

(For the case where $A \cap B = \varnothing$, this yields (1.22)).

This formula can be generalised to the case of $n (n > 2)$ sets. In the case of 3 sets, as the reader can check by way of an exercise, the expression becomes:

$$n(A \cup B \cup C) = n(A) + n(B) + n(C) - n(A \cap B) - n(A \cap C)$$
$$- n(B \cap C) + n(A \cap B \cap C). \qquad (1.25)$$

Note. — Similar formulae are used in the theory of probabilities. The probability of an event can in effect be defined as the "measure" of a set.

1.3 Relations

1.3.1 Relations between the Elements of a Number of Sets

Consider for example 2 sets A and B. A subset of the cartesian product $A \times B$ is known as a *relation* existing between the elements $x \in A$ and $y \in B$. In other terms, the word relation will be used here in a slightly different sense from its meaning in everyday language; a relation is a *subset R* of $A \times B$. To define a relation is to specify this subset, which is the set of pairs (x, y) between which the relation holds true, in its usual sense.

More generally, given n sets $E_1, E_2, \ldots, E_n$, a sub-set of $E_1 \times E_2 \times \cdots \times E_n$ is again known as a relation. In the case of 2 sets this relation is a binary relation, with 3 sets it is a ternary relation, etc.

Particular case: relation between the elements of a single set

In a similar manner, a sub-set of the product $A \times A$ is known as the R (binary) relation on A. Similarly a relation of the order n (n-ary) is defined as being a sub-set of A^n.

Examples

(1) Consider a set of points (x, y) in a plane.
— The set of points such that $x \geq y$ is a sub-set of this set and constitutes a relation.
— The set of points (x, y) such that $x^2 + y^2 \leq a^2$ is the relation situated either inside or on the circumference of a circle of radius a.
— The set of points (x, y) which gives $x + y = 1$ is a straight line.
(2) Next consider for example, the set of straight lines in a plane or in 3-dimensional space: $X \perp Y$, $X // Y$ are relations. These are the sets of pairs of straight lines which are respectively perpendicular and parallel.
(3) It will be noted that the 'belonging' relation $\in$ is a 1-term (unary) relation. It is defined by a sub-set (A) of the universal set.
(4) The 'inclusion' relation between sets $(A \subset B)$ is a binary relation of the set of parts of a set.

1.3.2 Notation

The symbolic representation xRy indicates that x and y are linked by the relation R, i. e. $(x, y) \in R$, where R is the 'relation' $(R \subset A \times B)$ in question.

1.3.3 Graphical Representation of a Binary Relation

One may start from the representations of a set A (on which the relation is defined) or from the set $A \times A$ (of which the relation R is a sub-set). In the first case a directed arc is drawn through the points a_i, a_j of a pair (a_i, a_j) for which one has:

$$a_i R a_j \tag{1.26}$$

In the second case a special mark will be used to indicate the pairs (a_i, a_j) for which the relation holds true such that

$$(a_i, a_j) \in R \tag{1.27}$$

Fig. 1.2 illustrates 3 variants of this procedure (matrix, 2 types of array) as well as the relation of this type of representation with the first.

Special case:

In the case where the relation is symmetrical (cf. 1.3.4.2) it suffices to use one half of the tables (1.2 b, c, d). For the first representation the arrows may be deleted.

Further example:

Consider the set of binary words of length n. Two words satisfy the relation R (the adjacency relation) if they differ in a single position.

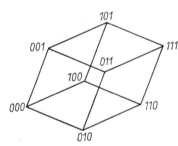

Fig. 1.3. Adjacency relation between binary words of length 3.

This relation is said to be *symmetrical*. By use of the first type of representation (on A) Figure 1.3 is obtained, and in which can be seen, with the shape given to the arcs (a_i, a_j), a cube in perspective. There will be a further opportunity of seeing this type of representation (planar representations of an n-dimensional cube) in Chapter 8.

1.3.4 Some Properties of Binary Relations

1.3.4.1 Reflexivity

A binary relation R on a set E is said to be *reflexive* if for every $x \in E$ xRx exists.

Examples: the relation $x = y$ for a set of real numbers is reflexive, since $x = x$ for every $x \in E$. Conversely, the relation $x < y$ is not reflexive.

The inclusion relation between sets is reflexive since, by the definition given $A \subset A$.

1.3.4.2 Symmetry

A binary relation R on a set E is said to be *symmetrical* if for every pair $(x, y) \in E \times E$, xRy implies yRx.

Example: the relation $x = y$ is symmetrical, the relation $x < y$ is not.

1.3.4.3 Transitivity

A binary relation R on E is said to be *transitive* if for every combination $(x, y, z) \in E \times E \times E$, xRy and yRz imply xRz.

Examples: for the set of real numbers the relations $x = y$ and $x < y$ are transitive, but the relation $x \neq y$ is not; neither is the relation $X \perp Y$ on the set of straight lines in space. The inclusion relation between sets is transitive.

1.3.5 Equivalence Relations

A binary relation R is known as an *equivalence relation* if it is simultaneously reflexive, symmetric and transitive. Given two elements x and y so as to have xRy, it can be said briefly that x is *equivalent* to y (through the relation R).

Examples:

(1) The relation $x = y$ for a set of real or complex numbers. It was shown (§ 1.3.4) that this relationship was reflexive, symmetric and transitive. The equality of numbers is thus an equivalence relation.

(2) The relation $X \parallel Y$ for a set of straight lines in space. It may be accepted that every straight line is parallel with itself, whence $X \parallel X$. Moreover, the property of parallelism is symmetric and transitive. The relation of parallelism between straight lines is thus an equivalence relation.

(3) The congruence of integers.

Consider the set Z of integers (of either sign). It is said that x and y are *congruent* with the modulus m (congruent modulo m), when:

$$x - y = km \text{ where } k \text{ is an integer.}$$

The following properties apply:

3.1 $x - x = 0\, m$

3.2 if $x - y = km,\ y - x = -km = (-k)m$ \hfill (1.28)

3.3 if $x - y = km,\ y - z = k'm,$ \hfill (1.29)

then

$$(x - z) = (k + k')m = k''m \tag{1.30}$$

0, $(-k)$, k'' are still integers. Thus the congruence relation is an equivalence relation. It is expressed as:

$$x \equiv y \ (\text{mod. } m) \tag{1.31}$$

(read as "x congruent with y modulo m", or "x equals y modulo m").

1.3.6 Equivalence Classes

For a given element x, the set $C(x)$ of elements y which are equivalent to x through the relation R, is known as the *equivalence class* containing x.

$$C(x) = \{y \mid xRy\}. \tag{1.32}$$

In the case of the relation of parallelism between straight lines, the equivalence class associated with a straight line $X \in D$ is the set of straight lines with which it is parallel.

*Theorem: An equivalence relation R separates (partitions) a set E into pairwise **disjoint** equivalence classes, and each element of S is precisely placed in a single equivalence class.*

In order to show that equivalence classes are disjoint, assume that the equivalence classes $C(x)$ and $C(y)$ associated with two elements x and y are different and not disjoint. There would exist an element z such that:

$$z \in C(x) \quad \text{and} \quad z \in C(y). \tag{1.33}$$

i.e. this would give:
$$zRx \quad \text{and} \quad zRy. \tag{1.34}$$

But, if zRx implies xRz (symmetry) then xRz, zRy produce xRy (transitivity). More generally, for every point x' of $C(x)$ we have:
$$x'Rx \quad \text{and} \quad xRy \tag{1.35}$$
whence (transitivity)
$$x'Ry \tag{1.36}$$
thus
$$C(x) \subset C(y). \tag{1.37}$$

But the demonstration is symmetric for both x and y such that:
$$C(y) \subset C(x) \tag{1.38}$$
and consequently:
$$C(x) = C(y) \tag{1.39}$$

which contradicts the hypothesis which states that the classes $C(x)$ and $C(y)$ are different.

It follows then that if two equivalence classes are different, they are necessarily disjoint.

Finally, every element $x \in E$ is squarely placed in an equivalence class since xRx (reflexivity) and thus $x \in C(x)$.

Rank of an equivalence relation

The rank of an equivalence relation R in a set E is the number of equivalence classes into which the relation R partitions the set E.

Examples:

(1) Consider for example, the set of straight lines in a plane and the relation $X \parallel Y$. Each class comprises the set of straight lines parallel with a single direction. There is thus an infinity of classes associated with the different directions in the plane. The rank is infinite.

It should be noted that all the classes are disjoint. No straight line can be parallel to two different directions.

(2) Consider a set of real numbers and the relation $x = y$. Again there is an infinity of classes associated with the different numbers. Here, each class contains only one number.

(3) Consider the set of whole numbers and the congruence relation $x \equiv y$ (mod. m).

For a given integer x, the class $C(x)$ comprises all the numbers x' such that:
$$x' = x - km \quad (k \text{ is any integer}) \tag{1.40}$$

Thus the equivalence class 0 is:

$$\{\cdots -2m, -m, 0, +m, +2m, +\cdots\}. \tag{1.41}$$

Similarly, those of $1, 2, 3, \ldots, m-1$ are:

$$C(1) = \{\ldots, -2m+1, -m+1, 1, m+1, 2m+1, \ldots\}$$
$$C(2) = \{\ldots, -2m+2, -m+2, 2, m+2, 2m+2, \ldots\} \tag{1.42}$$
$$C(3) = \{\ldots, -2m+3, -m+3, 3, m+3, 2m+3, \ldots\}$$

. . . .

$$C(m-1) = \{\ldots, -2m+(m-1), -m+(m-1), m-1,$$
$$m+(m-1), 2m+(m-1), \ldots\} \tag{1.43}$$

More generally:

$$C(r+km) = C(r). \tag{1.44}$$

In particular, for a given integer x, of either sign, it will be seen that a class $C(r)$ with $0 \leq r \leq m-1$ can be found, such that:

$$x \in C(r) \tag{1.45}$$

and then:

$$x - r = qm \tag{1.46}$$

$$x = qm + r \, (0 \leq r \leq m-1) \tag{1.47}$$

the symbol q is assigned to the quotient (to the nearest lower unit) of x by m, r is the *remainder of the division of x by m*, or the *residue modulo m* of x. Use is often made of the notations:

$$\begin{cases} q = \left[\dfrac{x}{m}\right] \\ r = |x|_m \end{cases} \tag{1.48}$$

Note:

In the case of the sum $x = a + b + c$ for example, it is said that:

$$|x|_m = |a+b+c|_m \tag{1.49}$$

is the sum modulo m of a, b, c.

Moreover, when $m = 2$, the following notation is also employed:

$$|a+b+c|_2 = a \oplus b \oplus c \tag{1.50}$$

used also in logic and in Boolean algebra to which further reference will be made later in this work.

Summarizing, the relation ≡ (mod. m) thus partitions the set of integers into m distinct equivalence classes, each containing terms of the form $(r + qm)$ $(0 \leq r \leq m - 1$, q any integer). The set of these classes is known as the *set of residual integers modulo m* — usually denoted by Z/m.

1.4 Univocal (many-to-one) Relations or Functions

It is said that a relation between A and B is *univocal* in y, if for every x there is at most one element y such that xRy (x determines y univocally). For such a relation, it can also be said that y is a *function* or *mapping* of x expressed as:

$$y = f(x) \quad \text{or} \quad y = f_R(x) \tag{1.51}$$

In another way, it can be said that a function (mapping) of a subset A into a set B is a correspondence rule, by means of which an element $x \in A$ can be associated with an (*unique*) element $y \in B$. y is said to be the image of x by the mapping f. A is the *domain* of the function f. The set of images of the points A is the *counter-domain* of f (Fig. 1.4).

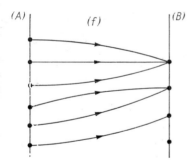

Fig. 1.4 Function of A into B.

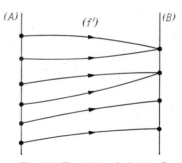

Fig. 1.5 Function of A onto B.

Note 1: In certain cases a function can be represented by a formula, as for example: $y = ax^2 + bx + c$, (x, y, a, b, c being real numbers). This is, however, not the case in general. For finite sets, a method of definition consists in giving all the pairs (x, y), where y represents the unique element of B associated with x. The set of these pairs (*graph f*) is a sub-set of $A \times B$ (Fig. 1.4).

Note 2: With an element $x \in A$ is associated a *single* element $y \in B$. It can, however, very well happen that several elements of A will produce the same element y of B by the mapping f.

Thus the function represented by the formula $y = \sin x$ is such that there is an infinity of arcs x having a given sine and an absolute value ≤ 1.

Note 3: It is equally possible that there may be elements $y \in B$ which are not the image of any element of A.

For example, if A and B are identical with the set of real numbers, there is no arc x for which the sine $y = 2$.

When, however, for every $y \in B$ there exists at least one $x \in A$ for which y is the image, this is said to be a mapping of A *onto* B (Fig. 1.5). The counter-domain of f is equal to B.

Fig. 1.6 A bi-univocal function.

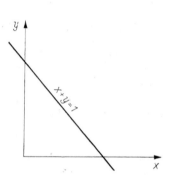
Fig. 1.7 Relation $x + y = 1$.

Special case: bi-univocal (one-to-one) functions

A mapping f of A onto B is said to be *bi-univocal* when an element $y \in B$ is the image of only one point x of A. In other terms, if

$$f(x_1) = f(x_2) \tag{1.52}$$

then

$$x_1 = x_2 \; (x_1, x_2 \in A). \tag{1.53}$$

This situation is represented symbolically in Fig. 1.6. It will be noted that, at the same time, it defines a bi-univocal mapping of B onto A.

1.5 Concept of an Algebraic Structure

From a given set E it is possible to define functions from $E \times E$ onto E: in other words 2 given elements x and y belonging to E can be made to correspond to an (unique) element z, itself also belonging to E, by means of a mapping $E \times E$ onto E: $z = f(x, y)$. This particular type of

1.5 Concept of an Algebraic Structure

function is known as an *operation* in E, and is expressed as $z = x \otimes y$ (the symbol $\otimes$ replacing $f(\ )$)[1].

Thus for the set of real numbers are to be found two operations: addition and multiplication, represented by the symbols $+$ and $\times$, for the set of square matrices an addition and a multiplication can be defined, for the set of vectors in a plane there is a vectorial addition, for the components of a set there are union and intersection, etc. An operation of the type $z = f(x, y)$ is said to be *binary* (having 2 terms). This is the case for the above operations. This concept can be generalised to define ternary (3 terms) ..., n-ary operations. Further examples will be encountered at a later stage in this work.

A set equipped with a certain number of operations is generally known as an *algebra*. The operations in question can be subjected to certain restrictions which give to the associated algebra a somewhat unusual structure. This algebra often takes a specific name (e. g. the algebra of groups, of rings, of fields, of lattices etc.).

Certain distinguished elements are known as constants.

Moreover, a number of different algebraic structures can be defined for a single set.

Consider for example the algebra of real numbers ("ordinary algebra") and the operation $\times$ (multiplication).

1. Two numbers (elements) a and b can be made to correspond with a number denoted by $a \times b$ which is also a real number.

2. For every number x:

$$x \times 1 = 1 \times x = x \qquad (1.54)$$

Thus there exists an element (1) which also plays a distinguished rôle.

3. For every number x, there exists a number y such that:

$$xy = yx = 1 \qquad (1.55)$$

(in practice y is denoted as $1/x$).

As a result of these 3 properties, the algebra of real numbers, with multiplication as operation and the number 1 as constant, constitutes a particular example of a *group*, which is, moreover, *commutative* ($a \times b = b \times a$). It should be noted that if limited to whole numbers, the third property would be defective. Finally it may be said that a set of real numbers with the 2 operators $\times$ and $+$, and the constants 0 and 1, also constitutes a *field*.

---

[1] also known as an *operator*.

At a later stage there will be opportunities for dealing with these diverse algebrae, whose remarkable characteristics will be defined in a series of axioms.

The word *algebra*, for which the everyday meaning is akin to "a set of rules for calculating" will, in the work which follows, take on a different meaning, that of a set equipped with operations.

Chapter 2

Numeration Systems. Binary Numeration

2.1 Numbers and Numeration Systems

The concept of a number is one of the fundamental concepts of arithmetic and of mathematics in general. At this stage, it is neither intended to trace back the history of this notion, nor to describe the evolution which has led to the perfectly axiomatic definitions currently in use. It will merely be pointed out that, in practice, numbers can be manipulated by means of representations which are usually decimal, although occasionally different (sexagesimal, roman ...). A method of representation constitutes what is known as a *numeration system*.

2.2 Decimal System of Numeration

This is by far the most commonly used system in everyday life. Consider a number such as:

$$N_{10} = 15\,703 \tag{2.1}$$

This actually means that N can be determined by compiling the sum:

$$N = 1 \cdot 10^4 + 5 \cdot 10^3 + 7 \cdot 10^2 + 0 \cdot 10^1 + 3 \cdot 10^0 \tag{2.2}$$

In other words, the ordinary way of writing numbers is to express them in terms of numbers from 0 to 9 and of powers of 10. This notation has the following properties:

— the position of a figure indicates the power of 10 of which it is the coefficient, as in the sum (2.2);
— where a power does not appear in the sum (2.2), the figure 0 is inserted in the corresponding position, which gives an unambiguous indication of the "spaces" required.
— it is not necessary to write in the zeros which would be found on the left of the last non-zero digit.

Because of these different properties, it is customary to say that the decimal numeration system is a *positional* one.

The number 10, whose powers are used to represent a given number N is known as the *base* (or radix) of the sytem. 10 separate symbols (0, 1, 2, ..., 9) are required to express a number in decimal form.

2.3 Numeration System on an Integer Positive Base

The use of the base 10 is not related to a logical or mathematical necessity, and probably originates from practical considerations: e.g one has ten fingers. In fact, many other counting systems have been used (with bases 12, 20, 60 in particular).

This notion of positional number system can be generalized as follows: given an integer b, greater than 1 and known as the base of the numeration system to be considered, and b distinct symbols x^0, x^1, ..., x^{b-1}, a number X may be expressed in the following way:

$$X_b = x_n x_{n-1} \ldots x_k \ldots x_0 \qquad (2.3)$$

which by convention signifies that X can be determined (for example in the decimal system), by the formula:

$$X = x_n b^n + x_{n-1} b^{n-1} + \ldots + x_k b^k + \ldots + x_0 b^0 \qquad (2.4)$$

i.e. in a shortened form:

$$X = \sum_{k=0}^{k=r} x_k b^k \qquad (2.5)$$

Each symbol x_n, ..., x_k, ..., x_0 is one of the b symbols used to represent the numbers from 0 to $b-1$. Given the representation $x_n x_{n-1} \ldots x_0$, the summation described in (2.5) uniquely defines X. Conversely, for a given integer X, it can be shown that the representation $x_n x_{n-1} \ldots x_0$ is, in fact, unique.

Notation:

1) To indicate the base to which a number is represented, it can be said that:

$$X_b = x_n x_{n-1} \ldots x_0 \qquad (2.6)$$

X_b then represents the sequence of symbols x_n, ..., x_0; and the value of the number X will be given by (2.5) with the base b shown as a sub-script.

2) To represent a number in an abstract manner without indication of the numeration system in which it is represented use is made of the notation X without sub-script. Where there is no ambiguity X will be written without a sub-script.

2.3 Numeration System on an Integer Positive Base

Note 1:

The names given to the most commonly used bases are as follows:

$b = 2$ binary numeration (or straight binary)
$b = 8$ octal numeration
$b = 10$ decimal numeration
$b = 12$ duodecimal numeration
$b = 16$ hexadecimal numeration
$b = 60$ sexagesimal numeration

Note 2:

If consideration again be given to formula (2.5) it will be seen that there is a need to specify the way in which the sum is carried out. It can be done with the decimal system, but could equally well be done in any other system:

1) Consider the decimal number $X_{10} = 15703$ amd carry out the sum:

$$(X_{10}) = 1 \cdot 10^4 + 5 \cdot 10^3 + 7 \cdot 10^2 + 0 \cdot 10^1 + 3 \cdot 10^0 \qquad (2.7)$$

the number obtained can be expressed as

$$X_{10} = 15703$$

2) If, however, the sum is carried out in another system the result will be expressed as X'. For example, use can be made of a positional numeration system to a different base b'.

In this case:

$$X' = X_{b'} = (x_n)_{b'} (b^n)_{b'} + (x_{n-1})_{b'} (b^{n-1})_{b'} + \ldots + (x_k)_{b'} (b^k)_{b'} + (x_0)_{b'} (b^0)_{b'} \qquad (2.9)$$

$$= \sum_{k=0}^{k=r} (x_k)_{b'} (b^k)_{b'} \qquad (2.10)$$

Example:

Consider the number with the octal representation:

$$X_8 = 3017$$

This number can be determined (represented) in decimal as follows:

$$X_{10} = 3 \cdot 8^3 + 0 \cdot 8^2 + 1 \cdot 8^1 + 7 \cdot 8^0 \qquad (2.12)$$
$$= 3 \cdot 512 + 8 + 7 \qquad (2.13)$$
$$X_{10} = 1551$$

Thus the number expressed as $X_8 = 3017$ in octal, is expressed as $X_{10} = 1551$ in decimal numeration.

Note 3:

The index k is known as the weight or rank: the lowest weight is 0, the highest is n.

Note 4:

It should be noted that a distinction has been made between the number and its representations. It is advisable to keep this distinction in mind, particularly with respect to the problem of conversions.

The notations X, X_b, (X_b) defined above permit the writing of:

$$X = (X_{10}) = (X_2) = (X_b) = (X_{b'}) = \cdots \qquad (2.15)$$

The above definitions correspond to the case of integers. More generally, expressions which include, as in the decimal system, an integer and a fractional integer can be considered and written as:

$$X_b = x_n x_{n-1} \ldots x_0, x_{-1} x_{-2} \ldots x_{-m} \ldots \qquad (2.16)$$

this expression representing a number N which is:

$$N = x_n b^n + x_{n-1} b^{n-1} + \cdots + x_0 b^0 + x_{-1} b^{-1} + x_{-2} b^{-2} + \cdots$$
$$+ x_{-m} b^{-m} + \cdots \qquad (2.17)$$

$$N = \sum_{i=-\infty}^{i=n} x_i b^i \qquad (2.18)$$

In order to separate the whole and fractional parts use is made, as in decimal, of a point.

2.4 Binary Numeration System

2.4.1 Binary Numeration

For various reasons of technology and reliability, digital computers make use of only 2 symbols represented by 0 and 1. They often (but not always: cf. Chapter 3) make use of the binary numeration system (known as "pure binary"). With this latter system a number is represented, in accordance with the general definition of § 2.3, by a sequence of 1's and 0's.

$$N_2 = x_n x_{n-1} \ldots x_0, x_{-1} x_{-2} \ldots x_{-m} \ldots \quad (2.19)$$

This expression signifies:

$$N = \sum_{i=-\infty}^{i=n} x_i 2^i \quad (x_i = 0 \text{ or } 1 \text{ for every } i) \quad (2.20)$$

Example:

$$N_2 = 1011{,}11 \quad (2.21)$$

$$N_{10} = 2^3 + 2^1 + 2^0 + 2^{-1} + 2^{-2} \quad (2.22)$$

$$= 8 + 2 + 1 + 0.5 + 0.25 \quad (2.23)$$

$$N_{10} = 11.75 \quad (2.24)$$

The symbols x_k are known as binary digits (or "bits").

Note: 10^n numbers can be represented with n decimal digits. To represent these numbers in the binary system, m binary digits are needed such that:

$$2^m \geq 10^n$$

$$m \geq n \log_2 10 \quad (2.25)$$

$$m \geq 3.32\, n \text{ approximately}$$

2.4.2 Binary Addition and Subtraction

2.4.2.1 Addition

For two given numbers

$$\begin{cases} X = x_n x_{n-1} \ldots x_i \ldots x_1 x_0 & (2.26) \\ Y = y_n y_{n-1} \ldots y_i \ldots y_1 y_0 & (2.27) \end{cases}$$

let $S = X + Y$. Preceding rank by rank, as in the decimal system, and commencing from rank 0 determine the sum S; at each rank it is necessary to determine:

— the sum modulo 2 of x_i, y_i and of the 'carry' obtained from the preceding rank, i. e. the digit s_i of the sum S.

— the 'carry' r_{i+1} which is to be added to the rank $i + 1$ at the next step.

These two operations are summarized in Table 2.1.

Table 2.1

$x_i y_i r_i$	$x_i + y_i + r_i$	s_i	r_{i+1}
0 0 0	0	0	0
0 0 1	1	1	0
0 1 0	1	1	0
0 1 1	10	0	1
1 0 0	1	1	0
1 0 1	10	0	1
1 1 0	10	0	1
1 1 1	11	1	1

$$\text{Example:} \begin{cases} R: 0\ 0\ 1\ 1\ 1\ 1\ 1\ 0 \\ X: 1\ 0\ 1\ 1\ 1\ 0\ 1\ 1\ 0 \\ Y: 0\ 0\ 0\ 0\ 0\ 1\ 0\ 1\ 1 \\ \overline{} \\ S: 1\ 1\ 0\ 0\ 0\ 0\ 0\ 0\ 1 \end{cases} \qquad (2.28)$$

2.4.2.2 Subtraction

Similarly, for each rank Table 2.2 gives the sum s_i and the 'carry' ("borrow") r_{i+1} for the subtraction $S = X - Y$.

Table 2.2

$x_i y_i r_i$	$x_i - (y_i + r_i)$	s_i	r_{i+1}
0 0 0	0	0	0
0 0 1	$-1 = -2+1$	1	1
0 1 0	$-1 = -2+1$	1	1
0 1 1	-2	0	1
1 0 0	1	1	0
1 0 1	0	0	0
1 1 0	0	0	0
1 1 1	$-1 = -2+1$	1	1

$$\text{Example:} \begin{cases} R: 0\ 1\ 0\ 1\ 1\ 0\ 0\ 0 \\ X: 1\ 0\ 1\ 1\ 0\ 1\ 1\ 1 \\ Y: 0\ 1\ 0\ 1\ 1\ 1\ 1\ 0 \\ \overline{} \\ S: 0\ 1\ 0\ 1\ 1\ 0\ 0\ 1 \end{cases}$$

2.4 Binary Numeration System

In practice, the complements method is often used (cf. Chapters 3 and 9), which reduces subtractions to additions.

Note:

It is interesting to try to express s_i and r_{i+1} algebraically in terms of x_i, y_i and r_{i+1}.:
Addition.

$$s_i = (1 - x_i)(1 - y_i) r_i + (1 - x_i) y_i (1 - r_i) + x_i (1 - y_i)$$
$$(1 - r_i) + x_i y_i r \qquad (2.30)$$

Indeed, it will be seen that:

1. Each of the variables x_i, y_i, r_i takes only the numerical values 1 or 0.
2. In these conditions each of the terms takes the value 1 for exactly 1 single combination of variables (thus $(1 - x_i)(1 - y_i) r_i$ equals 1 in just one case: $x_i = 0$, $y_i = 0$, $r_i = 1$)).
3. No two of these terms can take the value 1 at any one time.

The above expression can then be reduced to obtain:

$$s_i = x_i + y_i + r_i - 2(x_i y_i + y_i r_i + r_i x_i) + 4 x_i y_i r_i. \qquad (2.31)$$

It will also be seen that, s_i being equal to 1 or 0, is equal to its residue (modulo 2)

$$s_i = |s_i|_2 \qquad (2.32)$$

whence

$$s_i = |x_i + y_i + r_i|_2 \qquad (2.33)$$

Similarly, the following expression would be obtained for r_{i+1}:

$$r_{i+1} = (1 - x_i) y_i r_i + x_i (1 - y_i) r_i + x_i y_i (1 - r_i) + x_i y_i r_i \qquad (2.34)$$

i.e.:

$$r_{i+1} = x_i y_i + y_i r_i + r_i x_i - 2 x_i y_i r_i \qquad (2.35)$$

equally:

$$r_{i+1} = |r_{i+1}|_2 \qquad (2.36)$$

whence:

$$r_{i+1} = |x_i y_i + y_i r_i + r_i x_i|_2 \qquad (2.37)$$

Finally it will be seen that:

$$s_i + 2 r_{i+1} = x_i + y_i + r_i \qquad (2.38)$$

Subtraction.

The expression for s_i is the same as in the case of addition. The borrow is given by:

$$r_{i+1} = (1 - x_i) y_i r_i + (1 - x_i) y_i (1 - r_i)$$
$$+ (1 - x_i)(1 - y_i) r_i + x_i y_i r_i \qquad (2.39)$$
$$r_{i+1} = (1 - x_i) r_i + (1 - x_i) y_i - y_i r_i + 2 x_i y_i r_i \qquad (2.40)$$

or again:
$$r_{i+1} = (1 - x_i + (1 - x_i) y_i + y_i r_i + 2 (1 - x_i) y_i r_i \qquad (2.41)$$

A check will then show:

$$s_i = x_i + y_i + r_i + 2 [r_{i+1} - (r_i + y_i)] \qquad (2.42)$$
$$s_i - 2 r_{i+1} = x_i - (y_i + r_i). \qquad (2.43)$$

Other expressions closely akin to these will be encountered in Chapter 9 when Boolean equations for binary addition and subtraction will be established.

2.4.3 Binary Multiplication and Division

2.4.3.1 Multiplication

The basic method is similar to that used in decimal: dependent on whether the figure of the multiplier equals 1 or 0, the multiplicand is or is not added to the partial sum, by displacing it one rank to the left each time.

Example:

```
        1 0 1 1 1
      × 1 1 0 1
      ─────────
        1 0 1 1 1
      1 0 1 1 1
    1 0 1 1 1
    ─────────────
    1 0 0 1 0 1 0 1 1
```

2.4.3.2 Division

Here again, the basic method consists of an attempt, as in the decimal system, to subtract the divider, here multiplied by 1 (the only factor possible).

Example:

```
1 0 0 1 0 1 1 0 0    | 1 1 0 1
  1 1 0 1            |----------
  0 0 1 0 1 1 1        1 0 1 1 1
    1 1 0 1
    0 1 0 1 0 0
      1 1 0 1
      0 0 1 1 1 0
          1 1 0 1
          0 0 0 1
```

$\begin{cases} \text{quotient:} & 1\ 0\ 1\ 1\ 1 \\ \text{remainder:} & 0\ 0\ 0\ 1 \end{cases}$

For the four operations, the methods explained above are only basic procedures, from which a great many refinements have been derived, with the aim of increasing the speed of execution when they are implemented in a computer or a digital system.

2.4.4 Systems with Base 2^p ($p \geq 1$)

For $p = 1$, use is made of the pure binary system. For $p > 1$ there are systems (frequently: octal, hexadecimal) which are often used as abridged notations for the pure binary system, either with a view to being read by a human operator, or for certain other calculations. The conversions from base 2-to base 2^p and vice versa are effectively reduced to simple groupings of terms. It is clear that by the addition of a suitable number of zeros to the left of a binary number, the latter can always be written as:
$$N = X_k X_{k-1} \ldots X_i \ldots X_0 \qquad (2.46)$$
with
$$X_i = x_{(i+1)p-1} \ldots x_{(i+1)p-j} \ldots x_{ip}, \quad (i = 0, 1, \ldots, k) \qquad (2.47)$$

(i.e. by grouping the binary figures in groups of p commencing from the right). Whence:
$$N = \sum_{i=0}^{i=k} X_i 2^{ip} = \sum_{i=0}^{i=k} X_i (2^p)^i \qquad (2.48)$$

The groups X_i are, therefore, the representations in pure binary of the figures N in the system with base 2^p.

Conversely, for a given number N in a system with base 2^p, it suffices to represent each digit in binary and to juxtapose the resulting binary words to obtain the required number N in pure binary.

Example: octal code

$$\begin{cases} N_2 = 0\;0\;1 \quad 0\;1\;1 \quad 1\;0\;1 & (2.49) \\ N_8 = 1 \quad 3 \quad 5 & (2.50) \end{cases}$$

The octal code is frequently used as a compact notation for pure binary. In many cases an examination of the numbers in decimal would provide the user with nothing more. The octal code, by reason of the simplicity of the conversion procedure, saves machine-time (computers for example) and simplifies the means of conversion (visual-mechanical displays). It is convenient, too, for systems where insufficient programming capacity is available, and where the use of an autonomous (auxiliary) wired-in conversion device is necessary.

2.5 Passage from One Numeration System to Another

2.5.1 First Method

— *Whole numbers.* Consider a number N. By using two different bases b_1 and b_2 two expressions will be obtained for N to be denoted by N_{b_1} and N_{b_2}. Thus giving:

$$\begin{cases} N_{b_1} = x_n x_{n-1} \ldots x_0 \\ N_{b_2} = y_m y_{m-1} \ldots y_0 \end{cases} \quad (2.51)$$

with the relationships:

$$N = \sum_{i=0}^{i=r} x_i b_1 = \sum_{k=0}^{k=m} y_k b_2 \quad (2.52)$$

The problem of conversion is as follows: given the expression N_{b_1} in the system to base b_1, it is necessary to determine the expression N_{b_2} for the same number N in the system to the base b_2, and vice versa. For this it will be noted that y_0 is the remainder of the division of N by b_2. Effectively:

$$N = y_m b_2^m + y_{m-1} b_2^{m-1} + \cdots + y_1 b_2 + y_0 \quad (2.53)$$

which gives:

$$N = (y_m b_2^{m-1} + y_{m-1} b_2^{m-2} + \cdots + y_1) b_2 + y_0 \quad (0 \leq y_0 < b_2) \quad (2.54)$$

i.e.

$$N = N_1 b_2 + y_0 \quad (0 \leq y_0 < b_2) \quad (2.55)$$

2.5 Passage from One Numeration System to Another

Incidentally, the same thing can be said with respect to the number N_1: y_1 is the remainder of the division of N_1 by b_2, etc. Thus it can be seen that the successive digits of N_{b_2} can be obtained from consideration of the successive remainders:

$$\left.\begin{aligned} N &= N_1 b_2 + y_0 \\ N_1 &= N_2 b_2 + y_1 \\ N_2 &= N_3 b_2 + y_2 \\ &\vdots \\ N_k &= N_{k+1} b_2 + y_k \\ &\vdots \\ N_m &= N_{m+1} b_2 + y_m \end{aligned}\right\} \qquad (2.56)$$

$y_0, y_1, \ldots, y_m$ represent the remainders for the successive divisions.

Example:

In order to convert $N_{10} = 545$ to the system base 4, the successive dividends may be set out on the left and the successive remainders on the right of a vertical column.

Thus are obtained:

$$\text{Successive dividends} \left\{ \begin{array}{c|c} 545 & 4 \\ \hline 136 & 1 \\ 34 & 0 \\ 8 & 2 \\ 2 & 0 \\ 0 & 2 \end{array} \right\} \text{Successive remainders} \qquad (2.57)$$

Hence:

$$N_4 = 20\,201 \qquad (2.58)$$

It can further be verified that:

$$N = 1 \times 4^0 + 2 \times 16 + 2 \times 256 = 545 \qquad (2.59)$$

Note:

The method for the conversion between two bases b_1 and b_2 which has been expounded is, theoretically, equally applicable in both directions ($b_1 \to b_2$ and $b_2 \to b_1$).

In practice, it requires operations of division to be carried out in the system to the base b_1 (if it is required to pass from b_1 to b_2).

In 'pencil and paper' conversions this method will only be used for the conversion of a decimal number to a representation in a system to another base (as in the above example). However, when the operation of conversion must be mechanized by an automatic system, the above method, for various reasons (e.g. circuit complexity), may prove to be more advantageous in one direction than the other. An alternative method will be shown at a later stage.

— *Fractional numbers less than* 1.

Similarly, for a number N represented by N_{b_1} in the system to the base b_1 it is required to determine the representation of N in a system to the base b_2. For this it will be noted that the expression:

$$N_{b_2} = 0 . \; y_{-1} y_{-2} y_{-m} \ldots \tag{2.60}$$

signifies:

$$N = y_{-1} b_2^{-1} + y_{-2} b_2^{-2} + y_{-3} b_2^{-3} + \cdots + y_{-m} b_2^{-m} + \cdots \tag{2.61}$$

If N be multiplied by b_2 this will give:

$$N' = (N \cdot b_2)_{b_2} = y_{-1} \cdot y_{-2} y_{-3} \ldots \tag{2.62}$$

In other words the figure which in N' appears to the left of the point ("the overspill") is the first figure sought, i.e. y_{-1}.

Consideration can now be given to the number obtained by retaining only what appears to the right of the point and applying to it the same operation: it will be multiplied by b_2 and a check made on the overspill to the left which is y_{-2} etc.

Moreover this process may never end, as will be shown with examples. The representation of a number having a fractional part may well terminate in one numeration system and not in another.

Note:

As for the case of the conversion of whole numbers, it should be noted that the method given here for fractional numbers leads to the execution of the multiplications in the initial numeration system to the base b_1. Consequently, in practice, the method may not be equally convenient in both directions. In the case of 'pencil and paper' conversions, only the "decimal → base b" method will be used.

— *Numbers having a whole and a fractional part.*

These conversions will proceed in 2 stages: the method of (2.56) will be applied to the whole part, then the method of (2.62) will be applied to the fractional part. The number sought is that obtained from the juxtaposition of the newly acquired whole and fractional parts.

2.5.2 Second Method

This is the most natural method. Given a number represented as:

$$N_{b_1} = x_n x_{n-1} \ldots x_0, x_{-1} x_{-2} \ldots x_{-k} \ldots \qquad (2.63)$$

the figures $x_n, x_{n-1}, \ldots, x_0, x_{-1} \ldots$ are expressed in the numeration to the base b_2, as are the powers of b_1: $b_1{}^n, b_1{}^{n-1}, \ldots, b_1, b_1{}^{-1}, b_1{}^{-2}, \ldots$ etc., and the summation carried out to this same base b_2.

$$N_{b_2} = \sum_{-\infty}^{n} [x_i]_{b2} \cdot (b_1{}^i)_{b2} \qquad (2.64)$$

Note 1:

Here, in practice, the passage from b_1 to b_2 requires the summation to be effected in the system b_2. For paper conversions in particular, this method will be convenient for conversions from any other base to the base 10.

Note 2:

The two methods of conversion are complementary: method 1 leads to performing the operations in the "initial" system (base b_1), method 2 leads to the use of the terminal system (base b_2). For paper conversions between the decimal system and any other base b, use will be made of:

— method 1 (successive multiplications and/or divisions)
 in the direction $10 \to b$;
— method 2 (summation)
 in the direction $b \to 10$.

For "machine" produced conversions the problem becomes more complex for various reasons:

1. No matter what the bases utilised, the corresponding digits will be coded in binary.

2. In the case of a computer, the necessary conversions can be programmed (making use of the main arithmetic unit or a peripheral arithmetic unit) or by a specialized 'wired-in' programme network which, in its turn, is able to make use of certain elements (registers) or signals from the main arithmetic unit. In the latter case, the complexity and speed of the conversion equipment may favour one or another method or, possibly, a special method differing from the above general method. By way of an example consideration will be given to the particular case of decimal-binary and binary-decimal conversion.

2.6 Decimal-Binary and Binary-Decimal-Conversions

2.6.1 Conversion of a Decimal Number to a Binary Number

Consider the number
$$N_{10} = 17\,506,\,71\,875$$

— *Conversion of the whole (integer) part.*

This is written as:
$$N'_{10} = 17\,506$$

A search is then made for the successive remainders resulting from progressive divisions by 2. As has already been shown above, this procedure is presented in tabular form. Thus giving:

17 506	2
8 753	0
4 376	1
2 188	0
1 094	0
547	0
273	1
136	1
68	0
34	0
17	0
8	1
4	0
2	0
1	0
0	1

(2.67)

The 'whole' part of the number is thus expressed in binary as:

$$N'_2 = 1\ 0\ 0\ 0\ 1\ 0\ 0\ 0\ 1\ 1\ 0\ 0\ 0\ 1\ 0 \tag{2.68}$$

— *Conversion of the fractional part*

This is written as:
$$N''_{10} = 0.71875 \tag{2.69}$$

whence:

$$\begin{array}{r|l} 0.71875 & \\ 1.43750 & 1 \\ 0.87500 & 0 \\ 1.75000 & 1 \\ 1.50000 & 1 \\ 1.00000 & 1 \\ 0.00000 & 0 \end{array}$$

whence the fractional part:

$$N''_2 = 0.101\,1100\ldots \tag{2.70}$$

— *The complete number.*

This is expressed as:

$$N_2 = 100010001100010.10111 \tag{2.71}$$

2.6.2 Conversion of a Binary Number into a Decimal Number

In this instance the second general method (summation) is used. That is:
$$N_2 = 11011.101 \tag{2.72}$$
Here:

$$N_{10} = 1 \times 2^4 + 1 \times 2^3 + 0 \times 2^2 + 1 \times 2^1 + 1 \times 2^0 + 1 \times 2^{-1}$$
$$+ 0 \times 2^{-2} + 1 \times 2^{-3}$$
$$= 16 + 8 + 2 + 1 + 0.5 + 0.125 \tag{2.74}$$
$$N_{10} = 27.625 \tag{2.75}$$

2.6.3 Other Conversion Methods

2.6.3.1 Successive Subtraction Method (Binary-Decimal and Decimal-Binary)

— *Binary-decimal conversion.*

Given a binary N, an integer,

$$N_2 = B_n B_{n-1} \ldots B_k \ldots B_1 B_0 \tag{2.76}$$
$$N_{10} = D_m D_{m-1} \ldots D_l \ldots D_1 D_0$$

a search is made for the greatest number of the form $D_m 10^m$ which is less than or equal to N. This being determined in binary form $N^{(m)}$ then $N^{(m)} = N - D_m 10^m$. If $N^{(m)} = 0$ the conversion is terminated; if $N^{(m)} \neq 0$, the search is resumed for the largest term of form $D_{m-1} 10^{m-1}$ which is less than or equal to $N^{(m)}$ etc., until such time as $N^{(k)} = 0$ which, ultimately, will be found. ($k \geq 0$; $D_l = 0$ for $0 \leq l \leq k-1$ if $k > 0$). This method is applicable to whole, fractional or mixed numbers. It requires:

— a table of terms of the form $D_k 10^k$ in the field of numbers to be treated (or the means of recalculating them in every case);
— the comparison and subtraction operations to be made in binary.

Example:

$$N_2 = 1\,0\,0\,1\,0\,1\,1\,1 \quad (N_{10} = 151)$$

Giving:

$$0 \leq N_2 \leq 2^8 - 1, \quad D_m \leq 2, \quad m \leq 2 \qquad (2.77)$$

$$N_2^{(2)} = (N - 1.100)_2 = \quad 1\,0\,0\,1\,0\,1\,1\,1$$

$$- \quad 1\,1\,0\,0\,1\,0\,0 \qquad (2.78)$$

$$= \quad 0\,0\,1\,1\,0\,0\,1\,1\, (<1\,1\,0\,0\,1\,0\,0)$$

$$D_2 = 1$$

This then gives:

$$(N_2^{(1)} = N^{(2)} - 50)_2 = \quad 1\,1\,0\,0\,1\,1$$

$$- \quad 1\,1\,0\,0\,1\,0 \qquad (2.79)$$

$$\overline{0\,0\,0\,0\,0\,1}$$

$$D_1 = 5 \qquad (2.80)$$

and finally

$$D_0 = 1 \qquad (2.81)$$

whence:

$$D_2 D_1 D_0 = 151 \qquad (2.82)$$

Decimal-binary conversion

Given a decimal number

$$N_{10} = D_m D_{m-1} \ldots D_i \ldots D_1 D_0$$

a search is made for the greatest power of 2, i. e. 2^k, which is less than or equal to D. Then $D^{(k)} = D - 2^k$; etc., are formed following a procedure similar to that described above for binary-decimal conversions.

Example:

$$D = 151 \qquad (2.83)$$

2^7 is the greatest power of 2 contained in 151

$$D^{(7)} = 151 - 2^7 = 151 - 128 = 23 \qquad (2.84)$$

Thus giving:

$$23 - 2^4 = 23 - 16 = 7$$

$$7 - 2^2 = 3 \qquad (2.85)$$

$$3 - 2^1 = 1, \quad 1 - 2^0 = 0$$

whence:

$$N_2 = 1\,0\,0\,1\,0\,1\,1\,1 \qquad (2.86)$$

It is necessary to:
— have a table of decimal representations of the powers of 2 (or the means of calculating them in each case);
— make comparisons and subtractions in the decimal system.

2.6.3.2 *Mixed Method (Binary-Octal-Decimal)*

This method, quoted by J. E. Croy [40], proceeds via the octal code and effects the operations in decimal code. The method is based on the following considerations. Let:

$$N_2 = B_n B_{n-1} \ldots B_1 \ldots B_0$$

be the representation of a number N in pure binary.

If N_8 is the representation of this same number N in the octal system:

$$N_8 = A_m A_{m-1} \ldots A_k \ldots A_1 A_0 \qquad (2.87)$$

then:

$$N = \sum_{i=0}^{i=n} B_i 2^i = \sum_{k=1}^{k=m} A_k 8^k \qquad (2.88)$$

Now consider the number N' whose decimal representation N'_{10} is the word:

$$N'_{10} = A_m A_{m-1} \ldots A_k \ldots A_1 A_0 \qquad (2.89)$$

which gives:

$$N' = \sum_{k=0}^{k=m} A_k 10^k \qquad (2.90)$$

2. Numeration Systems. Binary Numeration

Proceeding in a similar manner are formed:

$$N^{(1)} = N' - \frac{2}{10} A_m 10^m \qquad (2.91)$$

$$= A_m 10 \cdot 10^{m-1} - A_m 2 \cdot 10^{m-1} + \sum_{k=0}^{k=m-1} A_k 10^k \qquad (2.92)$$

$$= 8 A_m 10^{m-1} + \sum_{k=0}^{k=m-1} A_k 10^k \qquad (2.93)$$

$$= (8 A_m + A_{m-1}) 10^{m-1} + \sum_{k=0}^{k=m-2} A_k 10^k \qquad (2.94)$$

Then

$$N^{(2)} = N^{(1)} - \frac{2}{10} \left(N^{(1)} - \sum_{k=0}^{k=m-2} A_k 10^k \right) \qquad (2.95)$$

$$= (8^2 A_m + 8 A_{m-1} + A_{m-2}) 10^{m-2} + \sum_{k=0}^{k=m-3} A_k 10^k \qquad (2.96)$$

Next

$$N^{(3)} = N^{(2)} - \frac{2}{10} \left(N^{(2)} - \sum_{k=0}^{k=m-3} A_k 10^k \right) \qquad (2.97)$$

$$= (8^3 A_m + 8^2 A_{m-1} + 8 A_{m-2} + A_{m-3}) 10^{m-3} + \sum_{k=0}^{k=m-4} A_k 10^k \qquad (2.98)$$

and more generally for $i < m$:

$$N^{(i)} = N^{(i-1)} - \frac{2}{10} \left(N^{(i-1)} - \sum_{k=0}^{k=m-1} A_k 10^k \right) \qquad (2.99)$$

$$N^{(i)} = (8^i A_m + 8^{i-1} A_{m-1} + 8^{i-2} A_{m-2} + \cdots + 8 A_{m-i+1} + A_{m-i})$$
$$+ \sum_{k=0}^{k=m-i-1} A_k 10^k \qquad (2.100)$$

When $i = m$; then $N^{(m)}$ is the *decimal* representation of the number

$$\sum_{k=1}^{k=m} 8^k A_k,$$

i.e. the number N.

The above considerations lead to the following rule:

Rule:

1. Convert the binary number into an octal number.
2. Take the first *octal* figure commencing from the left. Multiply it by 2 in *decimal*, displace the result one rank to the right and subtract it from the initial number.

2.6 Decimal-Binary and Binary-Decimal Conversions

3. Take the first two figures of the result obtained and multiply them by 2, displace them one rank to the right, subtract, then take the first three figures of the new result, etc. (at the k-th operation take the first k digits of the resulting number).

The last subtraction is made when the least significant digit of the number to be subtracted is below the digit with the weakest weight of the present number. The result of this last subtraction is the number required.

Example:

$$N_2 = 1 \; \underbrace{011} \; \underbrace{100} \; \underbrace{110} \; \underbrace{111} \quad (N_{10} = 5943)$$

$$\rightarrow N_8 = 1 \quad \; 3 \quad \; 4 \quad \; 6 \quad \; 7$$

$$N_8 = 1\,3\,4\,6\,7$$

$$2$$

$$\overline{1\,1\,4\,6\,7}$$

$$2\,2$$

$$\overline{0\,9\,2\,6\,7}$$

$$1\,8\,4$$

$$\overline{0\,7\,4\,2\,7}$$

$$0\,1\,4\,8\,4 \quad \text{(end)}$$

$$\overline{0\,5\,9\,4\,3} \tag{2.101}$$

2.6.3.3 Mixed Method (Binary-Hexadecimal)

By way of an exercise, study the following method, proposed by O. *Goering* [66], for the binary-decimal conversion of whole numbers. With this process it is first necessary to convert to the hexadecimal system, in which all subsequent operations are carried out.
This gives:

$$N_{16} = A_{l-1} A_{l-2} \dots A_i \dots A_0 \quad \text{(base 16)} \tag{2.102}$$

i.e.
$$N = \sum_{i=0}^{i=l-1} A_i (16)^i \qquad (2.103)$$

The method consists of a search for the successive remainders of the divisions by 10, as indicated for the general method, but in effecting these divisions in a special way. All these operations are effected in hexadecimal (or binary coded hexadecimal).

Determination of the remainder.

Having:

$$16^i = (6+10)^i \equiv (6)^i \ (\text{mod. } 10) \equiv 6 \ (\text{mod. } 10) \quad (i \geq 1). \qquad (2.104)$$

For the determination of the value of N (modulo 10):

$$N \equiv A_0 + \sum_{i=1}^{i=l-1} A_i (16)^i \qquad (\text{mod. } 10) \qquad (2.105)$$

$$N \equiv A_0 + 6 \sum_{i=1}^{i=l-1} A_i \qquad (\text{mod. } 10) \qquad (2.106)$$

Since
$$A_0 \equiv 11 A_0 \equiv 5 A_0 + 6 A_0 \qquad (\text{mod. } 10) \qquad (2.107)$$

then:
$$N \equiv 5 A_0 + 6 \sum_{i=0}^{i=l-1} A_i \qquad (\text{mod. } 10) \qquad (2.108)$$

Further, check that:

$$\text{if } A \text{ is even} \quad \begin{cases} 6A \equiv A & (\text{mod. } 10) \\ 5A \equiv 0 & (\text{mod. } 10) \end{cases}$$
$$\text{if } A \text{ is odd} \quad \begin{cases} 6A \equiv A+5 & (\text{mod. } 10) \\ 5A \equiv 5 & (\text{mod. } 10) \end{cases} \qquad (2.109)$$

which can be expressed as:

$$N = 5P + \sum_{i=0}^{i=l-1} A_i \qquad (\text{mod. } 10) \qquad (2.110)$$

$$\left(P = 0 \text{ if } \sum_{i=0}^{i=l-1} A_i \text{ and } A_0 \text{ have the same parity, otherwise } P=1\right)$$

2.6 Decimal-Binary and Binary-Decimal Conversions

The smallest integer ≥ 0 equal to N (mod. 10) is precisely the remainder of the division of N by 10. This provides a means for the determination of the remainder at each stage of the conversion.

Determination of the quotient

Since R is the remainder (obtained by the above method for example) it is necessary to determine $Q = (N - R)/10$. ($N - R$ is thus a multiple of 10). For this, use can be made of the following relationship:

$$\frac{N-R}{10} \equiv -(N-R)\left(\frac{3}{2}\right)(16^k - 1)\left(\frac{1}{15}\right) \quad (\text{mod. } 16^k) \quad (2.111)$$

which is a way of writing:

$$\frac{N-R}{10} \cdot 16^k \equiv 0 \quad (\text{mod. } 16^k) \quad (2.112)$$

Additionally:

$$\frac{16^k - 1}{15} = \frac{16^k - 1}{16 - 1} = \sum_{i=0}^{i=k-1}(16)^i = 111\ldots 11 \text{ (in hexadecimal)} \quad (2.113)$$

Then, for a given hexadecimal number

$$X = X_{k-1}X_{k-2}\ldots X_0 \quad (2.114)$$

the number

$$Y = X\left(\frac{16^k - 1}{15}\right) = X \cdot (111\ldots 11) \quad (2.115)$$

can be determined in various ways. By way of an exercise, the reader will be able to effect this operation by the summation of partial products (as in decimal), to see if, taking account of the form of the multiplier, he can deduce a simplified procedure.

Moreover given:

$$\frac{3}{2}(N-R) = \frac{3 \cdot 10}{2}Q = 15Q \quad (2.116)$$

we then have:

$$\frac{3}{2}(N-R)\left(\frac{16^k - 1}{15}\right) = Q(16^k - 1) \quad (2.117)$$

$$= 16^k Q - Q \quad (2.118)$$

But:

$$16^k Q - Q \equiv -Q \; (\text{mod. } 16^k) \equiv 16^k - Q \; (\text{mod. } 16^k)$$

The last expression, the true complement (mod. 16^k) of Q, can be obtained, digit by digit, commencing from the right — that on the far right being the complement to $16(16-x)$, the others being the complements to $15(15-x)$ of the digits of Q.

Whence a method for the determination of the quotient Q from $N-R$ onwards:

1. Multiply $N-R$ by $3/2$ to obtain $X = \dfrac{3}{2}(N-R)$.
2. Multiply X by $(16^k-1)/15 = (11 \ldots 1 \ldots 1)$.
3. Take the complement mod. 16^k.

Example. (Determination of a remainder and of a quotient.)

$$N_2 = \underbrace{0\,0\,0\,1}\ \underbrace{0\,1\,1\,1}\ \underbrace{0\,0\,1\,1}\ \underbrace{0\,1\,1\,1}\quad (N_{10} = 5943) \tag{2.119}$$

$N_{16} = 1.7.3.7.$ (the points . separate the hexadecimal figures).

1) *Remainder R.*

$$A_0 = 7,\ \sum_{i=0}^{i=k-1} A_i = 7+3+7+1 \equiv 8 \pmod{10} \tag{2.120}$$

$$P = 1.$$

Whence:
$$N \equiv 5 + 8 \equiv 3 \pmod{10} \tag{2.121}$$

i.e.:
$$R = 3. \tag{2.122}$$

2) *Quotient Q.*

Given that: $N - R = N - 3 = 1734$

$$1.7.3.4. \times 3 = 4.5.9.12 \tag{2.123}$$

$$\dfrac{4.5.9.12}{2} = 2.2.12.14$$

$(2.2.12.14) \cdot (1\,1\,1\,1) = 2.5.01.15.13.10.14 = 15.13.10.14 \pmod{16^4}$

Hence:
$$Q_{16} = 0\,2\,5\,2 \quad (Q_{10} = 5\,9\,4).$$

Note: The above method consists of a search for the remainder which is effected by way of the hexadecimal numeration system. Other methods are possible (cf. in particular: Chap 3, residue of the powers of integers). The search for the quotient is effected by a specialized method of division, taking due account of the fixed divisor (10) which, subsequently, can be more advantageous than the general methods (successive subtractions) from the point of view of conversion time and machine complexity. Finally it should be noted that the above two search operations for remainder and quotient have been expounded for the case of 'paper' conversions, by coding the hexadecimal figures greater than 9 in decimal. In the case of 'machine' conversions a binary coding is used for the 16 figures.

General note on conversion methods

The above conversion methods (the binary-decimal case, in particular) lend themselves to a number of variants and refinements. In the corresponding systems, from a technical viewpoint, the two most important parameters are the speed of conversion and machine complexity. Other factors such as the ease of multiplexing of several conversions etc., are also to be taken into consideration. The numerous systems in current use for the resolution of this problem (which mathematically speaking is relatively trivial) correspond to different speed-complexity 'trade-offs'.

Finally, it should be noted that for switching from system 1 to system 2, nothing a priori prevents the use of a third system differing from the first two if the conversions $1 \to 3$ and $3 \to 2$ can conveniently be effected. This third system could also be a weighted representation: cf. § 2.6.3.2—3. It is, however, not necessarily limited in this way: it can equally provide an analog representation. In practice this is the case for binary-decimal conversion systems operating in 2 stages: binary-analog and analog-binary conversions.

Chapter 3

Codes

3.1 Codes

Mention has already been made in Chapter 2 that, for considerations of reliability, electronic computers and digital systems in general make use of a *binary* internal representation of numbers and symbols.

For every number or symbol (e.g. letters, operational signs) there is a *binary word*, i.e. a sequence of 0's and 1's which serves to define a *binary code*. Generally speaking, when a 'one-to-one' (bi-univocal) correspondence has been established between the symbols of a set E_1 (the "set of messages to be coded") and those of a set E_2 (the "set of code-words") then, by definition, the elements of E_1 are *coded* with the assistance of those of E_2. In practice, a code is a correspondence (set of translation rules) between the sets E_1 and E_2, or the set of pairs: messages/code-words. This chapter will be devoted to the codes (essentially binary) used for the representation of numbers.

3.2 The Coding of Numbers

3.2.1 Pure Binary Coding

The binary number system, described in Chapter 2, constitutes one of the ways of representing a number (assumed to be written in decimal in the text which follows) in binary form. In a certain sense, this is merely a transposition, of what is usually done in decimal, to the system of base 2. However, as has been seen, binary-decimal conversions necessitate a relatively complex procedure and, consequently, to avoid or to simplify as much as possible the operation of the conversion of external data to internal representations, it is necessary in certain cases to have recourse to another method of binary coding — that provided by the "binary coded decimal" (BCD) type code.

3.2.2 Binary Coded Decimal Codes (BCD Codes)

3.2.2.1 Principle

For a given decimal number each decimal digit is encoded separately and the desired binary representation for the number is obtained by juxtaposition of the binary codes of the successive decimal digits.

3.2 The Coding of Numbers

Table 3.1 gives an example of the coding of decimal digits (in this instance the representation is given in pure binary, the minimum number (4) of binary symbols being employed).

Table 3.1. *Coding of decimal digits (pure binary)*

0	0 0 0 0
1	0 0 0 1
2	0 0 1 0
3	0 0 1 1
4	0 1 0 0
5	0 1 0 1
6	0 1 1 0
7	0 1 1 1
8	1 0 0 0
9	1 0 0 1

By the use of this coding, the (decimal) number 1358 will be represneted by:

$$\underbrace{0\,0\,0\,1}_{1}\ \underbrace{0\,0\,1\,1}_{3}\ \underbrace{0\,1\,0\,1}_{5}\ \underbrace{1\,0\,0\,0}_{8} \tag{3.1}$$

3.2.2.2 The Number of 4-Position BCD Codes

A minimum of 4 binary positions are required in order that the 10 figures 0 to 9 may be encoded in binary. This enables a total of $2^4 = 16$ combinations to be formed. If this minimum of 4 positions be used it will be seen that a particular coding can be defined if 10 of these 16 combinations are selected and assigned the figures 0, 1, ... 9 in a certain order. It will thus be seen that the number of codings for 10 decimal digits with 4 binary positions is simply the number (regardless of order) of the combinations of 16 items 10 × 10, i.e.

$$A_{16}^{10} = \frac{16!}{6!} \approx 2{,}9 \times 10^{10} \tag{3.2}$$

In fact not all of the codes formed in this way are really distinct in practice. Two codes which differ only by permutation of the variables, (which is equivalent to the crossing of two wires) are not different. Furthermore, in the case (frequent) where all of the variables are available in the two forms, complemented and non-complemented (e.g. the outputs from electronic flip-flops), two codes which differ only by the complementation of one or more of the 4 positions, can then be considered to be equivalent. However, notwithstanding this reduction, there remain several million truly distinct codings, as the reader can verify by way of an exercise.

3.2.2.3 4-Bit Weighted BCD Codes

The coding used in (Table 3.1) above was weighted: it was, in fact, simply the pure binary code. If

$$X = x_3 x_2 x_1 x_0 \qquad (3.3)$$

is the decimal figure, with x_0, x_1, x_2, x_3 as the binary symbols of the corresponding code-word, it follows that:

$$X = x_3 2^3 + x_2 2^2 + x_1 2^1 + x_0 2^0 = \sum_{i=0}^{i=3} x_i 2^i \qquad (3.4)$$

$$X = 8x_3 + 4x_2 + 2x_1 + x_0 \qquad (3.5)$$

The integers 8, 4, 2, 1 constitute the weights (coefficients affected to x_3, x_2, x_1, x_0, when X is "reconstituted" by use of the formula (3.5)). The code of Table 3.1, known as the (8421) code, is only one example, among others, of a 4-bit weighted BCD code. In such a code, each decimal digit from 0 to 9 is represented by an expression of the form:

$$\sum_{i=0}^{i=3} x_i{}^j P_i \quad (x_i = 0 \text{ or } 1; \quad i = 0, 1, 2, 3; \quad j = 0, 1, \ldots, 9) \qquad (3.6)$$

where the coefficients P_i are known as the weights typical of the code in question, and it is customary to designate the code by the ordered sequence of the P_i: thus reference is made to the 8421 code, the 2421 code etc. Five examples of positive weighted codes are to be found in Table 3.2.

Table 3.2.

	8421	7421	2421	4311	5121
0	0000	0000	0000	0000	0000
1	0001	0001	0001	0001	0001
2	0010	0010	0010	0011	0010
3	0011	0011	0011	0100	0011
4	0100	0100	0100	1000	0111
5	0101	0101	1011	0111	1000
6	0110	0110	1100	1011	1001
7	0111	1000	1101	1100	1010
8	1000	1001	1110	1110	1011
9	1001	1010	1111	1111	1111

For the given weights $P_i{}^j$, the 10×4 matrix $[x_i{}^j]$ of the 10 code-words

$$(x_3{}^j x_2{}^j x_1{}^j x_0{}^j) \quad (j = 0, 1, \ldots, 9, \quad x_i{}^j = 0 \text{ or } 1)$$

is usually not unique. Thus, for positive weights, there exist 17 weighted systems (*codings*) which correspond with the 225 actual *codes*, or matrices $[x_i{}^j]$. It can be shown (Beyer)[1] that for the 4-bit codes, the weights P_i are of necessity real and integers. 88 codings of this type exist, 17 of which have all-positive weights (*Richards*, *Weeg*) and 23 which correspond to a unique code.

It should be noted that in the weighted code (4 or more bit-positions), each binary coded decimal group can easily be converted to a 10-level analog signal: by summation of currents proportional to the weights is obtained a current proportional to the decimal figure in question. This is a useful property.

3.2.2.4 Non-Weighted 4-Bit BCD Codes

Not all codes are weighted. Among the 4-bit non-weighted codes, the one most widely used to date is the "excess 3" code. In Table 3.3 the representation of the decimal figure $x(0 \leq x \leq 9)$ corresponds with the pure binary representation of $x + 3$.

Table 3.3. *The Excess 3 code*

0	0	0	1	1
1	0	1	0	0
2	0	1	0	1
3	0	1	1	0
4	0	1	1	1
5	1	0	0	0
6	1	0	0	1
7	1	0	1	0
8	1	0	1	1
9	1	1	0	0

This code may be more difficult than the 8421 code for a human operator to read and lends itself less readily to a numerical digit to analog conversion of decimal figures. Conversely, from the point of view of the realisation of digital circuits, it possesses diverse remarkable properties of symmetry, in particular that of self-complementation, i.e. for every decimal digit $x = (x_3 x_2 x_1 x_0)$:

$$C_9(x) = 9 - x = [(1 - x_3), (1 - x_2), (1 - x_1), (1 - x_0)]$$

---

[1] Beyer, W. A.; Uniqueness of Weighted Codes Representations, II, III, IEEE Transactions on Electronic Computers, Vol. EC-12, N° 2, p. 137, April 1963; Vol EC-13, N° 2, pp 153—154, April 1964.

where $C_9(x)$ is the complement to 9 of x: in order to obtain the complement to 9 of a decimal digit, it suffices to "complement" in binary, digit by digit, the representation $x_3 x_2 x_1 x_0$ of x. (In the case of the 8421 code, for example, the binary digits $9 - x$ do not depend so simply on those of x). This property is useful for the subtraction operation. Moreover, the fact that the figures from 3 to 13 are situated symmetrically in relation to the transition 4-5, presents certain advantages for the addition operation (elaboration of the decimal 'carry').

3.2.2.5 BCD Codes Having More than 4 Positions

— *The "2 out of 5" Code*

In this *BCD* code, 5 binary positions are used to represent the 10 figures from 0 to 9. To this end are chosen $C_5^2 = 10$ binary combinations for which 2 of the 5 figures have the value 1. Table 3.4 gives an example of a code of this type corresponding to a given permutation of the 10 binary words in question. If the representation of the zero be disregarded, it will be seen that the "2 out of 5" code can also be considered as a weighted code with the 5 weights 74210. The presence of the weight 0 signifies that the 4 columns to the left are those of the 7421 weighted *BCD* code of Table 3.2 (except for the decimal 0), completed by the addition of a *5th* parity number (cf. § 9.4.5) of value 1 or 0 and, by definition, such that the total number of 1's in the 5 positions is even.

Table 3.4. *A "2 out of 5" code*

0	1	1	0	0	0
1	0	0	0	1	1
2	0	0	1	0	1
3	0	0	1	1	0
4	0	1	0	0	1
5	0	1	0	1	0
6	0	1	1	0	0
7	1	0	0	0	1
8	1	0	0	1	0
9	1	0	1	0	0

This code (including 0) is not weighted. It will be noted that any error which changes the weight (2) of the code-words causes a forbidden word to appear: a device capable of detecting the $2^5 - 10 = 22$ forbidden combinations can be designed for the checking of the arithmetic operations.

— *The bi-quinary code.*

This code is given in Table 3.5:

Table 3.5. *Biquinary code*

0	0 1 0 0 0 0 1
1	0 1 0 0 0 1 0
2	0 1 0 0 1 0 0
3	0 1 0 1 0 0 0
4	0 1 1 0 0 0 0
5	1 0 0 0 0 0 1
6	1 0 0 0 0 1 0
7	1 0 0 0 1 0 0
8	1 0 0 1 0 0 0
9	1 0 1 0 0 0 0

It will be seen that 7 digits are used; they can be divided into one group of 2 and another of 5 digits. The decimal figures to be encoded are divided into 2 groups of 5 figures: figures < 5 and figures ≥ 5. Five different code-words are required in order to code the 5 elements in each group. They are chosen as indicated in the table: each of the 5 values to be encoded is represented by a 5-digit word possessing a single digit equal to 1. The rank of this 1 in the codeword is equal to the rank of the decimal figure in ascending order within each group. Although a single digit would suffice to distinguish between the 2 groups, 2 digits are used such that the number of 1's in each codeword is constant and equal to 2. Looked at in this way, the code of Table 3.5 provides an example of the "2 in 7" code. It can also be considered as being a 5043210 weighted code.

Note. — The concept of a weighted code is easily generalised for the case of any number l of binary figures, as well as for certain qualities described above: thus, in order that a code be self-complementing, it is necessary (but not sufficient) that the sum of the weights be equal to 9.

Similarly, the concept represented by the 'excess 3' code can be extended. This is the way [Lippel][1] in which the 'excess E' codes are defined. In the case of codes having integer weights, it can be shown that with every weighted code having negative weights, can be associated an "excess E" code and that, conversely, with every 'excess E' code where the representation of zero is unique, can be associated a weighted code of mixed sign. It is thus possible to associate with the 88 weighted codes, 86 'excess' codes, engendered by only 33 weight systems, combined with suitable 'excess $E \geq 0$' codes.

---

[1] *Lippel, B.:* Negative-Bit-Weight Codes and Excess Codes, IEEE Transactions on Electronic Computers, Vol EC-13, N° 3, pp 304—306, June 1964.

3.3 Unit Distance Codes[1]

3.3.1 Definition (Binary Case)

A unit distance code is a binary coding of numbers which possesses the following property: the representations in this code of 2 consecutive numbers differ in a single position. If the Hamming distance $D(X, Y)$ of 2 code-words X, Y is defined as the number of positions by which they differ, it will be seen that for 2 given numbers x_{i+1} and x_i having the code words X_{i+1}, X_i such that $x_{i+1} - x_i = 1$, this property can be expressed as:

$$[(x_{i+1} - x_i) = 1] \Rightarrow [D(X_i, X_{i+1}) = 1]. \qquad (3.7)$$

It is easily verified that the pure binary number system does not possess this property. For example (Table 3.6):

Table 3.6

x	X
7	0 1 1 1
8	1 0 0 0

giving: $D(X_7, X_8) = 4$

3.3.2 Reflected Code (Gray Code, Cyclic Code)

3.3.2.1 Construction of a Reflected Code with n Binary Digits

This code, indicated for 4 binary positions in Table 3.7, is a special case of a unit distance code. It can be constructed as follows:
— commencing from the vertical block (b_1) (having 1 column and 2 lines), below this block mark out a block (b_1') symmetrical with respect to the arrow 1. In the binary rank to the left (rank 1) write in 0's opposite (b_1) and 1's opposite (b_1'). This results in a block (b_2) having 2 columns and 2^2 lines.
— below the block (b_2) mark out a symmetrical block (b_2') and complete the rank on the left (rank 3) with 0's opposite (b_2) and 1's opposite (b_2'). This results in a block (b_3) having 3 columns and 2^3 lines.
— repeat the operation step-by-step.

[1] Also known as "monostrophic codes" (i.e. single change) (*Russel* [135]).

3.3 Unit Distance Codes

This is clearly a unit distance code. In effect, if a block (b_n) is obtained in this way and possesses this property, then the block (b_{n+1}) possesses it too, since:

— the figure on the left changes only around the axis of symmetry (n) whereas the remainder of the word (the right hand portion) is unchanged.

— where the figure on the left does not change, the only changes must be within the block (b_n) which, by hypothesis, is a unit distance code.

(b_1) now possesses this property and, by repeated application of the above procedure, it is true for every (b_n).

3.3.2.2 Conversion Formulae

— *Reflected binary/pure binary conversion*

It will be noted that the completion of a *symmetric* block as indicated above is tantamount to the completion of the *complementary* block, i.e. that for which the elements equal 1 when the elements of the initial block equal 0 and vice versa, for the column to the left. The result is that each column k is composed of blocks of length 2^{k+1} of the type:

$$(V_k) \quad \overbrace{0\,0\,\ldots\,0}^{2^k} \quad \overbrace{1\,1\,\ldots\,1}^{2^k} \quad (0 \leq k \leq n) \tag{3.8}$$

and of complementary blocks, in an alternate manner, commencing with V_k at the top $(0 \leq k \leq n-1)$, and thus also of blocks formed of 0's and 1's of length 2^{k+1}; provided that the blocks $\frac{1}{2} V_k$ at the ends be joined.

The length of these blocks is therefore the same as that of the blocks of 0's plus that of the blocks of 1's appearing in the column of rank $k+1$ in the normal binary code. It is then clear from the nature of the block in column $k+1$ of the latter whether or not the block in column k in the Gray code is complemented. The blocks in the 2 codes in Fig. 3.7 reveal this type of correspondence.

Thus by making use of notations similar to those of Chapter 2 for addition and substraction it can be said that:

$$B_n = R_n \tag{3.9}$$

$$B_k = B_{k+1}(1 - R_k) + (1 - B_{k+1})R_k \tag{3.10}$$

$$= B_{k+1} + R_k - 2 B_{k+1} R_k \quad (0 \leq k < n) \tag{3.11}$$

Now:
$$|B_k|_2 = B_k \qquad (3.12)$$
hence:
$$B_k = |B_{k+1} + R_k|_2. \qquad (3.13)$$
Whence a rule:

Rule

The digit of rank $k < n$ in pure binary is equal to the sum modulo 2 of the reflected digit of the same rank and of the binary digit in the next higher rank. The digits of rank n are equal.

This rule is restated in slightly longer form and describes the equations (3.9) and (3.10):

Rule

The digit of rank $k < n$ in pure binary is equal to that of the same rank in reflected binary if the digit of rank $k+1$ in pure binary equals 0 and is equal to its complement if it equals 1. The digits of rank n are equal.

From the formulae (3.12) and (3.13) it can be deduced that:

$$\begin{aligned} B_n &= R_n \\ B_{n-1} &= |R_n + R_{n-1}|_2 \\ B_{n-2} &= |R_n + R_{n-1} + R_{n-2}|_2, \text{ etc.} \end{aligned} \qquad (3.14)$$

and in general:

$$B_k = \left| \sum_{i=k}^{i=n} R \right|_2 \quad \text{(for every } k\text{)} \qquad (3.15)$$

whence another expression for the conversion rule:

Rule

For every rank $k \leq n$, the pure binary digit of rank k is equal to the sum modulo 2 of the digits of rank $\geq k$ of the code-word in reflected binary.

— *Pure binary/reflected binary conversion*

From equation (3.13) it can be deduced that:

$$R_k = |B_{k+1} + B_k|_2. \qquad (3.16)$$

hence the rule:

3.3 Unit Distance Codes

Rule

The digit of rank $k\,(k < n)$ in reflected binary is the sum of the digits of ranks k and $k+1$ in pure binary.

This can also be expressed as:

Rule

The digit of rank $k\,(k < n)$ in reflected binary is equal to that of the same rank in pure binary if the digit of rank $k+1$ in pure binary is equal to 0 and to its complement if the latter equals 1.

It will be noted that for both types of conversion it is the *pure binary* digit of rank $k+1$ which, dependent on whether its value be 0 or 1, determines whether the digit of rank k of the code-word to be converted must be complemented or not.

Note:

The above conversion formulae show that the determination of the symbol of rank k takes account of the ranks higher than k, i.e. that the conversion is made by commencing with the highest significant digit. This peculiarity can be an embarrassment in practice, since for binary computers, by reason of the structure of the binary code, the conversions are usually effected by commencing with the least significant digit. The formula (3.15) can, however, be expressed in the following form which lends itself to a treatment by increasing rank:

$$B_0 = \left| \sum_{i=0}^{i=n} R_i \right|_2 \qquad (3.17)$$

$$B_1 = \left| \sum_{i=0}^{i=n} R_i \right|_2 = \left| \left| \sum_{i=0}^{i=n} R_i \right|_2 + R_0 \right|_2 \qquad (3.18)$$

$$B_2 = \left| \left| \sum_{i=0}^{i=n} R_i \right|_2 + |R_0 + R_1|_2 \right|_2 \qquad (3.19)$$

$$\cdots\cdots\cdots\cdots\cdots\cdots\cdots\cdots$$

$$B_k = \left| \left| \sum_{i=0}^{i=n} R_i \right|_2 + \left| \sum_{i=0}^{i=k-1} R_i \right|_2 \right|_2 \qquad (3.20)$$

In the case of a serial conversion, the formula can be mechanised very simply: it suffices to cause the reflected word to be circulated twice, the successive digits being applied to the input of a type-T flip-flop (§ 11.2.2) which performs the sum modulo 2 of these digits (with an initial state 0). The first circulation permits the determination of B_0, whilst, during the second circulation, the flip-flop determines the symbols B_k as successive outputs.

3.3.3 The Usefulness of Unit Distance Codes

Unit distance codes and, in particular, the reflected code, are very useful whenever transitions in more than one rank are to be avoided. This is the case for encoders, whose operating principles are briefly described below.

Disc type encoders.

Such an encoder, comprising a disc carrying concentric tracks divided into zones of two types for which detection is by means of brushes situated on each track, supplies a binary representation of the angular position of the disc. The disc can be made up of conducting and nonconducting zones which are in contact with the "brushes". Dependent on whether or not a current flows through a brush, one of the 2 binary states can be read. Other systems use transparent and opaque zones detected by means of photoelectric cells

	(B)					(R)		
0	0	0	0		0	0	0	0
0	0	0	1		0	0	0	(b_1) 1
0	0	1	0		0	0	1	(b_1') 1
0	0	1	1		0	0	(b_2) 1	0
0	1	0	0		0	1	(b_2') 1	0
0	1	0	1		0	1	1	1
0	1	1	0		0	1	0	1
0	1	1	1		0	(b_3) 1	0	0
1	0	0	0		1	(b_3') 1	0	0
1	0	0	1		1	1	0	1
1	0	1	0		1	1	1	1
1	0	1	1		1	1	1	0
1	1	0	0		1	0	1	0
1	1	0	1		1	0	1	1
1	1	1	0		1	0	0	1
1	1	1	1		(b_4) 1	0	0	0

a)   *b)*

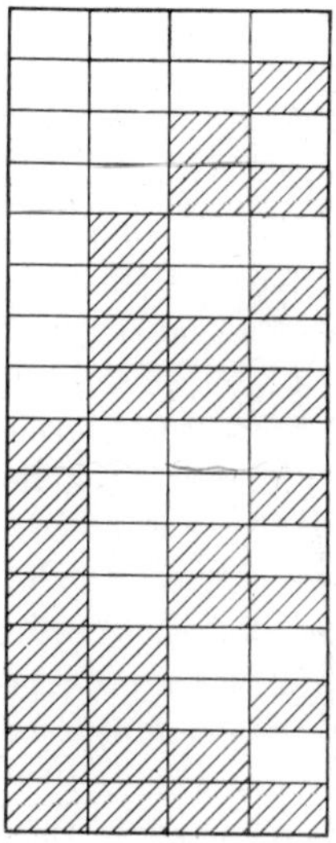

Fig. 3.7. Pure binary and reflected binary codes (4 digit case).

Certain of these encoders are in pure binary (Fig. 3.7b). (The disc is represented here in rectangular form for greater convenience.)

They cannot, however, function with a single row of brushes simply reading the state of the tracks. This, as was shown in § 3.3.1, is because the binary code is not a unit distance code: for certain transitions between successive integers (e.g. 0111 → 1000) there can be a *simultaneous* transition of 4 ranks. This is precisely what cannot be assured in practice: because of defects in the alignment of the brushes, or inaccuracies of contact points and of transitions between zones, the transition between 2 consecutive sectors will not be made *simultaneously* for every track, but generally in a *staggered* manner. This transition can occur in several ways and the intermediary values cannot be foreseen. In addition, it can easily be verified that the numerical difference between these parasitic intermediary values and the values between which the transition is made can be very much greater than 1, which is the normal transition value. If, for example, the simultaneous transition 0111 → 1000 is replaced by the staggered transition 0111 → 0011 → 0001 → 0000 → 1000, the errors with respect to the final value 1000 are respectively equal to 1 (normal), 5, 7, 8, 0. This phenomenon of "reading ambiguity" during the transition between numbers which differ in more than one rank is a defect of the pure binary system which, in the case of disc type encoders, has been avoided by two main methods: in the first method the pure binary code has been retained but a double row of brushes is used for all but the least significant rank, a method of which the V-scan encoder is a typical example. This type of coding will be studied in Chapter 9.

The other method consists in the suppression of this reading ambiguity by use of a unit distance code — in practice, the reflected code. There is thus only a single row of brushes, which reduces the friction and simplifies the construction of the encoder. In fact, in the case of encoders with 2 or 3 geared discs, it is necessary, for reasons of manufacturing precision, to use a 2-brush selection device for some tracks.

Finally it should be noted that it is often possible to preclude the reading of numbers during transition (as in the case of certain counters) and the use of the pure binary code presents no inconvenience, the parasitic phase being ignored by the remainder of the system. When this is not the case, the situation again becomes similar to that of the encoders, e.g. the case of asynchronous networks, where different parasitic phenomena (races) (§ 12.4.2.9) occur.

In practice, these are the two applications (encoders, asynchronous networks) where the unit distance codes are chiefly employed. From a digital point of view, a code such as the reflected code lends itself badly to the performance of the different operations, and when encoders are used in the reflected binary form and the information to be used is processed in a computer, the usual method is to reconvert the numbers into pure binary, using simple formulae wherever possible. There are, however, some exceptions: it will be shown in Chapter 16 that up/downcounters can be produced for the reflected code and for which the progression logic is simpler than that for the pure binary up/downcounters of the same capacity.

3.4 Residue Checks

3.4.1 Residue Checks

Electronic switching circuits, in particular those of digital computers, are not endowed with perfect reliability: errors may be produced due to various causes such as electro-magnetic parasitic interference, mains borne static, transient errors caused by a component about to exceed the permitted tolerances, or permanent error caused by a component failure, accidental failures, failures caused by wear. These errors can effect the transmission of coded messages or the digital operations to be performed. In all such cases the aim is to counter the problem by the addition of supplementary symbols to the messages to be processed (transmissions, digital operations), determined as functions of those of the original message and intended, by means of the *redundancies* thus introduced, to detect errors, or at any rate to detect those which are the most probable.

The methods used, although based on this common general concept of redundancy[1] differ in practice, depending on whether the emphasis is placed on the transmission of messages or on the execution of the digital operations. With respect to the transmission of messages, on the whole the easiest checks are those which relate rather to the *form* of the message (number of 1's, ranks of 1, etc.) than to the associated digital *magnitudes* (when numbers are concerned). Conversely, for the digital operations the methods best adapted appear to be those which cause the *arithmetic* properties of the numbers to intervene, and which can be defined independently of the particular number system utilised. Thus the redundant coding with the parity check (§ 9.4.5) and its generalization constituted by the linear codes (error correcting codes, self-detecting codes) is aimed at the protection of the transmission of messages. It will be seen (§ 9.4.6) that the use of the parity check is relatively more complex. On the other hand, the theory of integer congruence provides checks often used in practice and which are well adapted for the verification of digital operations, as will now be shown.

---

[1] For the theoretical basis (Shannon theorem in particular) the reference [128] may be consulted.

3.4 Residue Checks

For a given number X (in binary *for example*) the remainder R_i

$$R_i = |X|_{m_i} \tag{3.21}$$

of the division of X by one or more different numbers m_i is determined. For the integer m (cf. Chap. 1) this gives:

$$X = m \left[\frac{X}{m}\right] + |X|_m, \quad (0 \leq |X|_m < m) \tag{3.22}$$

where $[X/m]$ is the integer quotient of X by m, $|X|_m$ is the remainder. It is easily verified that this gives:

$$|X + Y|_m = \bigl||X|_m + |Y|_m\bigr|_m \tag{3.23}$$

$$|X - Y|_m = \bigl||X|_m - |Y|_m\bigr|_m \tag{3.24}$$

$$|X \cdot Y|_m = \bigl||X|_m \cdot |Y|_m\bigr|_m \tag{3.25}$$

These relations define the method of checking the operations $(+, -, \cdot)$, the residues, now known as residue checks, must satisfy the requirements of the above relations. In the case of division, a relation of the type:

$$X = AY + B \tag{3.26}$$

produces:

$$|X|_m = |AY + B|_m = \bigl|\bigl||A|_m \cdot |Y|_m\bigr|_m + |B|_m\bigr|_m \tag{3.27}$$

The verification of operations by a residue does not allow, as is easily checked, the detection of all errors: the most that can be said is that if the residues do not check, there is certainly an error present. Even if there is agreement, a doubt remains: the "check" on operations by this process is nothing more than "the luck of the draw" (a probability decision) between the two possibilities ("correct result, incorrect result").

3.4.2 Residue Check Computation

3.4.2.1 *Modulo m Residues of the Powers of an Integer*

Consider the residues $r_n = |b^n|_m$, $(n > 1)$, of the successive powers of an integer b. They make up a sequence:

$$r_1, r_2, \ldots, r_k, \ldots \tag{3.28}$$

for which all the terms are smaller than m. Consequently, a term r_n is necessarily repeated with a shift of $k \leq m$, thus giving:

$$r_{n+k} = r_n \qquad (k \leq m) \tag{3.29}$$

Then for every $l > 0$:

$$r_{n+k+l} = |b^{n+k+l}|_m = \Big||b^{n+k}|_m \cdot |b^l|_m\Big|_m = \Big||b^n|_m \cdot |b^l|_m\Big|_m = |b^{n+l}|_m = r_{n+l}. \tag{3.30}$$

Thus the successive residues of the powers, commencing with the least significant n, form a periodic sequence. The following paragraph provides a number of examples, and applies this property to the determination of the residue for a number in a weighted system to an integer base.

3.4.2.2 Examples:

Modulo 3 check (binary numbers).

$$1 \equiv 1 \qquad (\text{mod. } 3)$$

$$2 \equiv 2 \equiv -1 \quad (\text{mod } 3) \tag{3.31}$$

$$2^2 \equiv 1 \qquad (\text{mod. } 3)$$

etc.

$$2^{2p} \equiv 1 \qquad (\text{mod. } 3) \tag{3.32}$$

$$2^{2p+1} \equiv 2 \equiv -1 \quad (\text{mod. } 3)$$

For a given binary number X

$$X = x_{n-1}\, x_{n-2} \ldots x_k \ldots x_2 x_1 x_0 \tag{3.33}$$

then

$$|X|_3 = \left|\sum_{i=0}^{i=n-1} 2^i x_i\right|_3 \tag{3.34}$$

hence, by use of the residues:

$$|X|_3 = |x_0 + 2x_1 + x_2 + 2x_3 + \cdots|_3 \tag{3.35}$$

it may be convenient to use the negative values, giving:

$$|X|_3 = |x_0 - x_1 + x_2 - x_3 + \cdots|_3 \tag{3.36}$$

3.4 Residue Checks

— *Modulo 5 check (binary numbers)*

Thus:
$$\begin{aligned}
2^0 &\equiv 1 \equiv -4 &&(\text{mod. } 5) \\
2^1 &\equiv 2 \equiv -3 &&(\text{mod. } 5) \\
2^2 &\equiv 4 \equiv -1 &&(\text{mod. } 5) \\
2^3 &\equiv 8 \equiv 3 &&(\text{mod. } 5) \\
2^4 &\equiv 16 \equiv 1 \equiv -4 &&(\text{mod. } 5)
\end{aligned} \quad (3.37)$$

and step-by-step:
$$\begin{aligned}
2^{4p} &\equiv 1 \equiv -4 &&(\text{mod. } 5) \\
2^{4p+1} &\equiv 2 \equiv -3 &&(\text{mod. } 5) \\
2^{4p+2} &\equiv 4 \equiv -1 &&(\text{mod. } 5) \\
2^{4p+3} &\equiv 3 \equiv -2 &&(\text{mod. } 5)
\end{aligned} \quad (3.38)$$

For a given binary number X, this gives:

$$|X|_5 = |x_0 + 2x_1 + 4x_2 + 3x_3 + x_4 + 2x_5 + 4x_6 + 3x_7 + \ldots|_5. \quad (3.39)$$

and by alternating the signs:

$$|X|_5 = |x_0 + 2x_1 - x_2 - 2x_3 + x_4 + 2x_5 - x_6 - 2x_7 + \ldots|_5$$

— *Modulo 9 check (decimal numbers)*

Let N be any decimal number

$$N = D_{m-1} D_{m-2} \ldots D_k \ldots D_3 D_2 D_1 D_0 \quad (3.40)$$

$$N = \sum_{i=0}^{i=n-1} 10^i D_i \quad (3.41)$$

This now gives:
$$10^i \equiv 1 \quad (\text{mod. } 9) \quad (3.42)$$

It follows that:
$$N \equiv \sum_{i=0}^{i=m-1} D_i \quad (\text{mod. } 9) \quad (3.43)$$

and

$$|N|_9 = \left| \sum_{i=0}^{i=m-1} D_i \right|_9 \quad (3.44)$$

Thus the modulo 9 check is obtained by forming the sum of the digits of its decimal representation and removing therefrom the greatest multiple of 9.

Note: This last property may be generalized. To obtain the remainder of the division of a number by m it is necessary to form the sum of the digits of the number in a system to the base $m + 1$. Removing from this sum the greatest multiple of m contained therein then supplies the required remainder.

The following relation holds:

$$(1 + m)^n \equiv 1 \quad (\text{mod. } m) \tag{3.45}$$

as

$$(1 + m)^n = (1 + m)(1 + m) \ldots (1 + m). \tag{3.46}$$

Each factor $1 + m$ equals 1 mod. m and consequently the product of the n factors is still equal to 1 (mod. m).

The "proof by 9" (the casting-out of 9's) in the case of the decimal number system constitutes the most widespread illustration of this property. As an additional example, refer back to the case of a binary number and a modulo 3 check.

It is necessary to summate the digits of the representation of the number x in the number system to the base $3 + 1 = 4$. Here it suffices to cut the word X into groups of 2 digits, giving:

$$k = (x_0 x_1)(x_2 x_3)(x_4 x_5) \ldots (x_{2k} x_{2k+1}) \ldots \tag{3.47}$$

whence:

$$|X|_3 = \Big||x_0 x_1| + |x_2 x_3| + \cdots + |x_{2k} x_{2k+1}| + \cdots\Big|_3 \tag{3.48}$$

($|x_{2k} x_{2k+1}|$ is the numerical value of the number represented by $x_{2k} x_{2k+1}$).

By expressing

$$|x_{2k} x_{2k+1}| = x_{2k} + 2 x_{2k+1},$$

the expression (3.35) above can again be produced.

It should be noted that this method of finding the remainder of the division of a binary number by m is only of interest for a number such as $m + 1 = 2^k$, since the conversion to the base $m + 1$ follows immediately and is translated by a subdivision into blocks of length k.

3.4.3 Examples of Congruence

3.4.3.1 Problem

The following is a conventional problem involving the concept of residue:

Given that 1 January 1966 is a Saturday, what day of the week is 1 January 1970?

Let the day of the week be denoted by a variable x, counted from Saturday onwards, (0 for Saturday, 1 for Sunday, 2 for Monday, etc.). It then follows that:

$$x = |(1970 - 1966) \cdot (365) + 1|_7 \qquad (3.49)$$

(1 for the leap year)

$$x = |4 \times 365 + 1|_7 \qquad (3.50)$$

$$= \big||4|365|_7 + 1\big|_7 \qquad (3.51)$$

Hence:
$$|365|_7 = |52 \times 7 + 1|_7 = |1|_7 = 1 \qquad (3.52)$$

$$x = |4 \cdot 1 + 1|_7 = |5|_7 = 5 \qquad (3.53)$$

Answer: Thursday

It should be noted that $|365|_7 = 1$, $|366|_7 = 2$. In other words, each ordinary year introduces a shift of one day, each leap year a shift of 2. In practice it therefore suffices to take the residue modulo 7 of the number thus calculated:

$$x = |3 \times 1 + 1 \cdot 2|_7 = |5|_7 = 5. \qquad (3.54)$$

3.4.3.2 Counters

A binary counter is comprised of n memory elements ("stages") which are able to store one 'bit' of information, i.e. able to store one or another of the two binary symbols, denoted by 0 and 1, which form a *register* R. In addition it possesses: (Fig. 3.8)

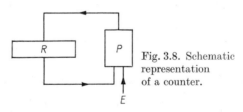

Fig. 3.8. Schematic representation of a counter.

— one input into which a signal can be fed indicating that 1 (a least significant digit) is added to the number contained in the register in binary form.
— a progression system P which, from the content N of the register, enables the unit presented at the input to be added in to replace N by $N + 1$, which thus becomes the new state of the counter.

This constitutes an overall description of the counter, which is sufficient at this stage. Counters will be studied in greater detail in Chapter 16.

This being so, let it be assumed that the content of the register is the pure binary representation of N. In these circumstances, it will be shown (Chapter 16) that the position N_k of rank $k (n-1 \geq k \geq 0)$, starting from the right, is or is not modified when 1 is added in a way which involves only the positions $i < k$. The progression system ignores the ranks superior to k when determining the new value of N. Thus for 2 counters with $n + l$ positions ($l > 0$) and n positions, the n digits on the right evolve in the same way. On arrival at the state formed uniquely of 1, representing the value $2^n - 1$ (Chapter 2), then the addition of an extra 1 moves the first counter into the state $2^n - 1 + 1 = 2^n$, and the second into the state 0 (Table 3.9)

Table 3.9

$$\begin{array}{r} 0 \\ + 1 \\ \hline = 1 \end{array} \qquad \begin{array}{r} 1\ 1\ 1\ 1 \\ 1 \\ \hline 0\ 0\ 0\ 0 \end{array} \qquad \begin{array}{r} 1\ 1\ 1\ 1 \\ + 1 \\ \hline 0\ 0\ 0\ 0 \end{array}$$

(a) \hspace{4em} (b)

$$\begin{array}{r} 1 \\ - \\ \hline = 0 \end{array} \qquad \begin{array}{r} 0\ 0\ 0\ 0 \\ 1 \\ \hline 1\ 1\ 1\ 1 \end{array} \qquad \begin{array}{r} 0\ 0\ 0\ 0 \\ - 1 \\ \hline = 1\ 1\ 1\ 1 \end{array}$$

(c) \hspace{4em} (d)

The content of the n-position counter after receipt of K units from an initial content K_0 is therefore:

$$C = |K_0 + K|_{2^n} \tag{3.55}$$

if $K_0 + K < 2^n$, then:

$$C = |K_0 + K|_{2^n} = K_0 + K \tag{3.56}$$

An n-position counter counts modulo 2^n. Here the modulus is the capacity 2^n of the register.

— *The case of a down-counter ("reverse counter")*

This resembles the usual counter, but with a different progression system. Assuming (frequent case) that for $N > 0$, N is replaced by $N - 1$, again for the determination of the digit values of rank k the associated device ignores the higher ranks. The comparison of counters with n and $n + l$ ($l > 0$) stages shows that from the "all zeros" state

3.4 Residue Checks

the counter moves to the state formed uniquely of 1's in the n positions. The application of other pulses leads to decreasing this number ($2^n - 1$). Consequently, if the down-counter receives K pulses, commencing with the content K_0, then for the n-position counter is obtained:

$$C = |K_0 - K|_{2^n} \qquad (3.57)$$

Special cases:

$$(K_0 - K \geqq 0), \qquad C = |K_0 - K|_{2^n} = K_0 - K \qquad (3.58)$$

$$(-2^n < K_0 - K < 0), \quad C = |K_0 - K|_{2^n} = \Big| -|K_0 - K| \Big|_{2^n}$$

$$= \Big| 2^n - |K_0 - K| \Big|_{2^n} = 2^n - |K_0 - K|. \qquad (3.59)$$

The quantity $2^n - |K_0 - K|$ is known as the true complement (or "to 2") of $|K_0 - K|$. Its most important properties are described in the following example.

Reversible counter

In a currently used unit having a 'sign' type flip-flop (e.g. 0 for $+$, 1 for $-$), the n other stages combine the above functions, which still gives (see 3.57)

$$C = |K_0 + K|_{2^n} \qquad (3.60)$$

Assuming that $0 \leqq |K_0 + K| < 2^n$ (normal condition in use)

$$\begin{cases} S = 0 & C = K_0 + K \\ S = 1 & C = 2^n - |K_0 + K| \end{cases} \qquad (3.61)$$

In another type the 'sign' feature is retained, but the progression system is such that the content of the n flip-flops always takes the absolute value $|K_0 + K|$ for $0 \leqq |K_0 + K| < 2^n$. Then, in a general way, the content of the n flip-flops is:

$$C = \Big| |K_0 + K| \Big|_{2^n} \qquad (3.62)$$

For $0 \leqq |K_0 + K| < 2^n$, as above, the content is:

$$\begin{cases} S = 0 & C = \Big| |K_0 + K| \Big|_{2^n} = |K_0 + K|_{2^n} = K_0 + K \\ S = 1 & C = |K_0 + K| \end{cases}$$

6 Chinal, Design Methods

3.4.3.3 True Complement

For a given integer $0 \leq X < 2^n$, this, by definition, is the quantity:

$$C_{2^n}(X) = 2^n - X \tag{3.63}$$

— *Application to subtraction.*

Let A and B be two integers such that:

$$\begin{cases} 0 \leq A < 2^n \\ 0 \leq B < 2^n \end{cases}$$

this then gives:

$$0 \leq |A - B| < 2^n \tag{3.64}$$

Two cases then present themselves:

1. $A - B \geq 0$.

   Can be expressed as:

   $$A - B = |A - B|_{2^n} = |A - B + 2^n|_{2^n} = |A + (2^n - B)|_{2^n}$$

2. $A - B < 0$.

$$|A - B|_{2^n} = 2^n - |A - B|. \tag{3.65}$$

Similarly:

$$2^n - |A - B| = |A - B|_{2^n} = |A - B + 2^n|_{2^n} \tag{3.66}$$
$$= |A + (2^n - B)|_{2^n}$$

In other words, the operation $A - B$ can be replaced by the operation $A + (2^n - B)$. As long as it is possible to formulate $2^n - B$ simply, the subtraction is reduced to addition. Effectively, this is the case and the use of the complement to 2^n to perform the subtraction is one of the most frequently used procedures, since additions and subtractions must generally be performed by means of the same arithmetic unit.

In order to distinguish between the two cases, it is sufficient to determine what the position n becomes in the operation $A + (2^n - B)$. If it takes the value 1 then $A - B \geq 0$, if not $A - B < 0$. It will be seen that in the first case the difference $A - B$ itself is obtained, and in the second case the complement of the absolute value: $2^n - |A - B|$. This property can be generalized to handle the numbers A and B with arbitrary signs.

Chapters 9 and 15 describe the determination of the true complement and the associated circuits.

3.5 The Choice of a Code

The choice of a code to represent numbers or other messages is in fact an extremely complex operation. Among the requirements (often contradictory) of which account must be taken, and between which a compromise is necessary, are the following which constitute the most important factors to be decided.

3.5.1 Adaptation to Input-Output Operations

1. The qualities of the code vis-à-vis the input-output operations necessary for the communication between the system and the exterior constitute an important factor to start off with.

Examples:
— digital-analog conversion: weighted codes are often advantageous
— digital-analog conversion by disc encoders: reflected code is of interest.
— counters: in the case of a counting of events, the code (and the structure of the counter) depend on the objective: for a simple display of results a *BCD* coding will be used. If calculations are to be made a pure binary counter may have to be employed. Finally if there are neither display nor calculations on the accumulated content, but simple detection of coincidences, then a reflected code counter (cf. Chap. 16) may prove to be more suitable.
— binary-decimal and decimal-binary conversions: these conversions between two digital representations are relatively complex: the detailed study of conversion systems, as might be expected, shows that the complexity of conversion equipment, or conversion time, or both at the same time, increase together with the size of the numbers to be converted.

3.5.2 Adaptation to Digital Operations

The qualities from the viewpoint of the construction of information processing circuits must also be taken into account.

For example, in the case of digital operations and the use of binary circuits, experience indicates that the pure binary code leads to simpler realisations than the decimal codes (binary codes).

3.5.3 Aptitude for the Automatic Detection or Correction of Errors

This is linked to the redundancy of the coding used. In the case of the 4-figure *BCD* codes, for example, the redundancy is due to the fact that 6 of the possible 16 combinations are not used. Thus their appearance in the circuits indicates an error: they represent the "illicit" combinations and thought may be given to the realisation of a system for their detection (cf. § 9.4.10). This process is applicable to the *BCD* codes with more than 4 digits: here the number of "illicit" combinations is even greater. Processes such as parity checks, residue checks by congruence, all possess the same property in the last analysis: the coding of N messages by a subset of a set E having $N' > N$ elements, the N code-words being pairwise at the maximum possible distance (in the Hamming sense). In order to detect (locate) all the errors affecting d positions at least, it is necessary and sufficient that the minimum mutual distance of the code-words be superior to or equal to $d + 1$. To correct (i.e. to localize) all errors at d positions it is necessary that this minimum mutual distance be superior to or equal to $2d + 1$. The verification of these two conditions is immediate.

Thus in the case of the parity digit appended to an n-rank word, there are $n + 1$ digits, corresponding to the 2^{n+1} combinations, for coding 2^n messages only. Similarly, in the case of a modulo 3 check, encoded with 2 binary digits and adjoined to an n-position word, there are $n + 2$ digits (capable of providing 2^{n+2} combinations) for 2^n distinct numbers.

In this respect it must be noted that the choice of a code is only one means among others of introducing a redundancy into a system: other methods such as the repetition of messages or of operations, the self-checking programmes in the case of computers, produce a similar result. It should also be noted that even within the limited hypothesis of a redundant coding, the quality of the code can only be measured in terms of probabilities (the probability of an undetected error, etc.) and depends on the characteristic statistics for *noise* which degrades the messages. In effect, it will be seen that the problem of the elimination of errors falls into a more general framework, that of the *reliability* of the system, which is again a concept of probability. It is only by adopting this overall standpoint and in the application of this concept at all levels: components, packaging techniques, codings, logic organisation etc., that there can be any hope of successful treatment, the choice of the code being only one aspect of the problem.

Chapter 4

Algebra of Contacts

4.1 Electromagnetic Contact Relays

Many electrical systems include electromagnetic contact relays in their control and actuation circuits. Essentially, such relays comprise an electromagnet and a moving armature carrying metallic blades (Fig. 4.1). Each blade is attached to the armature at one end. The other

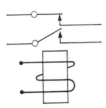

Fig. 4.1. Contact relay.

end is free and can, according to the position of the armature, either be free or in contact with a contact point mounted on a blade affixed to the relay. The assembly of the blade and the contact point associated with it constitutes what is known as a "contact" which may either be open or closed. Dependant upon whether or not the coil of the electromagnet be excited by the passage of a current, the moving armature takes up one of 2 positions: so long as no excitation current is applied the armature takes up the "rest" position; when an excitation current is applied the armature is attracted into a second position ("attracted" position). Thus it is possible, in general, to recognise two, principal types of contacts (Figs. 4.2 & 4.3):

Fig. 4.2. Normally open contact ("make contact").

Fig. 4.3. Normally closed contact ("break contact").

— contacts known as *normally open or* "make" contacts. They are in the open position so long as the armature remains at rest, and closed when the armature is attracted by the electro-magnet.
— contacts known as *normally closed* or "break" contacts. They are in the closed position so long as the armature is at rest and open when the armature is attracted by the electro-magnet.

There exists a third contact, the *transfer contact*, consisting of a moving blade and 2 contact points (Fig. 4.4). Dependent upon the position of the armature, the blade is in contact with one of these 2 contacts; functionally, this type of contact is the equivalent of pair of ordinary contacts, one of the normally open type, the other of the normally closed type. By consideration of the transfer contact of Fig. 4.4a it will be noted that in passing from one condition to the other the blade is positioned between the two contact points for a non-negligible period of time by reason of the various inherent inertias of the system and

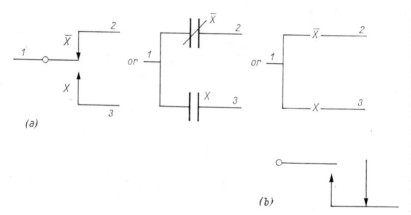

Fig. 4.4. a) Transfer contact   b) Break before make transfer contact.

that, with this type of contact, between the two permanent positions will exist a transitory state where both contacts will be open. Such a transfer contact is often known as a *break before make contact*. In certain applications it is desirable to avoid the simultaneous opening of both contacts and use is then made of another type of contact known as a *make before break contact* (Fig. 4.4b). With this type of contact there is, for every change-over condition, a very short transient phase during which both contacts are closed.

Apart from these fundamental types of relay, there exists a large variety of others (polarised relays, multi-coil relays, etc.) whose operation is somewhat more complex, to take account of particular working conditions.

In short, the relay of the type described possesses the following characteristics:

1. There is one input, the state of the induction coil, which may be either excited or non-excited.
2. There are a certain number of outputs, each of which is conditioned by the closed, or open state of the contacts.

The input-output relationship (i.e. the manner in which the relay controls the opening or closing of the contacts in accordance with the input signal) is a function of the nature of the contacts as summarised in Table 4.5 in which the behaviour of the armature is given by way of information.

Table 4.5

Condition of the Induction Coil	Position of the Armature	State of a normally open contact	State of a normally closed contact
Non-excited	At rest	Open	Closed
Excited	Attracted	Closed	Open

Table 4.6

Induction Coil		Armature		Contact	
Variable	E	Variable	A	Variable	X
Non-excited	E_0	At rest	A_0	Open	X_0
Excited	E_1	Attracted	A_1	Closed	X_1

4.2 Binary Variables Associated with a Relay

An examination of table 4.5 reveals that the behaviour of the relay as a switching device may be described by variables capable of taking 2 values, i.e. by *binary* variables. With the state of the coil, therefore, can be associated a variable E, with the state of the armature a variable A, with that of a normally open contact a variable X and with a normally closed contact the variable Y. If these states are coded in the manner shown in Table 4.6 then the Table 4.5 takes on the form of Table 4.7. In fact, it is advisable to do away with the name of the variable and to use in place of the various pairs of symbols A_0, A_1, B_0, B_1 etc., 2 symbols only, the same two symbols for all the variables. The first figures 0 and 1 may be utilised for this purpose, allocating them in an arbitrary manner to the states for each particular variable; for example:

$$E_0 = 0, \quad E_1 = 1, \quad A_0 = 1, \quad A_1 = 0, \quad X_0 = 1, \quad X_1 = 0$$

Among others, the following convention may be used:

$$\left.\begin{array}{ll}\text{Non-excited} & E_0 = 0 \\ \text{Excited} & E_1 = 1 \\ \text{Rest position} & A_0 = 0 \\ \text{Attracted position} & A_1 = 1 \\ \text{Open} & X_0 = 0 \\ \text{Closed} & X_1 = 1\end{array}\right\} \qquad (4.1)$$

i.e. the symbols used in Table 4.7 are replaced by their single indices. See Table 4.8.

Table 4.7

E	A	X	Y
E_0	A_0	X_0	X_1
E_1	A_1	X_1	X_0

Table 4.8

E	A	X	Y
0	0	0	1
1	1	1	0

4.3 Complement of a Variable

It will be seen that when a variable X has one given value, the variable Y has the other value.

It can thus be said that:

$$Y = \overline{X} \qquad (4.2)$$

the variable $\overline{X}$ being defined as a function of X as in Table 4.9. This is known as the "complement of X" (read as $Y = \text{bar } X$)

Table 4.9

X	$\overline{X}$
0	1
1	0

4.3 Complement of a Variable

Thus are established the relationships:

$$A = E \tag{4.3}$$

$$X = A \tag{4.4}$$

$$Y = \bar{X} = \bar{A} \tag{4.5}$$

It will be seen that, with these conventions, the variables X associated with a normally open contact is equal to E, while that associated with a normally closed contact is $\bar{E}$ and, therefore, an unique variable E will suffice to define the relay.

Hence the following new convention:

Notation: A relay is associated with a letter X. The value of the variable x represents the state of excitation of the relay. It is then convenient to designate by X the normally-open contacts and by $\bar{X}$ the normally-closed contacts of the relay.

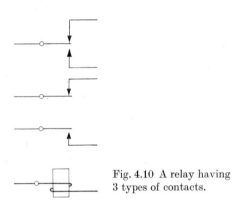

Fig. 4.10 A relay having 3 types of contacts.

In certain instances it is unnecessary to represent the coils, and only the contacts are retained. Otherwise, a capital letter is used for the excitation and a small letter for the contacts.

Thus the different letters appearing on a scheme drawing correspond to the different relays. For a given *letter X*, the allocations of the different *symbols X* and $\bar{X}$ correspond with the contacts of the relay, and the presence or absence of the bar above the letter indicates the nature of the contact (normally open or normally closed). The figures 4.2 to 4.4 give the symbols most commonly used.

The symbol on the right, in figures 4.2 to 4.4 will be used in this chapter. Fig. 4.10. gives are presentation of a simplified scheme for a relay with its excitation winding and 3 contacts. Finally Fig. 4.11 gives a representation of the same network with the aid of these three sets of symbols.

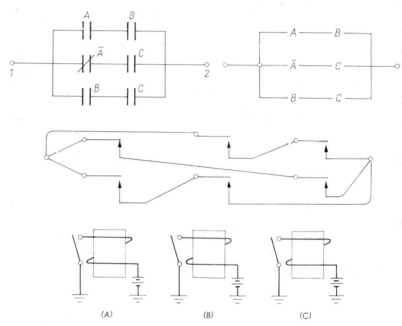

Fig. 4.11. Three representations of a relay contact network.

4.4 Transmission Function of a Dipole Contact Network

Consider a relay contact. It consists of a dipole network which can take 2 forms: open or closed. For this reason it has (above) been associated with a *binary* variable whose value indicates in which state it is to be found. In a general manner, a dipole circuit with interconnected contacts such as those of Fig. 4.11 can be made to define a *binary* variable T_{AB} which is known as the *transmission function* or, more simply, the *transmission* of the dipole with the convention:

$$\left. \begin{array}{l} T_{AB} = 1: \text{ the path } AB \text{ being closed} \\ T_{AB} = 0: \text{ the path } AB \text{ being open} \end{array} \right\} \quad (4.6)$$

4.4 Transmission Function of a Dipole Contact Network

The notation T_{AB} will be used so long as there is a need to define the extremities of the dipole under consideration, and more simply, T without suffix so long as there is no ambiguity.

4.4.1 Transmission of a Contact

Reverting to the case of the dipole consisting of a single contact (Fig. 4.12), the following relations become evident:

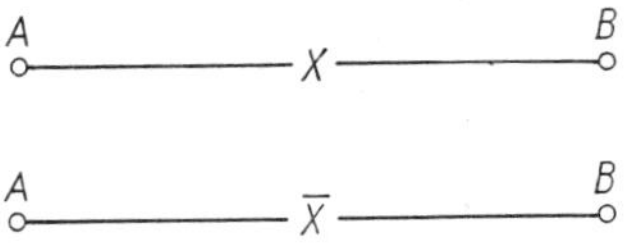

```
A ————— X ————— B

A ————— X̄ ————— B
```

Fig. 4.12. An elementary dipole: the contact.

— Contact normally open:

$$T = 1 \text{ when } X = 1$$
$$T = 0 \text{ when } X = 0 \tag{4.7}$$

which, algebraically, can be expressed as:

$$T = X \tag{4.8}$$

— Contact normally closed:

$$T = 1 \text{ when } X = 0$$
$$T = 0 \text{ when } X = 1 \tag{4.9}$$

which may be expressed as:

$$T = \overline{X} \tag{4.10}$$

4.4.2 Transmission of a Dipole Comprising 2 Contacts in Series

Consider now the closed dipole having contacts X and Y in series (for which the characteristics are as in Fig. 4.13). If both contacts X and Y

```
A ——— X ——— Y ——— B
```
Fig. 4.13

are closed, the series dipole is closed. On the other hand, if either of the two contacts is open, or, if both are open, then the path between A and B is open. This can be summarised by means of a table having, as inputs, the different possible combinations of the values of X and Y and, as outputs, the value of the transmission T_{AB} of the circuit between A and B. Thus is obtained Table 4.14.

Table 4.14

X Y	T
0 0	0
0 1	0
1 0	0
1 1	1

Thus, the dipole transmission binary is presented as a function $T = T(X, Y)$ of the two binary variables X and Y, or again, as an operation on the variables X and Y. It can be expressed:

$$T = X \cdot Y \quad \text{(read T is equal to XY)}$$

Thus with the operation of putting the 2 contacts X and Y in series is associated the operation ($\cdot$) on the algebraic values of the transmissions X and Y, known as the *product* of X and Y. This is usually expressed more briefly as:

$$T = X \cdot Y = XY$$

4.4.3 Transmission of a Dipole Consisting of 2 Contacts in Parallel

Consider now the dipole having 2 contacts X and Y (of any type) mounted in parallel (Fig. 4.15). If one of the two contacts X or Y is

Fig. 4.15

closed, or if both are closed, the path between A and B is closed. Alternatively, if both contacts are together in the open position, the path AB is open. The functioning of the dipole AB can then be summarised as in Table 4.16:

Table 4.16

X Y	T
0 0	0
0 1	1
1 0	1
1 1	1

4.5 Operations on Dipoles

Here again the transmission T_{AB} of the dipole is shown as a function of the 2 variables X and Y, or as an operation on X and Y, which may be expressed as:

$$T = X + Y \quad \text{(read: } T \text{ is equal to } X \text{ plus } Y\text{)} \qquad (4.12)$$

Thus the operation of putting the contacts X and Y in parallel is associated with the operation $(+)$ on the transmissions X and Y, and is known as the *sum* of X and Y.

Note — Examination of Table 4.14 reveals that the operation $X \cdot Y$ is identical, from the formal viewpoint, with the ordinary multiplication of X and Y, when considering 0 and 1 as figures. By coincidence the symbol $\cdot$ is retained by way of a mnemonic, since the symbols 0 and 1 have no arithmetic significance here and have been retained only for their convenience. On the other hand it will be seen from Table 4.16 that the symbol $+$ does not coincide with the ordinary addition of the numbers 0 and 1. It will be used here because of its simplicity of description for this operation — there being no fear of confusion with ordinary arithmetic addition, and the apparently surprising character of the relationship $1 + 1 = 1$ is allied only to this convention. Thus 3 contact operations can be defined as:

1. Complementation: i.e. the replacement of a contact of type X by a contact of another type $\overline{X}$; the new resultant transmission is $T = \overline{X}$.
2. Contacts in series: the resultant transmission is $T = X \cdot Y$.
3. Contacts in parallel: the resultant transmission is $T = X + Y$.

Table 4.17 summarises these three operations.

Table 4.17

Elementary dipoles	Operation	Transmission
1 contact X	Complementation	$T = \overline{X}$
2 contacts X & Y	In series	$T = X \cdot Y$
2 contacts X & Y	In parallel	$T = X + Y$

4.5 Operations on Dipoles

From the elementary contact dipoles discussed earlier have been developed other dipoles covering the complementary, series and parallel operations. These operations can be defined as $-$, $\cdot$, and $+$ for all dipoles.

4.5.1 Complementation

A dipole whose transmission

$$T_c = \overline{T} \tag{4.13}$$

is known as the complement of a dipole whose transmission is T.

4.5.2 Connections in Series

From the discussion above it will be seen that the transmission of the resultant dipole is

$$T_s = T_1 \cdot T_2 \tag{4.14}$$

where T_1 and T_2 are the transmissions of the 2 dipoles.

4.5.3 Connections in Parallel

Similarly, it will be seen that the transmission of the resultant dipole is:

$$T_p = T_1 + T_2 \tag{4.15}$$

It will, therefore, be seen that by the complementation, series and parallel operations, a procedure has been established for the construction of dipoles of ever-increasing complexity based on the elementary dipoles; for with each operation on the circuits it is possible to associate an algebraic operation with the transmission of the circuits. With the final circuit, therefore, can be associated an algebraic transmission which may be expressed as a function of the contacts, i.e. finally as a function of the excitation of the various relays. Before giving examples of the method it is necessary to demonstrate some properties of the operations $-$, $\cdot$ and $+$. To demonstrate these properties it is possible to proceed in 2 ways:

1. By direct reasoning on the circuits.
2. By the use of tables defining the operations.

4.5.4 Properties of the Operations $-$, $\cdot$ and $+$

Fig. 4.18 gives the equivalences between the circuits and the algebraic relationships which represent them. These relationships are designated by the symbols R_1 to R_{20}. Only one example is demonstrated, the task of justifying the others is left to the reader.

4.5 Operations on Dipoles

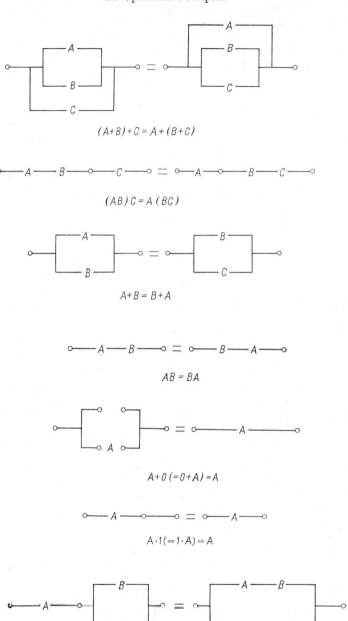

(R_1) $(A+B)+C = A+(B+C)$

(R_2) $(AB)C = A(BC)$

(R_3) $A+B = B+A$

(R_4) $AB = BA$

(R_5) $A+0 (=0+A) = A$

(R_6) $A \cdot 1 (=1 \cdot A) = A$

(R_7) $A(B+C) = AB + BC$

Fig. 4.18. Some remarkable properties, (R_1 to R_7).

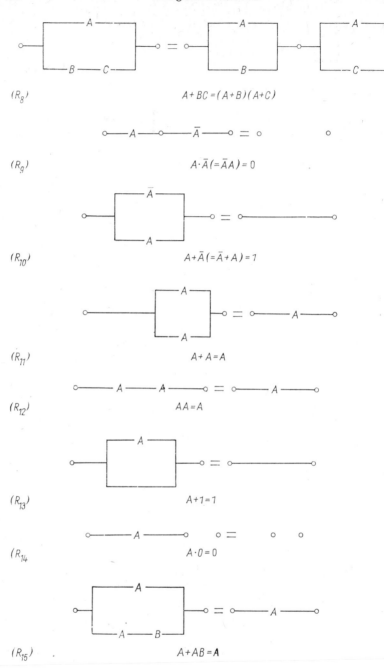

Fig. 4.18, (R_8 to R_{15}).

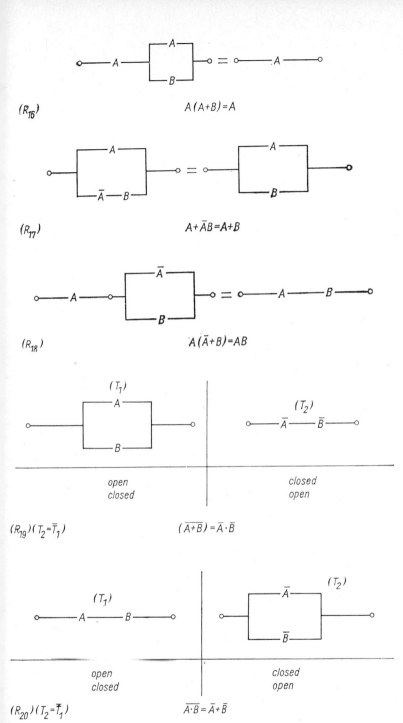

Fig. 4.18, (R_{16} to R_{20}).

7 Chinal, Design Methods

Example: relationship

$$A + BC = (A + B)(A + C) \tag{4.16}$$

First method: By direct reasoning on the circuits, it will be seen that if A is open the two circuits are closed only when B and C are closed and are open in all other cases. If A is closed, the two circuits are closed no matter what the condition of B and C. Thus the two circuits have the same function for the various combinations of A, B and C.

Second method: the use of tables.

There are $2^3 = 8$ combinations of values for A, B and C as shown on the left hand side of Table 4.19. The terms of the remaining columns are then calculated by use of the rules defining the operations of $-$, $\cdot$ and $+$.

The columns corresponding to $A + BC$ and $(A + B)(A + C)$ are identical showing that the two expressions under consideration have the same values for all combinations of the variables A, B and C. Hence the relationship (4.16).

Table 4.19

$A\ B\ C$	A	BC	$A+BC$	$(A+B)$	$(A+C)$	$(A+B)(A+C)$
0 0 0	0	0	0	0	0	0
0 0 1	0	0	0	0	1	0
0 1 0	0	0	0	1	0	0
0 1 1	0	1	1	1	1	1
1 0 0	1	0	1	1	1	1
1 0 1	1	0	1	1	1	1
1 1 0	1	0	1	1	1	1
1 1 1	1	1	1	1	1	1

Note 1 — Relation (4.16) expresses the 'distributivity' of addition with respect to multiplication: this property does not exist in the algebras of real or complex numbers. There are, in addition to the above, other relationships which differ from those of ordinary algebra. This matter will further be discussed later in this work.

Note 2 — By means of the expressions

$$A + (B + C) = (A + B) + C = A + B + C \qquad (4.17)$$

$$A(BC) \quad = (AB)C \quad = ABC \qquad (4.18)$$

the operations for 3 variables, denoted by $A + B + C$ and ABC, have been defined. In an analogous manner the quantities $A_1 + A_2 + \cdots + A_n$, $A_1 \cdot A_2 \cdot \cdots \cdot A_n$ for n variables can be defined. These definitions will be dealt with in a more general way later in this work.

4.5.5 2 Variable Functions

The two functions for 2 variables, denoted by $X \cdot Y$ and $X + Y$ have already been defined by the use of a table giving the value of the function for each combination of values of the 2 variables X and Y. This procedure can be used in defining other functions having 2 variables: here again a list of input combinations has been established, each of which has been allocated the value 1 or the value 0.

There are 4 input combinations, to each of which can be ascribed 2 corresponding values. There are, therefore, $2^4 = 16$ possible functions of 2 variables (Table 4.20).

(It will be noted that the combinations of 0 and 1 in the columns correspond to the binary representations of the indices allocated to the different transmission functions.)

Table 4.20. *Functions of 2 Variables*

X	Y	T_0	T_1	T_2	T_3	T_4	T_5	T_6	T_7	T_8	T_9	T_{10}	T_{11}	T_{12}	T_{13}	T_{14}	T_{15}
0	0	0	1	0	1	0	1	0	1	0	1	0	1	0	1	0	1
0	1	0	0	1	1	0	0	1	1	0	0	1	1	0	0	1	1
1	0	0	0	0	0	1	1	1	1	0	0	0	0	1	1	1	1
1	1	0	0	0	0	0	0	0	0	1	1	1	1	1	1	1	1

It will now be shown that these diverse functions can be expressed by means of the operations $\cdot$, $+$ and that of complementation.

1. Consider for example the function T_1:

T_1 has the value 1 when both X and Y are zero, i.e. when both $\overline{X}$ and $\overline{Y}$ have the value 1, and 0 in the other cases; thus

$$T_1 = \overline{X} \cdot \overline{Y} \qquad (4.19)$$

In a more detailed way, this relation may be checked by the use of Table 4.21:

Table 4.21

X Y	T_1	$\overline{X}$	$\overline{Y}$	$\overline{X} \cdot \overline{Y}$
0 0	1	1	1	1
0 1	0	1	0	0
1 0	0	0	1	0
1 1	0	0	0	0

It will be seen that the columns corresponding to T_1 and $\overline{X} \cdot \overline{Y}$ have the same value for each combination of the values of X and Y.

2. *Further example:*

$$T_6 = X\overline{Y} + \overline{X}Y \tag{4.20}$$

Similarly, the values of $T_6, \overline{X}, \overline{Y}$ and $X\overline{Y} + \overline{X}Y$ can be listed for all combinations of values of X and Y (Table 4.22).

Table 4.22

X Y	T_6	$\overline{X}$	$\overline{Y}$	$X\overline{Y} + \overline{X}Y$
0 0	0	1	1	0
0 1	1	1	0	1
1 0	1	0	1	1
1 1	0	0	0	0

3. In a general way the reader will be able to check, by means of the tables, that:

$$\begin{aligned}
T_0 &= 0 \\
T_1 &= \overline{X}\overline{Y} \\
T_2 &= \overline{X}Y \\
T_3 &= \overline{X} \\
T_4 &= X\overline{Y} \\
T_5 &= \overline{Y} \\
T_6 &= X\overline{Y} + \overline{X}Y \\
T_7 &= \overline{X} + \overline{Y} \\
T_8 &= XY
\end{aligned} \tag{4.21}$$

$$T_9 = XY + \overline{X}\overline{Y}$$
$$T_{10} = Y$$
$$T_{11} = \overline{X} + Y$$
$$T_{12} = X$$
$$T_{13} = X + \overline{Y}$$
$$T_{14} = X + Y$$
$$T_{15} = 1$$

4.5.6 Examples of Application

Example 1.

Consider the network (R) of Fig. 4.23. It consists of 3 relays A, B, C and of 6 contacts. It will be shown that it is possible to find an equivalent network (R') in the sense that it is closed when R is closed, open when R is open and which, however, comprises fewer contacts.

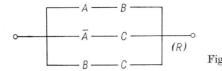

Fig. 4.23

The network (R) possesses 3 branches in parallel, each branch comprising 2 contacts in series.

The branches have as transmissions:

$$AB,\ \overline{A}C\ \text{and}\ BC \tag{4.22}$$

from which the transmission T of network R

$$T = AB + \overline{A}C + BC \tag{4.23}$$

This is the algebraic expression of the transmission of the network. In substituting $A, B, C, \overline{A}$ by their numerical values for each of the $2^3 = 8$ combinations of A, B, C the transmission value can be obtained in each case.

The following operations can be carried out (the properties utilised are indicated on the right hand side of each line):

$$AB + \overline{A}C + BC = (AB + \overline{A}C) + BC \tag{4.17}$$
$$= (AB + \overline{A}C) + (\overline{A} + A)BC \quad (R_{10},\ R_6)$$
$$= (AB + \overline{A}C) + (\overline{A}BC + ABC) \quad (R_7)$$

$$= AB + [\bar{A}C + (\bar{A}BC + ABC)] \quad (R_1)$$
$$= AB + [(\bar{A}C + \bar{A}\dot{B}C) + ABC] \quad (R_1)$$
$$= AB + [(\bar{A}C + \bar{A}CB) + ABC] \quad (R_4)$$
$$= AB + [(\bar{A}C + (\bar{A}C)B) + ABC] \quad (4.18)$$
$$= AB + (\bar{A}C + ABC) \quad (R_{15})$$
$$= AB + (ABC + \bar{A}C) \quad (R_3)$$
$$= (AB + ABC) + \bar{A}C \quad (R_1)$$

from which, finally (R_{15})

$$AB + \bar{A}C + BC = AB + \bar{A}C \quad (4.24)$$

This expression $AB + \bar{A}C$ represents the transmission of a new network, represented in Fig. 4.24 and which is equivalent to the first. However, this time the network possesses only 4 contacts.

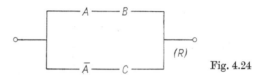

Fig. 4.24

The reader could retrace, step by step, the algebraic equations leading to the simplified expression $AB + \bar{A}C$ and, for each case, draw the circuit by using a procedure analogous to that which was used for the relations R_1 to R_{20}. The example above shows how the circuit operations can be replaced by algebraic operations. In this instance it is possible to arrive at a simplification of the circuit.

Example 2.

Consider the network of Fig. 4.25. When is it open?

Fig. 4.25

4.5 Operations on Dipoles

By analogy, making use of the principal properties involved, it may be shown that:

$$(4.25) \begin{cases} T = \bar{A}\,\bar{B}\bar{C} + \bar{C} + \bar{A}C + AC & \\ \quad = (\bar{A}\,\bar{B}\bar{C} + \bar{C}) + (\bar{A}C + AC) & (4.17) \\ \quad = \bar{C} + (\bar{A}C + AC) & (R_{15}) \\ \quad = \bar{C} + (\bar{A} + A)C & (R_{7}) \\ \quad = \bar{C} + 1C & (R_{10}) \\ \quad = \bar{C} + C = 1 & (R_{6}, R_{10}) \\ T = 1 & \end{cases}$$

The circuit is, therefore, always closed and is the equivalent of a simple connection.

This example shows how direct reasoning from the circuits, often laborious, can be replaced by simple algebraic manipulations to determine when the circuit is closed.

In short: wherever there are networks of relay contacts the following operations can be effected:

1) With each dipole circuit can be associated a transmission function T.
2) With every operation on the circuit (complementation, series, parallel) are associated operations on the transmissions.
3) Consequently, with every *dipole circuit* can be associated an *algebraic expression* as a function of the contacts. This expression consists of nothing more than another representation of the circuit.

It provides the following advantages:

1. It is a compact and *linear* representation, replacing the 2 dimensional drawing.
2. It can be transformed algebraically, by use of the rules R_1 to R_{20}, to give other expressions for as many circuits as desired.

Operations on the circuits are, then, replaced by algebraic manipulations. Hence there is no need for re-drawing the intermediary variants.

Thus has been defined for the range of dipole contact circuits a kind of algebra hereafter referred to as the "contact network algebra", or simply "contact algebra". This gives, in the particular case of contact relays, an example of *a switching algebra*. At a later stage other switching elements, principally electronic and magnetic elements, will be introduced. These can be organised into complex networks to be studied with the help of an algebra which will be seen to have the same structure as the algebra of contacts.

Note — So far only dipole circuits have been considered. The above ideas can be extended to cover a multipole circuit by the introduction of a *connection matrix* between the various pairs of poles. Alternatively, there exist dipole networks known as "bridge networks" which cannot be obtained merely by the series or paralleling operations. These can be more economical than the series-parallel networks. For these problems the reader is recommended to consult the bibliography (p. ex.: [26, 36]).

Chapter 5

Algebra of Classes. Algebra of Logic

5.1 Algebra of Classes

Consider a set I which, in the text which follows, will play the role of universal set. The different subsets of I form a set on which may be defined three operations: union, intersection and complementation. Endowed with these, the set of subsets of I then constitutes an '*algebra*', usually known as an "algebra of sets" or an "algebra of classes". From the definitions of set, subset and those of the three operations above the diverse properties of this algebra can be demonstrated. The more important of these properties will be stated later. They will be justified with the aid of Venn diagrams. However, by way of an exercise the reader can carry out rigorous abstract demonstrations based on the definitions in Chapter 1. It will be shown by an example (relationship E_9 below) which type of demonstration can then be made.

Among others, the following relationships can be demonstrated:

E_1) $$(A \cup B) \cup C = A \cup (B \cup C) \qquad (5.1)$$

This is the property known as the "*associativity of union*" $\cup$ (Fig. 5.1a). The above identity involves:

- 3 sets A, B and C written in the same order in each member and playing the role of variables.
- the operation $\cup$ applied twice.
- parentheses defining a grouping of variables and consequently the train of operations leading to one or other of the 2 expressions. This train of operations is either $(A \cup B)$, $(A \cup B) \cup C$, or $B \cup C$, $A \cup (B \cup C)$. The two expressions on the left and on the right of the sign = in (E_1) denote, therefore, the 2 ways (the only 2) of grouping the terms A, B and C in order twice to apply the operation $\cup$, which, it is to be noted, is an operation on *two* sets. The identity E_1 expresses, therefore, that the result of the double application of $\cup$ is independent of grouping. It will be *convenient* to call this unique result, the *union* of A, B and C denoted, *by definition* as $A \cup B \cup C$. Giving:

$$A \cup B \cup C = (A \cup B) \cup C = A \cup (B \cup C) \qquad (5.2)$$

The expression $A \cup B \cup C$ must, therefore, be considered, from theoretical viewpoint, as a *new notation*, denoting the result of a *equally new operation* with 3 variables: the union of A, B and C, and the result can be calculated by one (chosen) of the two procedures denoted by $(A \cup B) \cup C$ and $A \cup (B \cup C)$.

Consequently, from a *practical* viewpoint this time, the expression $A \cup B \cup C$ can be considered as an *abridged notation* where the parentheses "float" and are located to suit the convenience of the user.

The above considerations also apply in the case where there are variables instead of just 3, ($n \geq 3$, finite). Given n sets, $A_1, A_2 \ldots A$ in this order, it is possible by repeated application of the operation $\cup$ to obtain an unique set known as the union of $A_1, A_2, A_3 \ldots A_n$ and denoted by:

$$A_1 \cup A_2 \cup A_3 \ldots \cup A_n$$

In this case the law of associativity (the "generalised law of associativity") states that the grouping of the variables has no influence on the result, for a given order of groups $A_1, A_2, \ldots A_n$. These, for example translate for $n = 4$ into relationships of the following type:

$$A_1 \cup A_2 \cup A_3 \cup A_4 = (A_1 \cup A_2 \cup A_3) \cup A_4 = (A_1 \cup A_2) \cup (A_3 \cup A_4)$$
$$= A_1 \cup (A_2 \cup A_3 \cup A_4) = A_1 \cup (A_2 \cup A_3) \cup A_4$$

(5.?)

This generalised law of associativity (with n variables) can easily be deduced from that for three variables. The proof is left to the reader by way of an exercise.

(E_2) $\qquad\qquad (A \cap B) \cap C = A \cap (B \cap C) \qquad\qquad$ (5.?)

This, in an analogous manner, is the *law of associativity for the operation of intersection* $\cap$ (Fig. 5.1b). It is now possible to define the *intersection* of A, B and C, denoted by $A \cap B \cap C$ by the (generalised) equation also having n variables.

$$A \cap B \cap C = (A \cap B) \cap C = A \cap (B \cap C) \qquad (5.?)$$

(E_3) $\qquad\qquad A \cup B = B \cup A \qquad\qquad$ (5.?)

The *law of commutivity of the operation of union* $\cup$ which springs immediately from the definition $A \cup B$: Evidently the result does not depend upon the order of the two 'operands' A and B. In fact, when the operation of union was defined (chap I) immediate use was made of this property: indeed it was agreed to read the symbol $A \cup B$ (unsymmetri

5.1 Algebra of Classes

xpression) "union of A and B (symmetric expression) and not, for example, "A united with B", "A increased by B" etc., non-symmetric xpressions which could only be provisional on account of E_3.

$$E_4) \qquad A \cap B = B \cap A \qquad (5.7)$$

The *law of commutativity of the operation of intersection* $\cap$ calls into play the same observations as for (E_3).

$$E_5) \qquad A \cup \emptyset = A \qquad (5.8)$$

ince $\emptyset$ contains no point (element) its union with a set A is none other than A itself.

$$E_6) \qquad A \cap I = A \qquad (5.9)$$

ince A is included in I, (by definition), the points common to A and I ll belong to A.

$$E_7) \qquad A \cap (B \cup C) = (A \cap B) \cup (A \cap C) \qquad (5.10)$$

The *law of distributivity of intersection with respect to union*. In ordinary algebra (that of real or complex numbers) there is an analogous property: a, b and c being 3 numbers, gives:

$$a \times (b + c) = a \times b + a \times c \qquad (5.11)$$

This is the well-known property of the distributivity of multiplication with respect to addition.

$$E_8) \qquad A \cup (B \cap C) = (A \cup B) \cap (A \cup C) \qquad (5.12)$$

The *law of the distributivity of union with respect to intersection* Fig. 5.1c).

Taking up the correspondence (albeit arbitrary) earlier established between $\cup$, $+$ on the one hand and $\cap$, $\times$ on the other, the equation E_8) has no counterpart in ordinary algebra which, in respect of the laws of distributivity, does not, therefore, possess the symmetry of the algebra of sets.

$$E_9) \qquad A \cap \bar{A} = \emptyset \qquad (5.13)$$

A set A and its complement $\bar{A}$ have nothing in common, this is a consequence of the definition of A.

$$E_{10}) \qquad A \cup \bar{A} = I \qquad (5.14)$$

The union of an assembly A and of its complement $\bar{A}$ is always the universal set. This is a second consequence of the definition of $\bar{A}$.

The complement $\bar{A}$ of A thus satisfies the two equations E_9 and E_{10}. Conversely, if a set B satisfies both equations:

$$(E'_9)\ A \cap B = \emptyset \qquad (E'_{10})\ A \cup B = I \qquad (5.15)$$

it can be seen that B is effectively the complement of A or that $B = \bar{A}$. In fact, if an element $x \in B$, (E'_9) ensures that $x \notin A$ and (E'_{10}) that $x \in I$, then x belongs to $I - A = \bar{A}$, $x \in \bar{A}$. Therefore $B \subset \bar{A}$. But every element of $\bar{A}$ otherwise checks with the expressions E_9 and E_{10} so that if $x \in \bar{A}$, $x \in B$, hence $\bar{A} \subset B$. Thus $\bar{A} = B$. In other terms the two equations (E'_9) and (E'_{10}) have $\bar{A}$ as their solution and, in practice, are often used for defining $\bar{A}$ directly.

(E_{11}) $$A \cup A = A \qquad (5.16)$$

The *law of idempotence for union* $\cup$. In a general way this results in

$$\underbrace{A \cup A \ldots \cup A}_{n \text{ times}} = A \qquad (5.17)$$

(E_{12}) $$A \cap A = A \qquad (5.18)$$

The *law of idempotence for intersection* $\cap$. By analogy it may be deduced that:

$$\underbrace{A \cap A \ldots \cap A}_{n \text{ times}} = A \qquad (5.19)$$

The laws (E_{11}) and (E_{12}) thus offer a marked contrast with those of ordinary algebra: there being neither powers nor co-efficients in the algebra of sets.

(E_{13}) $\qquad\qquad\qquad A \cup I = I \qquad\qquad\qquad (5.20)$

(E_{14}) $\qquad\qquad\qquad A \cap \emptyset = \emptyset \qquad\qquad\qquad (5.21)$

(E_{15}) $\qquad\qquad\qquad A \cup (A \cap B) = A \qquad\qquad (5.22)$

First law of absorption.

The intersection $E = A \cap B$, which can only be a subset of A, the ultimate union of E with A can only produce A.

(E_{16}) $\qquad\qquad\qquad A \cap (A \cup B) = A \qquad\qquad (5.23)$

Second law of absorption.

A, being contained in $E = A \cup B$, $A \cap E = A$

(E_{17}) $\qquad\qquad A \cup (\bar{A} \cap B) = A \cup B \qquad\qquad (5.24)$

(E_{18}) $\qquad\qquad A \cap (\bar{A} \cup B) = A \cap B \qquad\qquad (5.25)$

(E_{19}) $\qquad\qquad \overline{A \cup B} = \bar{A} \cap \bar{B}. \qquad\qquad (5.26)$

5.2 Algebra of Logic (Calculus of Propositions)

The first theorem of De Morgan for sets is often quoted as: "the complement of the *intersection* is the *union* of the complements".

E_{20})
$$\overline{A \cap B} = \overline{A} \cup \overline{B} \qquad (5.27)$$

The second theorem of De Morgan states: "the complement of the *intersection* is the *union* of the complements".

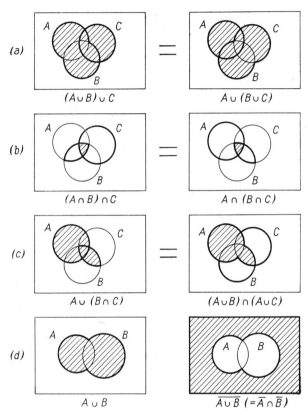

Fig. 5.1. Venn Diagrams — Some identities.

5.2 Algebra of Logic (Calculus of Propositions)

5.2.1 Propositions. Logic Values of a Proposition

An examination of the language in daily use or, more particularly, that which serves as a vehicle for scientific reasoning, reveals units, *propositions* from which can be derived theorems, proofs of theorems, definitions etc. The manipulation of language in the form of proposition is, moreover, one of the fundamental characteristics of human language. Hence the expressions:

"11 is a prime number"
"Paris is the capital of France"
"The triangle ABC is rectangular"

Each of these propositions can be examined from the point of view of its *truth* or *falsity*.

For example, the proposition "11 is a prime number" is a true proposition. On the other hand, the other two propositions can be examined from this same point of view and it must be possible to state whether they are true or false. In what follows we will study proposition and suppose that they can only be either true or false.

5.2.2 Operations on Propositions

Pursuing the examination of ordinary language it is possible to produce in evidence a whole class of words or even more complex expressions which are not propositions but *links* between propositions. Such, in particular, is the case for the so-called conjunctions of co-ordination: AND, OR, NEITHER, BUT, FOR, THEREFORE and also of expressions such as "leading to", "implies that", "equivalent to saying that" etc. These words and expressions allow the combining of propositions, "elements" in a certain sense, into more complex propositions i.e. to carry out a sort of operation which, based on the constituent propositions, results in a new "compound" proposition.

So long as one adheres to current language, these operations of liaison of propositions, as will be seen, are not always completely devoid of ambiguity. Nevertheless, the notion of proposition and of the elements of liaison as encountered provide a sort of framework of analogous mathematical concepts, which can be endowed, this time, with a rigorous definition.

In a way these concepts are "models" and play a role analogous to that of models for physics, social or other sciences.

5.2.2.1 Operation AND (Logical Product)

Consider, for example, the two propositions:
"X is a prime number"
"The triangle ABC is rectangular"
With the aid of the conjunction AND it is possible to state a new pro-

5.2 Algebra of Logic (Calculus of Propositions)

sition: "X a is prime number AND the triangle ABC is rectangular". hen is this true? If one adheres to current language this new propo-ion will be true so long as the two component propositions are both ιe, and false in all other cases. It is as well to note that two other pro-sitions could have been used such as, for example:

"Paris is the capital of France"
"The climate of France is moderate"

d, in the same way, to formulate a new proposition: "Paris is the pital of France AND the climate of France is moderate". This pro-sition will be recognised as being true if the two component proposi-ιns are true and only in this case. It is thus seen that the operation ND has produced a new proposition made up from two component opositions in a way that essentially determines whether or not it is lse, knowing the falsity or veracity of the first of the propositions, but ,sically is independent of the particular sense and nature of the pro-sitions.

The character of the operation AND can be demonstrated by using, stead of the two propositions quoted above, abstract variables, e.g. and Q. These variables will, therefore, represent the propositions, by aying a role analogous to that of the variables in ordinary algebra. The wly obtained proposition P AND Q will be denoted by uniting the opositions P and Q by means of the conjunction AND. At this point, by aking use of abstract models for the propositions (mathematical enti-es capable of being substituted for P and Q and, of necessity, endowed th one or other of the two exclusive qualities *truth* or *falsity*) it remains ly to complete the definition of the operation AND which, hence-rth, will be used in logic: with P it is necessary to associate a variable ₽, and with Q a variable X_Q, the variables X_P and X_Q being *binary*; ch of these variables can have but two values, the *values in logic*, noted by T (truth) and F (falsity) in the text which follows. The eration AND based on P and Q gives the proposition $P \wedge Q$ as shown the following table.

Table 5.2. Operation AND

X_P	X_Q	$X_{P \wedge Q}$
F	F	F
F	T	F
T	F	F
T	T	T

Thus has been defined one of the operations in the mathematic logic. Henceforth, this will be known as the *operation* AND or the *logic product* denoted by the symbol ∧, reminiscent of the ∩ used in the theor of sets. The operation ∧ has, in literature, also been given other name *conjunction*, *co-incidence* and *intersection*. A return to these and oth symbols to be used will be made later.

A table, such as that of Table 5.2 is known as a *truth table*: it d scribes for all possible combinations the value (truth or falsity) of th compound proposition as a function of the values of the componer propositions.

5.2.2.2 Operation OR (OR-Inclusive)

Consider the statement "The product $X \cdot Y$ (in ordinary algebra) nil when $X = 0$ or when $Y = 0$". For greater precision, it could add tionally be said "or again when X and Y are both zero". But if the phras is left alone it remains sufficiently clear for the majority of readers an will retain precisely the significance: "XY is nil when $X = 0$, $Y \neq$ or $X \neq 0$, $Y = 0$ or, even, $X = 0$, $Y = 0$". The proposition $(X = 0$ or $(Y = 0)$ thus furnishes an example of a new operation, which ca rigorously be defined by the aid of a table. For two given propositior P and Q the proposition P OR Q, denoted by $P \lor Q$, is known as th compound proposition, defined by means of the following truth tab (Table 5.3)

Table 5.3. *Operation OR*

X_P	X_Q	$X_{P \lor Q}$
F	F	F
F	T	T
T	F	T
T	T	T

The operation ∨ thus defined is also known as *logical sum*, *disjunctio* and *union*. Other names and symbols sometimes used in place of ∨ wi be discussed later. It should again be noted that current language ha in general, the sense given here to the operation ∨ (OR-inclusive). Th is, however, not always the case — witness the use of the abbreviatio "OR/AND". This possesses a second signification on whose basis can b defined a different operation, the "OR-*exclusive*) as will be shown late Before proceding to this it will be necessary to define the negation of proposition.

5.2.2.3 Negation of a proposition

Starting from a proposition such as "it is raining today", it is possible to produce "it is not raining today". It is clear that if the first is true, the second is false, and vice versa. Moreover, it is possible to proceed in the same way with any proposition.[1] Once again it is possible to make use of a symbolic variable P, representing any particular proposition, on which can be defined an operation of a mathematical character: *negation*. The negation of a proposition P denoted by $\sim P$ is defined by means of the truth Table 5.4, where X_P and $X_{\sim P}$ are variables taking the values T (truth) or F (falsity).

Table 5.4. *Negation*

X_P	$X_{\sim P}$
F	T
T	F

The symbol $\sim$ ("tilde") is, therefore, the equivalent in formal logic of the diverse expressions upon which can be constructed in current language the negation of a proposition. It is always placed before the symbol for the proposition. Summarizing, at this stage three fundamental operations have been defined which, for various reasons, will be better understood from what follows: i.e. the operations AND, OR and negation. It should be noted that one of the operations — negation — acts on one variable and that the two others, the operations AND and OR combine two operations. The negation is, therefore, in this sense a *unary* operation, the others are *binary* operations.

5.2.3 Compound Propositions and Functions of Logic

Use has already been made of two types of variables: the variables P and Q on the one hand, the variables X_P and X_Q on the other. To simplify the notations it will, henceforth, be convenient to use the same variables P and Q at the same time for the propositions and for the logic variables which express their truth or falsehood. As a counterpart, for the future it will be necessary to bear in mind that they possess a double sense: that of variables replacing the propositions or that of logic variables associated therewith.

Assume we are given an algebraic expression such as:

$$A = [(P \wedge Q) \vee P] \wedge (\sim R) \qquad (5.28)$$

---

[1] Thanks to various expressions, amongst which in particular is to be found the negative NOT.

representing either a complex proposition constructed with the aid of P, Q and R, or an algebraic expression in logic variables, which will permit, by simple substitution and the rules for calculation set out in the tables, the determination of the logic value of A. The first notion is that of a *compound proposition*, the second that of a *logic function*. Given a compound expression E having n variables $P_1 \ldots P_n$, replacement of these variables by the possible combinations of T or F allows the value (T or F) of the expression E to be derived for every case. In this sense the expression E generates a function from $\{T, F\}^n$ into $\{T, F\}$ known as a logic function of n variables. The above convention will be retained with respect to the variables as well as the theoretical distinction between the compound proposition and logic function for the other operations and functions to be defined at a later stage. In practice, either expression will be used indifferently in all cases where such usage does not lend itself to confusion.

5.2.4 Numeric Notation of Logic Values

In conformance with general usage, the two logic values of the variables will be represented, not by the symbols T and F used so far, but by the symbols 0 and 1 — identical to those of the figures 0 and 1. It must be understood that this is merely a matter of convention, a practice to which arithmetic is foreign from the theoretical point of view. There being two possible conventions, the one to be adopted here, the most widely used, is as follows:[1]

T	1
F	0

The truth tables 5.2, 5.3 and 5.4 thus become:

Table 5.5
AND

P Q	P∧Q
0 0	0
0 1	0
1 0	0
1 1	1

Table 5.6
OR (*inclusive*)

P Q	P∨Q
0 0	0
0 1	1
1 0	1
1 1	1

Table 5.7
Negation

P	∼P
0	1
1	0

---

[1] This is reminiscent of the calculation of probabilities: probability 1 for an event which is certain; 0 for an event which is impossible.

5.2.5 Other Logic Functions

5.2.5.1 2 Variable Functions

The method of defining the operations AND and OR may be generalised. In fact $2^4 = 16$ truth tables of 2 variables can be constructed in order to define as many operations or logic functions numbered from 0 to 15 and for which the list is to be found in Table 5.8.

In addition to the functions AND, OR and negation already defined, it contains 14 other functions the most important of which have been named and are, briefly, to be discussed at this time.

5.2.5.2 Function "Exclusive OR" $(P \oplus Q)$

This is the name given to the function F_6, sometimes also called *mutual exclusion*, *dilemma* or *non-equivalence*. It will be denoted by the symbol $\oplus$ and is reproduced, by way of comparison, in the two truth tables $P \oplus Q$ and $P \vee Q$ of Table 5.9.

Table 5.9. *OR-Exclusive and OR-Inclusive*

P Q	$P \oplus Q$	$P \vee Q$
0 0	0	0
0 1	1	1
1 0	1	1
1 1	0	1

Briefly, $P \oplus Q$ has the logic value 1 so long as P and Q have different logic values, otherwise the logic value is 0, while the proposition $P \vee Q$ is still true for $P = 1$, $Q = 1$. The OR of current language sometimes takes the sense of "exclusive OR" as, for example, in the sentence "the product of algebraic variables (not zero) A and B is negative when (or else) A or else B is negative". However, it is only by the use of the words "or else ... or else" that it is possible really to avoid any ambiguity, and it is this form of OR which, in current language, has the value "exclusive". OR alone, according to the context, can have the two values of inclusive OR or exclusive OR.

5.2.5.3 Equivalence $(P \leftrightarrow Q)$

With the aid of the operand $\leftrightarrow$ can be obtained a new proposition denoted as $P \leftrightarrow Q$, true so long as P and Q have the same logic value but otherwise false, known as the *equivalence* of P and Q or the *reciprocal*

Table 5.8 The 16 logic functions of 2 variables

P	Q	F_0	F_1 $P\downarrow Q$	F_2	F_3 $\sim P$	F_4	F_5 $\sim Q$	F_6 $P\oplus Q$	F_7 $P\vert Q$	F_8 $P\wedge Q$	F_9 $P\leftrightarrow Q$	F_{10} Q	F_{11} $P\to Q$	F_{12} P	F_{13} $Q\to P$	F_{14} $P\vee Q$	F_{15}
0	0	0	1	0	1	0	1	0	1	0	1	0	1	0	1	0	1
0	1	0	0	1	1	0	0	1	1	0	0	1	1	0	0	1	1
1	0	0	0	0	0	1	1	1	1	0	0	0	0	1	1	1	1
1	1	0	0	0	0	0	0	0	0	1	1	1	1	1	1	1	1

5.2 Algebra of Logic (Calculus of Propositions)

implication of P and Q. It will suffice to compare the truth table (Table 5.10) with that for OR exclusive:

Table 5.10. *Equivalence and OR exclusive*

P Q	$P \leftrightarrow Q$	$P \oplus Q$
0 0	1	0
0 1	0	1
1 0	0	1
1 1	1	0

to see that the proposition $P \leftrightarrow Q$ and $P \oplus Q$ are the negation the one of the other. This justifies the respective names of equivalence and non-equivalence that have been given to them. As for the second name for $P \leftrightarrow Q$, its justification is to be found in the following operation.

5.2.5.4 Implication $(P \rightarrow Q)$

The function F_{11} whose truth table is that of Table 5.11 translates, approximately, the sense, sometimes vague, of the notion of implication in ordinary language which can often be summarized thus: So long as P

Table 5.11. *Implication*

P Q	$P \rightarrow Q$
0 0	1
0 1	1
1 0	0
1 1	1

is true, Q is also true: therefore $P \rightarrow Q$ is true so long as $P = 1$, $Q = 1$ and false when $P = 1$ and $Q = 0$. Furthermore, when P is false nothing prevents Q from having an arbitrary value 1 or 0 and $P \rightarrow Q$ is considered to be true in the 2 cases where $P = 0$. Briefly: "Q cannot be false so long as P is true". The function F_{13} represents in an analogous manner the function $Q \rightarrow P$. The symbol $P \rightarrow Q$ can be read in various ways: P *implies* Q "implication of Q by P" and sometimes also "P is a condition sufficient for Q", "Q is a necessary condition for P".

Consider the proposition $(P \to Q) \wedge (Q \to P)$ from which can be established a truth table (5.12) as follows:

Table 5.12

P Q	$P \to Q$	$Q \to P$	$(P \to Q) \wedge (Q \to P)$
0 0	1	1	1
0 1	1	0	0
1 0	0	1	0
1 1	1	1	1

The right hand column is the same as that for the operation $P \leftrightarrow Q$ which, at the same time, justifies the symbol and the name of reciprocal implication defined in the preceding paragraph.

It is necessary once more to emphasise that the definition of implication given above does not always correspond with the intuitive notion often experienced, in fact, in certain cases it leads to diverse paradoxes which will not be dealt with here. The reader should refer to the table of references at the end of this work for details on this problem.

5.2.5.5 Sheffer Functions (P/Q) "Function NAND" (Sheffer Stroke)

This is function F_7 of Table 5.8

Table 5.13. *Sheffer Function and function AND*

P Q	P / Q	P∧Q
0 0	1	0
0 1	1	0
1 0	1	0
1 1	0	1

The proposition P/Q is false so long as P and Q are both true and true is all other cases. This justifies, in particular, the name also given to it of the *incompatibility* of P and Q. Additionally it will be seen that $P/Q = \overline{P \wedge Q}$ which has led to the other names ascribed to it, i.e. "*function* NOT-AND" and "NAND-*function*". Further reference will be made to this important function later.

5.2 Algebra of Logic (Calculus of Propositions)

5.2.5.6 Peirce Function $(P \downarrow Q)$ ("Function NOR"; Peirce stroke)

This is the function F_1

Table 5.14. *Peirce function and OR function*

P Q	$P \downarrow Q$	$P \vee Q$
0 0	1	0
0 1	0	1
1 0	0	1
1 1	0	1

The proposition $P \downarrow Q$ is therefore true so long as neither P nor Q is true. Hence the name *"function* NOR" often ascribed to it. Here $P \downarrow Q = \overline{P \vee Q}$.

5.2.6 Tautologies — Contradictions

Of all the logic functions of n variables $(n \geq 1)$ which can be defined, for example, by the use of a truth table to n variables, two among them present a distinguished character: the function invariably nil and the function always equal to 1, which will be denoted as F_0 and F_{2^n-1}. In table 5.8 this role is played by the functions F_0 and F_{15}. But, a function $f(P_1, P_2 \ldots, P_n)$ can be considered as defining (by the intermediary of the logic values) an operation on n propositions, resulting in a compound proposition.

The functions F_0 (always nil) and F_{2^n-1} (always equal to 1) are, therefore, associated with 2 types of proposition which are, respectively, always false or always true, no matter what the logic values of the component propositions. Generally, it may be stated that by definition:

— a compound proposition which is false for all possible logic values of the n component propositions, is a *contradiction:*
— a compound proposition, which is true for all the logic values of the n component propositions is a *tautology.*

However, one of the aims of mathematical logic is the determination of relations between propositions of a mathematical character, this in a manner independent, not only of the particular significance of the propositions, but also of their particular logic values. This is nothing more than a search for the tautologies and contradictions.

Examples of tautologies and contradictions:

These, together with their justifications, are given in the truth tables which follow.

1. *Principle of double negation:*

$$P \leftrightarrow \overline{\overline{P}} \quad \text{(tautology)} \tag{5.29}$$

Table 5.15

P	$\overline{P}$	$\overline{\overline{P}}$	$P \leftrightarrow \overline{\overline{P}}$
0	1	0	1
1	0	1	1

The compound proposition (with 1 variable) $P \leftrightarrow \overline{\overline{P}}$ is, therefore, always true: this is a tautology which can be expressed as:

"Every proposition is the equivalent of the negation of its own negation"

2. $(P \leftrightarrow Q) \leftrightarrow [(P \to Q) \wedge (Q \to P)]$ (tautology) $\tag{5.30}$

Table 5.16

$P\ Q$	$P \leftrightarrow Q$	$P \to Q$	$Q \to P$	$(P \to Q) \wedge (Q \to P)$	$(P \leftrightarrow Q) \leftrightarrow [(P \to Q) \wedge (Q \to P)]$
0 0	1	1	1	1	1
0 1	0	1	0	0	1
1 0	0	0	1	0	1
1 1	1	1	1	1	1

The equivalence of P and Q is the same as the conjunction of the implication of P by Q and that of Q by P.

3. *Principle of identity:*

$$P \to P \quad \text{(tautology)} \tag{5.31}$$

Table 5.17

$P\ P$	$P \to P$
0 0	1
1 1	1

Every proposition implies itself"

4. *Principle of the excluded middle:*

$$P \vee \bar{P} \quad \text{(tautology)} \tag{5.32}$$

Table 5.18

P	$\bar{P}$	$P \vee \bar{P}$
0	1	1
1	0	1

"Every proposition is either true or false".

5.
$$\overline{(P \wedge \bar{P})} \quad \text{(contradiction)} \tag{5.33}$$

"A proposition cannot at the same time be true and false".

6.
$$[(P \to Q) \wedge (Q \to R)] \to [P \to R] \quad \text{(tautology)} \tag{5.34}$$

Table 5.19

P Q R	$P \to Q$	$Q \to R$	$(P \to Q) \wedge (Q \to R)$	$P \to R$	$[(P \to Q) \wedge (Q \to R)]$ $\to [P \to R]$
0 0 0	1	1	1	1	1
0 0 1	1	1	1	1	1
0 1 0	1	0	0	1	1
0 1 1	1	1	1	1	1
1 0 0	0	1	0	0	1
1 0 1	0	1	0	1	1
1 1 0	1	0	0	0	1
1 1 1	1	1	1	1	1

5.2.7 Some Properties of the Functions AND, OR and Negation

In the following passage are found a certain number of algebraic identities which relate the three operations AND, OR, negation ($\wedge$, $\vee$, $-$). They are presented in the form of tautologies or contradictions where it is convenient to denote them as $\dot{P}$ or $\mathring{P}$ respectively[1] without reference to the number n ($n = 2$ or 3) variables.

---

[1] Notation as used by J. *Chauvineau* [31].

5. Algebra of Classes. Algebra of Logic

(L_1) $(P \vee Q) \vee R \leftrightarrow P \vee (Q \vee R)$ (associativity of the operation OR)
(L_2) $(P \wedge Q) \wedge R \leftrightarrow P \wedge (Q \wedge R)$ (associativity of the operation AND)
(L_3) $P \vee Q \leftrightarrow Q \vee P$ (commutativity of the operation OR)
(L_4) $P \wedge Q \leftrightarrow Q \wedge P$ (commutativity of the operation AND)
(L_5) $P \vee \mathring{P} \leftrightarrow P$
(L_6) $P \wedge \mathring{P} \leftrightarrow P$
(L_7) $P \wedge (Q \vee R) \leftrightarrow (P \wedge Q) \vee (P \wedge R)$
  (distributivity of AND with respect to OR)
(L_8) $P \vee (Q \wedge R) \leftrightarrow (P \vee Q) \wedge (P \vee R)$
  (distributivity of OR with respect to AND)
(L_9) $\sim [P \wedge (\sim P)]$ (principle of contradiction)
(L_{10}) $P \vee (\sim P)$ (principle of excluded third)
(L_{11}) $P \vee P \leftrightarrow P$ (idempotence of OR)
(L_{12}) $P \wedge P \leftrightarrow P$ (idempotence of AND)
(L_{13}) $P \vee \mathring{P}$
(L_{14}) $\sim (P \wedge \mathring{P})$
(L_{15}) $P \vee (P \wedge Q) \leftrightarrow P$ (absorption)
(L_{16}) $P \wedge (P \vee Q) \leftrightarrow P$ (absorption)
(L_{17}) $P \vee [(\sim P) \wedge Q] \leftrightarrow P \vee Q$
(L_{18}) $P \wedge [(\sim P) \vee Q] \leftrightarrow P \wedge Q$
(L_{19}) $\sim (P \vee Q) \leftrightarrow [(\sim P) \wedge (\sim P)]$ ⎫
(L_{20}) $\sim (P \wedge Q) \leftrightarrow [(\sim Q) \vee (\sim Q)]$ ⎬ De Morgan theorems

5.2.8 Examples and Applications

(1) *Nullity of an ordinary algebraic product.*

In ordinary algebra, in order that the product $m = a \times b \times c$ be $\neq 0$, it is necessary and sufficient the $a \neq 0$, $b \neq 0$ and $c \neq 0$. Thus giving the following equivalence, (placing the propositions in parentheses)

$$(a \times b \times c \neq 0) \leftrightarrow (a \neq 0) \wedge (b \neq 0) \wedge (c \neq 0) \qquad (5.35)$$

Conversely, in order that $m = 0$, it is necessary and sufficient that one or several factors be nil, and expressed as:

$$(a \times b \times c = 0) \leftrightarrow (a = 0) \vee (b = 0) \vee (c = 0) \qquad (5.36)$$

which can also be obtained by the complementation of (5.35)

$$\overline{(a \times b \times c \neq 0)} \leftrightarrow \overline{(a \neq 0) \wedge (b \neq 0) \wedge (c \neq 0)}$$

$$\overline{(a \times b \times c \neq 0)} \leftrightarrow \overline{(a \neq 0)} \vee \overline{(b \neq 0)} \vee \overline{(c \neq 0)}$$

$$(a \times b \times c = 0) \leftrightarrow (a = 0) \vee (b = 0) \vee (c = 0) \qquad (5.37)$$

5.2 Algebra of Logic (Calculus of Propositions)

(2) *Rule for signs in ordinary algebra.*

Consider the product $m = a \times b$.
First case: $a \neq 0$, $b \neq 0$
The rule of the signs can be expressed as

$$(a \times b > 0) \leftrightarrow [(a < 0) \wedge (b < 0)] \vee [(a > 0) \wedge (b > 0)]$$

$$(a \times b > 0) \leftrightarrow [(a < 0) \leftrightarrow (b < 0)] \tag{5.38}$$

or again

$$(a \times b > 0) \leftrightarrow [(a > 0) \leftrightarrow (b > 0)]. \tag{5.39}$$

Conversely

$$(a \times b < 0) \leftrightarrow [(a < 0) \oplus (b < 0)]$$

or

$$(a \times b < 0) \leftrightarrow [(a > 0) \oplus (b > 0)] \tag{5.40}$$

or

$$(a \times b < 0) \leftrightarrow \overline{[(a < 0) \leftrightarrow (b > 0)]}, \text{ etc.}$$

Second case: a, b can be positive, negative or nil.
It may, for example, be expressed

$$a \times b \geqq 0) \leftrightarrow (a \times b > 0) \vee (a \times b = 0)$$

$$\leftrightarrow (a \neq 0) \wedge (b \neq 0) \wedge [(a > 0) \leftrightarrow (b > 0)] \vee (a = 0)$$

$$\vee (b = 0) \tag{5.41}$$

$$\leftrightarrow \overline{[(a = 0) \vee (b = 0)]} \wedge [(a > 0) \rightarrow (b > 0)] \vee [(a = 0)$$

$$\vee (b = 0)] \tag{5.42}$$

$$\leftrightarrow [(a > 0) \leftrightarrow (b > 0)] \vee [(a = 0) \vee (b = 0)] \quad \big(\text{from } (L_{17})\big) \tag{5.43}$$

$$a \times b \geqq 0) \leftrightarrow [(a > 0) \leftrightarrow (b > 0)] \vee (a = 0) \vee (b = 0), \text{ etc.} \tag{5.44}$$

5.2.9 Algebra of Logic (Propositional Calculus)

So far a certain number of operations have been defined for one or two propositions and, in addition, it has been shown how this procedure can be generalised to the case of n propositions thanks to the "truth tables" which relate the logic values of the compound propositions to those of the component propositions: for each of the 2^n combinations of logic values of the latter, a truth table indicates the value of a logic function, thus defining an operation on n propositions. All the operations thus defined have one character in common: they permit the manipulation of the propositions independently of their internal structure (the propositions so far used and denoted by P, Q etc., are such). Thus they provide the instruments of a kind of algebra known as the *algebra of logic, propositional calculus* or, again, *propositional logic* which deals only with operations *on* propositions. This algebra comprises only one part of the mathematical logic which, itself, is not limited to the study of propositions. However, the study of propositions does not stop at the operations of logic so far examined — in fact, the study of their internal structure can equally form the object of a mathematical treatment: this is the aim of what is known as *functional logic* which will now be defined.

5.2.10 Functional Logic

5.2.10.1 Propositional Functions

Consider expressions such as:

"x is less than y"
"x is between y and z"

These are not, properly speaking, true propositions since nothing is known of x, y and z and the value (true or false) cannot be decided: x, y and z play the role of variables susceptible to replacement by particular numeric values. If this replacement be made (for example $x = 2$, $y = 3$, $z = 4$) then for the propositions as produced: the first is true, the second is false. Such propositions containing variables are known as *propositional functions* or *relations*. When the relation consists of only one variable it is often given the name of *property*. When replacing the variables by particular values, taken within their sphere of definition, the propositional function is transformed into an ordinary proposition.

5.2.10.2 Operations on Propositional Functions

Consider propositional functions such as:

"x is less than y «AND» z is even"
"the triangle x is equilateral OR x is less than y"
"x is not less than y"

If the variables were replaced by particular values, all the component functions would become propositions and, in addition, the compound functions would take the form of compound propositions, obtained by means of the usual operations AND, OR and negation.

5.2 Algebra of Logic (Calculus of Propositions)

Thus the operations $\wedge$, $\vee$ ($\sim$) on the propositional functions can be defined. These operations are formally identical to the operations on the propositions and have the significance indicated above.

5.2.10.3 Quantifiers

It has been seen that from a given propositional function can be derived a proposition by giving to the variables particular values taken from their sphere of definition. This is the so-called method of *specialization* of propositional functions. But is also possible to formulate propositions from propositional functions by using what are known as quantifiers (*quantification*).

— *Universal quantifier.*

With a relation $f(x, y, z)$ (having 3 variables) in associated the proposition:

"For any x, any y and any z, a relation $f(x, y, z)$ holds, which is expressed as

$$(\forall x)(\forall y)(\forall z) f(x, y, z) \tag{5.45}$$

The symbol $\forall$ is known as the "universal quantifier".

— *Existential quantifier.*

With a relation $f(x, y, z)$ is associated the proposition: "There exists at least one x, at leat one y, at least one z which will check the relationship $f(x, y, z)$". This is expressed as:

$$(\exists x)(\exists y)(\exists z) f(x, y, z) \tag{5.46}$$

The symbol $\exists$ is known as the "existential quantifier".

5.2.11 Binary Relations and Associated Propositions

For a given binary relationship $f(x, y)$ it is possible for each variable:
— to be replaced by a constant.
— to be universalised, i.e. endowed with the symbol $\forall$.
— to be extentialised, i.e. endowed with the symbol $\exists$.

Thus can be formulated $3^2 = 9$ types of proposition:

1. Proposition of the type $f(x_i, y_j)$
2. Proposition of the type $(\forall x) f(x, y_k)$
3. Proposition of the type $(\forall y) f(x_l, y)$

4. Proposition of the type $(\exists x) f(x, y_k)$
5. Proposition of the type $(\exists y) f(x_l, y)$
6. Proposition of the type $(\forall x)(\forall y) f(x, y)$ "universal" proposition
7. Proposition of the type $(\exists x)(\exists y) f(x, y)$ "existential" proposition

8a. Proposition of the type $(\forall x)(\exists y) f(x, y)$
which reads "for any x, there exists at least one y such that $f(x, y)$ holds true".

8b. Proposition of the type $(\forall y)(\exists x) f(x, y)$

} Universalised existential propositions

9a. Proposition of the type $(\exists x)(\forall y) f(x, y)$
which reads "there exists at least one x so that for any y $f(x, y)$ holds true".

9b. Proposition of the type $(\exists y)(\forall x) f(x, y)$

} Universal existentialised propositions

Note: case of a finite domain of definition.

It is possible to express the propositions containing the quantifier as a function of the specialised propositions, by the use of the operation OR and AND.

Examples:

1. *Relation with one variable (property)*

$$(\forall x) f(x) \leftrightarrow \bigwedge_i f(x_i)$$
$$(\exists x) f(x) \leftrightarrow \bigvee_i f(x_i) \qquad (5.47)$$

In effect, to say that the relationship $f(x)$ is true for any x, is to say that it is true for x_1 AND x_2 AND x_3 ... AND x_n. To say that there exists at least one x such that the relation $f(x)$ holds, is to say that at least one of the propositions $f(x_1), f(x_2)$ etc., must be true, i.e. that $f(x_1)$ OR $f(x_2)$ OR ... OR $f(x_n)$ is true.

2. *Binary relation*

One obtains, for example, with the obvious notations:

$$(\forall x)(\forall y) f(x, y) \leftrightarrow \bigwedge_{ij} f(x_i, y_j)$$
$$(\forall x)(\exists y) f(x, y) \leftrightarrow \bigwedge_i \left[\bigvee_j f(x_i, y_j)\right] \qquad (5.48)$$
$$(\exists x)(\forall y) f(x, y) \leftrightarrow \bigvee_i \left[\bigwedge_j f(x_i, y_j)\right]$$
$$(\exists x)(\exists y) f(x, y) \leftrightarrow \bigvee_{ij} f(x_i, y_j), \quad \text{etc.} \quad (i = 1, ..., I; \ j = 1, ..., J)$$

5.2 Algebra of Logic (Calculus of Propositions)

Application: To demonstrate in the case of a finite domain of definition the following implications:

$$(\forall x)(\forall y) f(x, y) \to (\exists x)(\forall y) f(x, y) \to (\forall y)(\exists x) f(x, y)$$
$$\to (\exists x)(\exists y) f(x, y) \qquad (5.49)$$

P_{ij} will be the proposition for $f(x_i, y_j)$.

The preceding equivalences can now be used and it will be necessary to show that:

$$\bigwedge_{ij} P_{ij} \to \bigvee_i \left(\bigwedge_j P_{ij}\right) \to \bigwedge_j \left(\bigvee_i P_{ij}\right) \to \bigvee_{ij} P_{ij} \qquad (5.50)$$

1. $\bigwedge_{ij} P_{ij} \to \bigvee_i \left(\bigwedge_j P_{ij}\right) \qquad (5.51)$

Using the following notations:

$$\bigwedge_{ij} P_{ij} = A, \quad \bigvee_i \left(\bigwedge_j P_{ij}\right) = B, \quad C = A \to B \qquad (5.52)$$

and:

$$(A \to B) \leftrightarrow (\bar{A} \vee B)$$

gives:

$$C \leftrightarrow \left\{ \overline{\bigwedge_{ij} P_{ij}} \vee \left[\bigvee_i \left(\bigwedge_j P_{ij}\right)\right]\right\} \leftrightarrow \left\{\bigvee_i \overline{\left(\bigwedge_j P_{ij}\right)} \vee \left[\bigvee_i \left(\bigwedge_j P_{ij}\right)\right]\right\}$$

$$\leftrightarrow \bigvee_i \left\{ \overline{\bigwedge_j P_{ij}} \vee \left(\bigwedge_j P_{ij}\right)\right\} \leftrightarrow \bigvee_i \dot{P} \leftrightarrow \dot{P}$$

(5.51) is, therefore, a tautology.

2. $\bigvee_i \left(\bigwedge_j P_{ij}\right) \to \bigwedge_j \left(\bigvee_i P_{ij}\right) \qquad (5.53)$

again with

$$\bigvee_i \left(\bigwedge_j P_{ij}\right) = A, \quad \bigwedge_j \left(\bigvee_i P_{ij}\right) = B \qquad (5.54)$$

and

$$C = A \to B. \qquad (5.55)$$

gives;

$$\bigwedge_j \left(\bigvee_i P_{ij}\right) \leftrightarrow \bigvee_{\substack{k_1, k_2, \ldots, k_J \\ (i, k_1, k_1, \ldots, k_J = 1, \ldots, I;\, j = 1, \ldots, J)}} \left(\bigwedge_j P_{k_j j}\right) \leftrightarrow \bigvee_i \left(\bigwedge_j P_{ij}\right) \vee S$$

By the use of the easily verified relationship

$$(B \leftrightarrow A \vee S) \leftrightarrow (A \to B)$$

it will be seen that C is a tautology.

3. $\bigwedge_j \left(\bigvee_i P_{ij} \right) \to \bigvee_i \left(\bigvee_j P_{ij} \right)$ (5.56)

Re-stating

$$\bigwedge_j \left(\bigvee_i P_{ij} \right) = A, \quad \bigvee_i \left(\bigvee_j P_{ij} \right) = B \qquad (5.57)$$

and

$$C = A \to B. \qquad (5.58)$$

gives

$$\bar{A} \vee B \leftrightarrow \left\{ \bigvee_j \overline{\left(\bigvee_i P_{ij} \right)} \vee \left(\bigvee_j \bigvee_i P_{ij} \right) \right\} \leftrightarrow \bigvee_j \left[\overline{\left(\bigvee_i P_{ij} \right)} \vee \left(\bigvee_i P_{ij} \right) \right] \leftrightarrow \bigvee_j \dot{P} \leftrightarrow \dot{P},$$

which demonstrates (5.56)

The different implications (5.50) have an obvious geometric interpretation if the 2 dimensional representations (the matrices of Chap. 1 for example) of the four relations appearing in (5.49) are considered.

5.2.12 Classes Associated with a Property or with a Relationship

Given a property $f(x)$, where x represents the elements belonging to a certain set defined as I, the class E_f associated with $f(x)$ is known as the set of the elements of I for which the property $f(x)$ holds. In other words, we have the equivalence:

$$\bigvee x \left[(x \in E_f) \leftrightarrow f(x) \right]. \qquad (5.59)$$

It should be noted that, in this respect, the procedure has been the converse of that used in Chap. 1 for defining a set as a collection of objects possessing a common property.

In the case of any relation having several variables, for example $f(x, y, z)$, the set of definition I is the set of all combinations of the elements x_i, y_j, z_k.

The class E_f associated with f is the set of combinations (x_i, y_j, z_k) for which the relationship $f(x, y, z)$ holds true. The class associated with a relation is, therefore, a subset of the set of definition I. This gives the following particular cases:

— there is no combination x, y, z such that the relationship $f(x, y, z)$ holds. The class E_f is the empty set: $E_f = \emptyset$;
— the relationship $f(x, y, z)$ holds for all combinations of the set of definition I. The class E_f coincides with the set I which, here, is the universal set.

5.2.13 Operations on Propositions and on Classes

Consider, for example, two properties: $f(x)$ and $g(x)$ with which are associated the classes E_f and E_g.

It will be seen that:
— the class associated with the relationship $f(x) \wedge g(x)$ is the class $E_f \cap E_g$,
— the class associated with the relationship $f(x) \vee g(x)$ is the class $E_f \cup E_g$,
— the class associated with $\overline{f(x)}$ is the class $\overline{E_f}$.

These properties are equally true for relations of any number of variables and their generalisation is immediate.

Thus we have a correspondence that exists between the propositional functions and the classes: any property causing the propositional functions to intervene can be expressed in the form of a property of the associated classes.

5.3 Algebra of Contacts and Algebra of Logic

Consider the propositional function:

(the path AB is closed)

By definition of the transmission:

(the path AB is closed) $\leftrightarrow (T_{AB} = 1)$
(the contact X is closed) $\leftrightarrow (X = 1)$

Hence the chain of equivalences, AB having 2 contacts X, Y in series:

$(T_{AB} = 1) \leftrightarrow (AB$ is closed$) \leftrightarrow ($contact X is closed$) \wedge ($contact Y is closed$)$

$$\leftrightarrow (X = 1) \wedge (Y = 1)$$

$(T_{AB} = 0) \leftrightarrow \overline{(T_{AB} = 1)} \leftrightarrow \overline{(X = 1) \wedge (Y = 1)} \leftrightarrow \overline{(X = 1)} \vee \overline{(Y = 1)}$

$$\leftrightarrow (X = 0) \vee (Y = 0).$$

In this way are obtained the conditions for the opening or closing of the series assembly.

Proceeding in an analogous manner for the parallel assembly and for the complementation. In this sense the algebra of contacts defined earlier is merely an application of formal logic to the network of relay contacts.

5.4 Algebra of Classes and Algebra of Contacts

Consider a contact network R. It possesses input variables which are the relay excitations and an output which is its transmission. To establish the idea consider the case of a network having 3 input variables A, B and C. There are $2^3 = 8$ combinations of (ABC) for these 3 variables. Let $(T = 1)$ denote the combinations which bring about the closing of the dipole, and $(T = 0)$ the combinations which brings about the opening of the dipole. These combinations form a set C having 8 elements.

Let C_1 be the set of the combinations of closing, and C_0 that of the combinations of opening. Then

$$\begin{cases} C_1 \cup C_0 = C \\ C_1 \cap C_0 = \emptyset \end{cases} \tag{5.61}$$

hence, in particular, $C_1 = \overline{C}_0$ (the dipole cannot be other than open or closed).

Thus, the network complementary to R, call it R_1, is such that

$$\begin{aligned} C_1' &= C_0 \\ C_0' &= C_1 \end{aligned} \tag{5.62}$$

Now consider 2 dipole networks R_1 and R_2 comprising the sets C_1^1, C_1^2, C_0^1, C_0^2. It is assumed that they possess the same input variables. (For this it will suffice to take as the set of input variables the union of the sets of inputs variables of R_1 and R_2 and to assume that R_1 and R_2 do not depend in any effective manner on certain of these variables).

— *Utilisation of R_1 and R_2 in parallel.*
For an input combination C:

$$\begin{aligned} T = 1 &\leftrightarrow (C \in C_1^1) \vee (C \in C_1^2) \\ &\leftrightarrow (C \in C_1^1 \cup C_1^2) \end{aligned} \tag{5.63}$$

resulting in:

$$\begin{aligned} T = 0 &\leftrightarrow \overline{(T=1)} \leftrightarrow \overline{(C \in C_1^1)} \wedge \overline{(C \in C_1^2)} \\ &\leftrightarrow (C \notin C_1^1) \wedge (C \notin C_1^2) \\ &\leftrightarrow (C \in C_0^1 \cap C_0^2) \end{aligned} \tag{5.64}$$

Therefore, in order that the combination of R_1 and R_2 in parallel is closed it is necessary and sufficient that the input combination should belong to the union of C_1^1 and C_1^2.

$$T = 1 \leftrightarrow (C \in C_1^1 \cup C_1^2) \tag{5.65}$$

5.4 Algebra of Classes and Algebra of Contacts

Utilisation in series:

$$T = 1 \leftrightarrow (C \in C_1^1) \wedge (C \in C_1^2) \leftrightarrow C \in C_1^1 \cap C_1^2 \tag{5.66}$$

resulting in:

$$T = 0 \leftrightarrow \overline{(T = 1)} \leftrightarrow \overline{(C \in C_1^1)} \vee \overline{(C \in C_1^2)} \tag{5.67}$$

or

$$(T = 0) \leftrightarrow (C \in C_0^1) \vee (C \in C_0^2) \leftrightarrow (C \in C_0^1 \cup C_0^2) \tag{5.68}$$

Thus in order that the combination of R_1 and R_2 in series is closed it will be necessary and sufficient that the input combination belongs to the intersection of C_1^1 and C_1^2.

Example (Fig. 5.2a, b)

$$\begin{cases} (R_1) : T_1 = (A\bar{B} + \bar{A}B)C \\ (R_2) : T_2 = A(B\bar{C} + \bar{B}C) \end{cases}$$

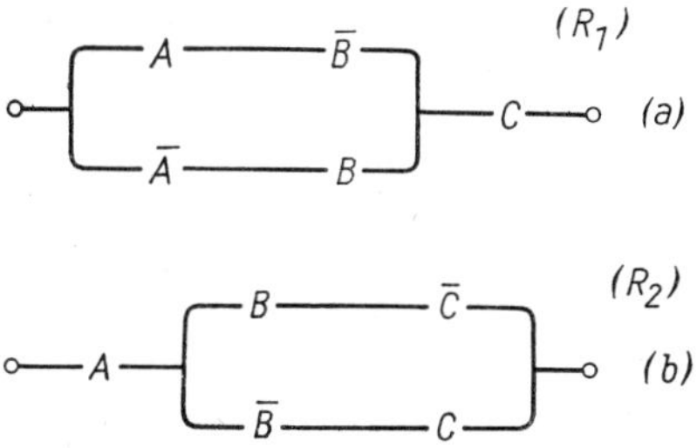

Fig. 5.2

Table 5.20 gives the conditions for closing of the different networks R_1, R_2, R_1 and R_2 in parallel and R_1 and R_2 in series.

Table 5.20

A B C	T_1	T_2	$T_1 + T_2$	$T_1 T_2$
0 0 0	0	0	0	0
0 0 1	0	0	0	0
0 1 0	0	0	0	0
0 1 1	1	0	1	0
1 0 0	0	0	0	0
1 0 1	1	1	1	1
1 1 0	0	1	1	0
1 1 1	0	0	0	0

Here are to be found

$$C_1^1 = \{0\ 1\ 1,\ 1\ 0\ 1\}, \quad C_1^2 = \{1\ 0\ 1,\ 1\ 1\ 0\}$$

$$C_1^1 \cup C_1^2 = \{0\ 1\ 1,\ 1\ 0\ 1,\ 1\ 1\ 0\} \qquad (5.69)$$

$$C_1^1 \cap C_1^2 = \{1\ 0\ 1\}$$

(sets of the closing combinations for operation in parallel and in series respectively).

5.5 Concept of Boolean Algebra

In the two preceding chapters diverse algebrae have been introduced the algebra of contacts, the algebra of classes and the algebra of propositions in logic. The first permits the rapid determination of the conditions governing the opening or closing of a dipole contact relay.

Evidence has now been produced for the advantage of an algebraic symbolism for the study of a circuit; once in possession of a certain number of theorems and algebraic identities established by reasoning in respect of the circuits, these can be abandoned and the work can proceed to the study of networks of considerably greater complexity with the aid of purely algebraic operations. Here, where intuition is no longer a help, or where reasoning applied to a circuit — always possible — becomes too delicate, the algebra of contacts provides an analytical tool.

In an analogous manner, the algebra of sets, apart from its theoretical interest, furnishes an algebraic method of resolving such diverse problems of sets as, for example, those of enumeration and of classification. Finally, the algebra of propositions is part of the logic thanks to which the truth or falsity of a compound proposition, however complex, can be determined in a systematic manner. It also provides the means of detecting the equivalence of different algebraic propositions by a comparison of their truth tables or, again, by application of known identities in an attempt to convert them into a common form. The reader will not have missed the striking resemblances between these three types of algebra. In fact, the relationships (R_1 to R_{20}), (E_1 to E_{20}) and (L_1 to L_{20}) are almost of the same form for each index $i (1 \leq i \leq 20)$. This means that the three algebrae examined, although different in the nature of their elements and of the associated operations, represent in a way three "versions" of the same algebraic structure. This abstract algebra is known as "Boolean algebra" and, at the same time as this is defined in the following chapter, it will be shown in a more precise manner how the three algebrae of contacts, sets and propositions can be considered as particular cases. The study will have the following advantages:

— it will systematically make clear the differences between ordinary algebra and that of Boole and will thus constitute an introduction to the particular rules for calculation of the latter.

— it will be independent of any particular application as well as of any technological support. Consequently, when it is desired to study commutation networks realised by the use of a certain technology, it will suffice to verify the possibility of defining a switching algebra taking the form of a Boolean algebra and to derive all the known properties thereof without the necessity for a step by step demonstration of each theorem for a particular application.

Chapter 6

Boolean Algebra

6.1 General. Axiomatic Definitions

The preceding chapter focussed attention on the advantages of an axiomatic structure which, in a way, would summarize identical rules for calculation encountered in widely differing circumstances, such as the study of contact networks, algebrae of sets and the propositional calculus. To this end, and in conformity with the general definition of an algebra, consideration will be given to a '*set*' for which some *operations* will be defined in the sense that, in the future application of the algebra to a particular situation, they might be replaced indifferently by contact networks, statements and, perhaps, other quantities. For this set of undefined elements (to which no meaning will be given here, other than that of a sort of 'mould' ultimately destined to receive a particular content), some operations will be defined. In their turn these operations, by reason of the undefined character of the elements of the set, are abstract and only have the meaning indicated in Chapter 1. They, too, are undefined and will have no particular significance until such time as the elements of the set themselves take on a particular meaning.[1]

Starting with this set and its operations we shall a priori, set forth a certain number of properties known as *axioms*, whose purpose it is to endow these operations with some properties worthy of note. Such an axiomatic presentation has often been given for diverse established mathematical theories; for example it is possible to demonstrate all the theorems of ordinary geometry by means of a small number of first concepts and axioms with respect to geometrical entities.

In order define a Boolean algebra — the object of this chapter — there exist a large number of axiomatic systems whose equivalence can be demonstrated. Furthermore, it is possible to pose the problem of the determination of a system of axioms which, from a certain point of view, are minimal. These problems of equivalence and minimality will not be discussed here. This work will be confined to a system of axioms which, although not the most compact possible, has, at least, the advantage of symmetry.

---

[1] *Cf. Liouville:* Classical algebra was the art of performing known operations on unknown quantities; modern algebra consists of performing unknown operations on unknown quantities.

6.2 Boolean Algebra

6.2.1 Set. Operations

A *Boolean algebra* is the name given to an algebra denoted here by:

$$B = \langle E, +, \cdot, 0, 1 \rangle$$

and possessing the following characteristics:
1. The set E, on which it is defined, contains at least two elements.
2. Two internal operations are defined for this set:
   — an operation denoted as $+$ and known as the "addition" or "sum" which, for every pair of elements A and B belonging to E, determines an element denoted as $A + B$ (read as A plus B) also belonging to E.
   — an operation denoted as $\cdot$ and known as the "multiplication" or "product" which, for every pair of elements A and B belonging to E, determines an element denoted as $A \cdot B$ or, more briefly, AB (read "AB") also belonging to E.
3. For E there exists an *equivalence* relation, denoted $=$ ("equal").

It possesses, therefore, the properties:

$$\left.\begin{array}{ll} 1.\ \forall x, \forall y \quad (x=y) \leftrightarrow (y=x) & \text{(symmetry)} \\ 2.\ \forall x \quad\quad\quad x=x & \text{(reflectivity)} \\ 3.\ \forall x, \forall y, \forall z\ (x=y) \wedge (y=z) \to (x=z) & \text{(transitivity)} \end{array}\right\} \quad (6.1)$$

6.2.2 Axioms for a Boolean Algebra [70, 37, 134];

$$\left.\begin{array}{l}(A_1)\ \forall A, \forall B, \forall B\ (A, B, C \in E) \\ \quad \text{whence: } (A+B)+C = A+(B+C) \\ (A_2)\ \forall A, \forall B, \forall C\ (A, B, C \in E) \\ \quad \text{whence: } (AB)C = A(BC) \end{array}\right\} \text{(associativity)}$$

$$\left.\begin{array}{l}(A_3)\ \forall A, \forall B\ (A, B \in E) \\ \quad A+B = B+A \\ (A_4)\ \forall A, \forall B\ (A, B \in E) \\ \quad AB = BA \end{array}\right\} \text{(commutativity)}$$

(A_5) There exists an element denoted as 0 which, for every $A \in E$, gives:

$$A+0 = 0+A = A$$

(A_6) There exists an element, denoted 1, which, for every $A \in E$, gives:
$$A \cdot 1 = 1 \cdot A = A$$

(A_7) $\forall A, \forall B, \forall C \ (A, B, C \in E)$
$$A(B + C) = AB + AC$$

(A_8) $\forall A, \forall B, \forall C \ (A, B, C \in E)$
$$A + BC = (A + B)(A + C)$$

} distributivity

(A_9) For every A, there exists an element, denoted $\bar{A}$, and known as the "complement of A", such that:
(A_{9a}) $A\bar{A} = 0$
(A_{9b}) and $A + \bar{A} = 1$

6.3 Fundamental Relations in Boolean Algebra

The preceding axioms now suffice to prove the diverse theorems which, as will be seen, have the same form as those encountered during the study of the algebra of contacts, the algebra of classes and of propositions. To simplify the presentation of the proofs, it will be convenient to indicate on the right of each identity or equation, the axiom(s) or theorem(s) contributing to the proofs. We have the following theorems:

(T_1) $A + A = A$

Indeed this can be expressed:

$$\begin{aligned}
A + A &= (A + A) \cdot 1 & (A_6) \\
&= (A + A) \cdot (A + \bar{A}) & (A_9) \\
&= A + A \cdot \bar{A} & (A_8) \\
&= A + 0 & (A_9) \\
&= A & (A_5)
\end{aligned}$$

Step by step (by recurrence) it can be demonstrated that, in a more general manner:

$$\underbrace{A + A + \cdots + A}_{n \text{ terms}} = A \qquad (6.2)$$

(T_2) $\quad A \cdot A = A$

$$\begin{aligned}
A \cdot A &= A \cdot A + 0 & (A_5) \\
&= A \cdot A + A \cdot \bar{A} & (A_9) \\
&= A \cdot (A + \bar{A}) & (A_7) \\
&= A \cdot 1 & (A_9) \\
&= A & (A_6)
\end{aligned}$$

6.3 Fundamental Relations in Boolean Algebra

In an analogous manner, this result can be extended as follows:

$$\underbrace{A \cdot A \cdots A}_{n \text{ terms}} = A \qquad (6.3)$$

(T_1) and (T_2) constitute the properties known as those of *idempotence*. They signify that in Boolean algebra there are neither multiples nor powers.

$T_3)$ $\qquad A + 1 = 1$

This may be expressed:

$$\begin{aligned}
A + 1 &= (A + 1) \cdot 1 & (A_6) \\
&= (A + 1) \cdot (A + \bar{A}) & (A_9) \\
&= A + 1 \cdot \bar{A} & (A_8) \\
&= A + \bar{A} & (A_6) \\
&= 1 & (A_9)
\end{aligned}$$

$T_4)$ $\qquad A \cdot 0 = 0$

$$\begin{aligned}
A \cdot 0 &= (A \cdot 0) + 0 & (A_5) \\
&= A \cdot 0 + A \cdot \bar{A} & (A_9) \\
&= A \cdot (0 + \bar{A}) & (A_7) \\
&= A \cdot \bar{A} & (A_5) \\
&= 0 & (A_9)
\end{aligned}$$

$T_5)$ $\qquad A + A \cdot B = A \qquad$ (law of absorption)

$$\begin{aligned}
A + A \cdot B &= A \cdot 1 + A \cdot B & (A_6) \\
&= A \cdot (1 + B) & (A_7) \\
&= A \cdot (B + 1) & (A_3) \\
&= A \cdot 1 & (T_3) \\
&= A & (A_6)
\end{aligned}$$

$T_6)$ $\qquad A \cdot (A + B) = A \qquad$ (law of absorption)

$$\begin{aligned}
A \cdot (A + B) &= A \cdot A + A \cdot B & (A_7) \\
&= A + A \cdot B & (T_2) \\
&= A & (T_5)
\end{aligned}$$

$T_7)$ $\qquad A + \bar{A} \cdot B = A + B$

$$\begin{aligned}
A + \bar{A} \cdot B &= (A + \bar{A}) \cdot (A + B) & (A_8) \\
&= 1 \cdot (A + B) & (A_9) \\
&= A + B & (A_6)
\end{aligned}$$

(T_8) $\quad A \cdot (\bar{A} + B) = A \cdot B$
$$A \cdot (\bar{A} + B) = A \cdot \bar{A} + A \cdot B \qquad (A_7)$$
$$= 0 + A \cdot B \qquad (A_9)$$
$$= AB \qquad (A_5)$$

(T_9) The complement $\bar{A}$ of an element A is unique.

Let it be assumed that an element of A possesses 2 complements $\bar{A}_1$ and $\bar{A}_2$. By virtue of axiom A_9 the following relations exist:

$$\begin{cases} A + \bar{A}_1 = 1 \\ A \cdot \bar{A}_1 = 0 \end{cases}$$

and

$$\begin{cases} A + \bar{A}_2 = 1 \\ A \cdot \bar{A}_2 = 0 \end{cases}$$

giving:

$$\bar{A}_1 = \bar{A}_1 \cdot 1 \qquad (A_6)$$
$$= \bar{A}_1 (A + \bar{A}_2) \qquad (A_9)$$
$$= \bar{A}_1 A + \bar{A}_1 \bar{A}_2 \qquad (A_1)$$
$$= 0 + \bar{A}_1 \bar{A}_2 \qquad (A_9, A_4)$$
$$= A \bar{A}_2 + \bar{A}_1 \bar{A}_2 \qquad (A_9)$$
$$= \bar{A}_2 A + \bar{A}_2 \bar{A}_1 \qquad (A_4)$$
$$= \bar{A}_2 (A + \bar{A}_1) \qquad (A_7)$$
$$= \bar{A}_2 \cdot 1 \qquad (A_9)$$
$$\bar{A}_1 = \bar{A}_2 \qquad (A_6)$$

Thus, $\bar{A}_1$ and $\bar{A}_2$ can only be equal. In other words, $\bar{A}$ *is* unique. *Complementation* is the name given to the operation which associates the element $\bar{A}$ with the element A:

(T_{10}) (Involution) $\quad \bar{\bar{A}} = A$

In considering an element $\bar{A}$ whose complement is to be determined and expressed as $\bar{\bar{A}} (= (\bar{A}))$, by virtue of the axiom (A_9) it can be said that:

$$\begin{cases} \bar{A} + \bar{\bar{A}} = 1 & (6.4) \\ \bar{A} \cdot \bar{\bar{A}} = 0 & (6.5) \end{cases}$$

6.3 Fundamental Relations in Boolean Algebra

But, on the other hand, the definition of $\bar{A}$ gives:

$$\begin{cases} A + \bar{A} = 1 \\ A\bar{A} = 0 \end{cases}$$

or again

$$\begin{cases} \bar{A} + A = 1 \\ \bar{A}A = 0 \end{cases} \qquad (6.6) \\ (6.7)$$

It will thus be seen that both A and $\bar{A}$ satisfy the same equations (6.4) (6.5) and (6.6) (6.7). Theorem T_6 states, however, that the complement is unique. Thus

$$\bar{\bar{A}} = A \qquad (6.8)$$

$T_{11})$
$$\overline{A + B} = \bar{A} \cdot \bar{B} \quad \text{(De Morgan Theorem)}$$

For this we will now determine the complement of $\bar{A} \cdot \bar{B}$ and demonstrate that this is $A + B$. Consider, therefore, the quantity $A + B + \bar{A}\bar{B}$:
Here:

$$\begin{align*}
(A + B) + \bar{A} \cdot \bar{B} &= [(A + B) + \bar{A}][(A + B) + \bar{B}] & (A_8) \\
&= [A + (B + \bar{A})][A + (B + \bar{B})] & (A_1) \\
&= [A + (\bar{A} + B)][A + (B + \bar{B})] & (A_3) \\
&= [(A + \bar{A}) + B][A + (B + \bar{B})] & (A_1) \\
&= (1 + B) \cdot (A + 1) & (A_9) \\
&= 1 \cdot 1 = 1 & (T_3, A_3, T_1)
\end{align*}$$

In an analogous manner we have:

$$\begin{align*}
(A + B) \cdot (\bar{A} \cdot \bar{B}) &= (\bar{A} \cdot \bar{B}) \cdot (A + B) & (A_4) \\
&= (\bar{A} \cdot \bar{B}) \cdot A + (\bar{A} \cdot \bar{B}) \cdot B & (A_7) \\
&= (\bar{A} \cdot A) \cdot \bar{B} + \bar{A} \cdot (\bar{B} \cdot B) & (A_4, A_2) \\
&= 0 \cdot \bar{B} + \bar{A} \cdot 0 & (A_9) \\
&= 0 + 0 = 0 & (T_4, A_4, A_5)
\end{align*}$$

Thus $X = A + B$ and $Y = \bar{A} \cdot \bar{B}$ verifies both equations:

$$\begin{cases} X + Y = 1 \\ X \cdot Y = 0 \end{cases}$$

i.e. that $A + B$ is the complement (unique, it must be remembered) of $\bar{A} \cdot \bar{B}$, hence:

$$A + B = \overline{(\bar{A}\bar{B})} \quad \text{and consequently} \quad \overline{A + B} = \overline{\overline{(\bar{A}\bar{B})}} = \bar{A} \cdot \bar{B}$$

$$\overline{A + B} = \bar{A} \cdot \bar{B} \quad \text{(first De Morgan Theorem)}$$

$(T_{12}) \quad \overline{AB} = \bar{A} + \bar{B} \quad \text{(second De Morgan theorem)}$

Next consider the second member and take its complement:

$$\overline{\bar{A} + \bar{B}} = \bar{\bar{A}} \cdot \bar{\bar{B}} \qquad (T_{11})$$
$$= A \cdot B \qquad (T_{10})$$

Taking next the complement of both members:

$$\overline{\overline{\bar{A} + \bar{B}}} = \overline{A \cdot B}$$
$$\bar{A} + \bar{B} = \overline{AB} \qquad (T_{10})$$

results in the theorem.

The theorems T_{11} and T_{12} (*De Morgan* theorems) are very important. They take the same form as the theorems of the same name discussed in Chapter 5 for sets and calculus of propositions.

6.4 Dual Expressions. Principle of Duality

6.4.1 Dual Expressions

By reverting to the axioms A_1 to A_9, it will be seen that they can be grouped together in pairs: $(A_1 A_2)$, $(A_3 A_4)$, $(A_5 A_6)$, $(A_7 A_8)$ and $(A_{9a} A_{9b})$. Two axioms of any one of these pairs correspond in the manner shown in Table 6.1, where X describes every variable appearing in the axioms. By this procedure, from one axiom of a pair of two axioms the other may be deduced. For example, from (A_1):

$$A + (B + C) = (A + B) + C$$

can immediately be written (A_2):

$$A(BC) = (AB)C.$$

It is said that the 2 axioms of such a pair are *dual*, or again, that the identities by which they can be translated are *dual*. In a general manner two algebraic expressions which can be deduced, the one from the other

6.4 Dual Expressions. Principle of Duality

Table 6.1. *Rules for dualisation*

+	·
·	+
0	1
1	0
X	X
$\bar{X}$	$\bar{X}$

in accordance with the rules of Table 6.1 (*"dualisation rules"*) are known as *dual expressions*.

Examples of dual expressions:

$$
\begin{array}{l|l}
A + BC & A(B+C) \\
A\bar{B} + \bar{A}B & (A+\bar{B})(\bar{A}+B) \\
AB + BC + CA & (A+B)(B+C)(C+A), \text{ etc.}
\end{array} \qquad (6.9)
$$

6.4.2 Principle of Duality

That the axioms of a Boolean algebra can be grouped in dual pairs already constitutes an interesting peculiarity. Also remarkable, in fact is the consequence to be derived therefrom, in the form known as the *principle of duality*, and which can be justified as follows. Consider, therefore, a theorem which holds true in a Boolean algebra. For its proof it is necessary to apply a succession of operations which, in the final analysis, can be reduced to the application, in a certain number of stages, of the various axioms (A_1 to A_9). If to each of these intermediate operations is now associated the dual operation it is clear that the final result will, itself, be dual (this is what has already been done for the theorems T_1 to T_4). On the basis of this reasoning, it is now possible, in each instance, to dispense with the proof, henceforth superfluous, for a dual theorem: it will suffice merely to apply the above principle (the *principle of duality*) which can be summarised thus:

"With every theorem, true in a Boolean algebra, there corresponds a dual theorem, which is also true."

Example: within a Boolean algebra, for a given identity:

$$AB + \bar{A}C + BC = AB + \bar{A}C \qquad (6.10)$$

which the reader can establish by following, step-by-step, the example given in chapter 4 for the particular case of the algebra of contacts making use of the axioms (A_1) to (A_9) and of the theorems (T_1) to (T_{12}). Once (6.10) has been accomplished the following identity can, immediately, be deduced by duality, namely:

$$(A + B)(\bar{A} + C)(B + C) = (A + B)(\bar{A} + C) \qquad (6.11)$$

N.B. — as has already been seen from the foregoing, the principle of duality could have served to prove half of the theorems (T_1) to (T_{12}). Intentionally, this has not been used in order to bring to light the symmetry of the expressions (e.g. $T_1 - T_2$) or to provide a new proof $(T_6$ and $T_{12})$. From this it follows that, in practice however, the principle of duality will be applied whenever possible, thus effecting an economy of one proof in every two.

6.4.3 Notation

For a given expression E, the dual expression is denoted as $\tilde{E}$ (E tilde). With the help of examples it is easy to verify that, in general E differs from $\tilde{E}$ in the sense that there exist combinations of variables which give the value 1 to one expression and the value 0 to the other.

6.5 Examples of Boolean Algebra

6.5.1 2-Element Algebra

Consider a set E with 2 elements, denoted by 0 and 1 and define for this set $\{0, 1\}$ the 3 operations given in Table 6.2 a, b, c which follows:

Table 6.2

$E = \{0, 1\}$,

(+)

B\A	0	1
0	0	1
1	1	1

(a)

(.)

B\A	0	1
0	0	0
1	0	1

(b)

A	$\bar{A}$
0	1
1	0

(c)

The set $\{0, 1\}$ thus equipped with the operations $+$, $\cdot$, and $-$ is a Boolean algebra.

$$B = \langle \{0, 1\}, +, \cdot, -, 0, 1 \rangle$$

For the justification, it suffices to show that the axioms A_1 to A_9 are satisfied. To this end, use will be made of the tables defining the operations. For the axioms (A_5), (A_6) and (A_9) which involve the quantificator ∃, it is easy to verify that the elements 0 and 1 above are none other than the 0 and 1 of the axioms, that they completely satisfy (A_5) and (A_6) and, moreover, that the element $\bar{A}$ defined in table 3.2 is indeed the complement in the sense of the axiom (A_9). As for the remaining axioms, which take the form of *universal* propositions, it is sufficient to establish their validity and, for each identity, to compile a table containing *all the combinations of variables*, the value of each member and to ensure that they are equal. This is the method already adopted on many occasions in Chapters 4 and 5. As an example only one single identity will be demonstrated here. The task of verifying the others by the same procedure is left to the reader.

Example:
$$(A+B)+C = A+(B+C) \qquad (A_1)$$

which leads to the following table:

Table 6.3

A B C	A+B	(A+B)+C	B+C	A+(B+C)
0 0 0	0	0	0	0
0 0 1	0	1	1	1
0 1 0	1	1	1	1
0 1 1	1	1	1	1
1 0 0	1	1	0	1
1 0 1	1	1	1	1
1 1 0	1	1	1	1
1 1 1	1	1	1	1

6.5.2 Algebra of Classes

Consider an set E and the set of its subsets I_E. The operations $\cap$, $\cup$, − and the constants $\varnothing$, I_E, confer on I_E the structure of a Boolean algebra, denoted:

$$B_E = \langle I_E, \cap, \cup, -, \varnothing, I_E \rangle$$

The relations of chapter 5 provide verification of axioms (A_1) to (A_9). The set of subsets E is, therefore, a Boolean algebra. Note that, E is not necessarily finite, but that when it is finite and contains $n (\geq 2)$ elements, the Boolean algebra B_E contains 2^n elements.

6.5.3 Algebra of Propositions

Let L be a set of propositions, with operations $\wedge$, $\vee$, $\sim$, such that $P \in L$, $Q \in L$, then $\sim P \in L$ and $P \wedge Q \in L$. Then $P \vee Q \in L$ als‍ In these circumstances, the axioms (A_1) to (A_9) take the form of t‍ relations $(L_1$ to $L_{10})$ which have been verified; and it will be seen th‍ a set of propositions L, closed under the operations $\sim$ and $\wedge$ (or $\vee$) a Boolean algebra, denoted:

$$B_L = \langle L, \wedge, \vee, -, \mathring{P}, \dot{P} \rangle$$

with the reciprocal implication ($\leftrightarrow$) as the equivalence relation.

6.5.4 Algebra of n-Dimensional Binary Vectors

Consider a set of binary vectors of n dimensions. A binary vector is an ordered set of variables (components) which can take on one 2 values:

$$X = (x_1 x_2 \cdots x_i \cdots x_n) = (x_i) \quad x_i \in \{0, 1\} \quad (i = 1, 2 \cdots n) \quad (6.1‍$$

This set comprises

$$\underbrace{2 \times 2 \times 2 \cdots \times 2}_{n \text{ factors}} = 2^n \text{ elements}$$

The following operations can be introduced:

1. *Product*

For 2 given vectors X and Y:

$$X = (x_1 x_2 \cdots x_i \cdots x_n) = (x_i)$$
$$Y = (y_1 y_2 \cdots y_i \cdots y_n) = (y_i)$$
(6.1‍

the Boolean product of X and Y, denoted as $X \cdot Y$ (or XY) is t‍ following vector:

$$X \cdot Y = (x_1 \cdot y_1, x_2 \cdot y_2, \ldots, x_i \cdot y_i, \ldots, x_n \cdot y_n) = (x_i \cdot y_i) \quad (6.1‍$$

2. *Sum*

The sum of the 2 given vectors X and Y, described above and denote‍ as $X + Y$, is the vector:

$$X + Y = (x_1 + y_1, x_2 + y_2, \ldots x_i + y_i, \ldots x_n + y_n) = (x_i + y_i) \quad (6.1‍$$

3. Complement

Given 1 vector $X = (x_1, x_2, ..., x_i, ..., x_n)$, the complement of X, denoted $\overline{X}$, is the vector:

$$\overline{X} = (\overline{x}_1, \overline{x}_2, ..., \overline{x}_i, ..., \overline{x}_n) = (\overline{x}_i) \tag{6.16}$$

The operations $+, \cdot, -$, on the variables x_i, y_i are those defined in Tables 6.2 a, b, c.

4. Null vector

This, by definition, is the vector uniquely formed from zeros.

$$0 = (00 \cdots 0 \cdots 0) \quad (x_i = 0, \quad i = 1, 2, ..., n) \tag{6.17}$$

5. Unit vector

$$I = (1\ 1 \cdots 1 \cdots 1) \quad (x_i = 1, \quad i = 1, 2, ..., n) \tag{6.18}$$

In these conditions, the set of n-dimensional binary vectors subjected to these 3 operations is a Boolean algebra with 2^n elements, and having as constant elements the vectors 0 and 1.

$$B = \langle \{X\}, +, \cdot, -, 0, I \rangle \tag{6.19}$$

This can be shown in the following manner:

First method

If the axioms (A_1) to (A_9) are verified for the *vectors*, this must apply to all *components*. This is the case since this is the Boolean algebra already encountered in 6.5.1. Thus to verify the axiom (A_1), for example:

$$[(X + Y) + Z = X + (Y + Z)] \leftrightarrow [\forall_i \in \{1, 2, ..., n\}\ (x_i + y_i) + z_i$$
$$= x_i + (y_i + z_i)] \tag{6.20}$$

$$\leftrightarrow \bigwedge_{i=1}^{i=n} [(x_i + y_i) + z_i = x_i + (y_i + z_i)] \tag{6.21}$$

But for every i, the proposition within the square brackets is true (algebra of § 6.5.1). Thus the left hand proposition is also true: the axiom (A_1) has, therefore been properly verified. In an analogous manner it can be shown that the other axioms have all been satisfied.

Second method

This is given by an indirect proof: let E be a set having n distinct elements, denote them $e_1, e_2, \ldots, e_i, \ldots, e_n$.

$$E = \{e_1, e_2, \ldots, e_i, \ldots, e_n\} \tag{6.22}$$

A subset A of E ($A \subset E$) is none other than a choice of elements e_i ($\in A$). It can be defined by 'earmarking' those of the orderly sequence $e_i, \ldots, e_n$ to be retained for the construction of A or, just as easily, those to be rejected. In other words, A can be determined through the intermediary of a vector X_A:

$$X_A = (x_1, x_2, \ldots, x_i, \ldots, x_n) \tag{6.23}$$

subject to the one-to-one correspondence of Table 6.4.

Table 6.4

x_i	$e_i \in A$
1	yes (true)
0	no (false)

We shall call X_A the *characteristic vector* associated with the pair of sets (A, E). Thus is obtained the correspondence, indicated in Table 6.5, between sets and characteristic vectors (where E plays the part of the universal set).

Table 6.5

$A \cup B$	$X_{A \cup B} = X_A + X_B$
$A \cap B$	$X_{A \cap B} = X_A \cdot X_B$
$\bar{A}$	$X_{\bar{A}} = \bar{X}_A$
$\varnothing$	$X_\varnothing = 0$
E	$X_E = I$

Summarizing:

1. With every finite set $A \subset E$ there corresponds a vector X_A (in one-to-one fashion).
2. With the operation $\cup$ corresponds the operation $+$ for the vectors.
3. With the operation $\cap$ corresponds the operation $\cdot$ for the vectors.
4. With the operation $-$ corresponds the operation $-$ for the vectors.
5. With the constants $\varnothing$ and E, correspond the vectors null (0) and unity (1) respectively.

If, in addition, the letter X (which appears throughout) be deleted in order to retain only the indices, it will be seen that every law which is valid for sets will correspond formally with the same law for vectors apart from slight differences in *notations* for the operations and constants. In particular, this applies to the axioms, which are thus adequately verified for the vector algebra being considered.

6.6 Boolean Variables and Expressions

6.6.1 Boolean Variables

These are variables whose values belong to a Boolean algebra. Thus, in the axioms and theorems so far considered, the letters used have represented such variables. It can also be said that the algebra discussed constitutes the 'scope' of the variable.

6.6.2 Boolean Expressions

The term *Boolean expression*, or *Boolean form*, is applied to every algebraic expression formed from variables and constants belonging to a Boolean algebra, formed by a *finite number* of applications of the operations $+$, $\cdot$ and $-$.

More precisely it may be said that:

1. 0 and 1 are expressions.
2. Every variable X is an expression.
3. If A is an expression, then so too is $\bar{A}$.
4. If A, B are expressions, then $A + B$ and AB are also expressions.
5. An expression can be obtained only by application of the rules 1, 2, 3 and 4 a finite number of times.

Examples:

1. 0
2. A (6.24)
3. $A + BC$

(obtained, for example, from the following "programme":

$$\begin{cases} B, C \to BC = X_1 \\ A, X_1 \to A + X_1 = A + BC \end{cases} \quad (6.25)$$

4. $A + \bar{B}\left[\overline{CD + \left(\overline{A+B} + \overline{B+C}\right)}\right] + FG$

A possible 'programme'

$$
\begin{array}{lll}
C, D & \to CD & = X_1 \\
X_1 & \to \overline{X}_1 & = X_2 \\
A, B & \to A + B & = X_3 \\
B, C & \to B + C & = X_4 \\
X_3 & \to \overline{X}_3 & = X_5 \\
X_4 & \to X_4 & = X_6 \\
X_5, X_6 & \to X_5 + X_6 & = X_7 \\
X_2, X_7 & \to X_2 + X_7 & = X_8 \\
X_8 & \to \overline{X}_8 & = X_9 \\
B & \to \overline{B} & = X_{10} \\
X_{10}, X_9 & \to X_{10} \cdot X_9 & = X_{11} \\
A, X_{11} & \to A + X_{11} & = X_{12} \\
F, G & \to F \cdot G & = X_{13} \\
X_{12}, X_{13} & \to X_{12} + X_{13} : \text{expression (6.26).}
\end{array}
$$
(6.27)

6.6.3 Values for a Boolean Expression. Function Generated by an Expression

If the variables of a Boolean expression $G(X_1, X_2, \ldots X_n)$ be replaced by particular elements of the algebra E, which serves as a field of operation, the outcome is an element of E. Any combination $(a_1, a_2, \ldots, a_n)$ of particular values attributed to the variables $(X_1, X_2, \ldots X_n)$ is known as an assignment (of values to variables). The element obtained on the basis of $G(X_1, X_2, \ldots X_n)$ is thus the value of the expression for the assignment under consideration. Then if the variables $X_1, \ldots X_n$ be given all possible combinations of values, and the values of the expressions be calculated in each case, it will be seen that, through the intermediary of the expression (which is thus a means for calculation), is defined a function $f_G(x_1, \ldots, x_n)$ of E^n in E. It can be said that the expression G generates the function in question and that, conversely, G is an *expression* or *representation* of f_G. In this respect it should be noted that an expression defines an unique function, but, conversely, for a given function of E^n in E, there exists an infinity (by reason, for example, of the laws of idempotence) of expressions which give rise to it. Thus in the algebra

6.6 Boolean Variables and Expressions

$\langle\{0, 1\}, +, \cdot, - \rangle$ the expressions:

$$A\bar{B} + \bar{A}B, A\bar{B} + \bar{A}B + \bar{A}\bar{B}, (A + B) \cdot (\bar{A} + \bar{B}), \overline{(A + B)(\bar{A} + B)} \quad (6.28)$$

among others, represents the same function, defined in the following table:

Table 6.6

A B	$f(A, B)$
0 0	0
0 1	1
1 0	1
1 1	0

6.6.4 Equality of 2 Boolean Expressions

It is said that two expressions are equal (or equivalent) if they take the same value for each assignment of values, i.e. if they generate the same function; the equality of the 2 expressions E and F is denoted by the sign $= (E = F)$.

The reader should verify that it really does constitute an equivalence relation in the normal sense, i.e. with properties of symmetry, reflexivity and transivity.

6.6.5 Canonical Forms Equal to an Algebraic Expression

Every Boolean algebraic expression can be reduced to one or other of 2 distinguished forms, known as canonical forms, whose existence can be demonstrated as follows:

6.6.5.1 Expressions Having 1 variable

Every single-variable expression $F(X)$ can be expressed in the form:

$$F(X) = F(1)X + F(0)\bar{X} \quad (6.29)$$

(the Shannon theorem for one variable)

Effectively, the following cases can be considered:

a) $F(X) = \text{Constant} = A$.
This can be expressed as:

$$A = A \cdot 1 = A(X + \bar{X}) = AX + A\bar{X} = F(1)X + F(0)\bar{X} \quad (6.30)$$

b) $F(X) = X$.
Here:
$$F(X) = X = 1 \cdot X + 0 \cdot \overline{X} = F(1) X + F(0) \overline{X} \qquad (6.31)$$

c) $F(X) = \overline{X}$.
In an analogous manner:
$$F(X) = \overline{X} = 0 \cdot X + 1 \cdot \overline{X} = F(1) X + F(0) \overline{X}$$

d) Assuming that the Shannon theorem (6.29) be true for 2 functions $F(X)$ and $G(X)$; it will be shown to be true for $F(X) \cdot G(X)$, $F(X) + G(X)$, and $\overline{F(X)}$.
Here:

$$\begin{aligned} H(X) = F(X) \cdot G(X) &= \big[F(1) X + F(0) \overline{X}\big]\big[G(1) X + G(0) \overline{X}\big] \\ &= F(1) G(1) X + F(0) G(0) \overline{X} \qquad (6.32) \\ &= H(1) X + H(0) \overline{X} \end{aligned}$$

$$\begin{aligned} K(X) = F(X) + G(X) &= F(1) X + F(0) \overline{X} + G(1) X + G(0) \overline{X} \\ &= [F(1) + G(1)] X + [F(0) + G(0)] \overline{X} \qquad (6.33) \\ &= K(1) X + K(0) \overline{X} \end{aligned}$$

$$\begin{aligned} N(X) &= \overline{F(X)} \\ &= \overline{[F(1) X + F(0) \overline{X}]} \\ &= \overline{F(1) X} \cdot \overline{F(0) \cdot \overline{X}} \\ &= \big[\overline{F(1)} + \overline{X}\big] \cdot \big[\overline{F(0)} + X\big] \\ &= \overline{F(1)}\,\overline{F(0)} + \overline{F(0)}\, \overline{X} + \overline{F(1)} X + X\overline{X} \\ &= \overline{F(1)}\,\overline{F(0)} + \overline{F(0)}\, \overline{X} + \overline{F(1)} X \qquad (6.34) \\ &= \overline{F(1)}\,\overline{F(0)} (X + \overline{X}) + \overline{F(0)}\, \overline{X} + \overline{F(1)} X \\ &= \big[\overline{F(1)}\,\overline{F(0)} + \overline{F(1)}\big] X + \big[\overline{F(1)}\,\overline{F(0)} + \overline{F(0)}\big] \overline{X} \\ &= \overline{F(1)} X + \overline{F(0)}\, \overline{X} \\ &= N(1) X + N(0) \overline{X}. \end{aligned}$$

But an expression is constructed in a finite number of steps, as described in § 6.6.2. At every step, the operation that can be effected, preserves the Shannon property. This, in short, is true for every Boolean expression. This is equivalent to saying that every expression can be reduced ("equality") to a linear form (in Boolean algebra), homogeneous with respect to the variables X and $\overline{X}$.

$$f(X) = AX + B\overline{X}, \quad \text{with} \quad \begin{cases} A = F(1) \\ B = F(0) \quad \text{(constants)} \end{cases}$$

It should be noted that this expression involves only 2 constants $F(1)$ and $F(0)$ which are 2 elements of the algebra in question.

6.6.5.2 Expressions Having n Variables

Consider next an expression in n variables $F(X_1, X_2, ..., X_n)$ and assume that $n-1$ of these variables are given constant values. The expression $F(X_1, X_2, ..., X_n)$ is now none other than a single-variable expression which can be subjected to the Shannon theorem — (6.29) to yield the expression:

$$F(X_1, X_2, ..., X_n) = F(1, X_2, ..., X_n) X_1 + F(0, X_2, ..., X_n) \overline{X}_1 \quad (6.35)$$

If their freedom be restored to the variables, the terms $F(1, X_2, ... X_n)$ and $F(0, X_2, ..., X_n)$ again become expressions with $n-1$ variables of which $(n-2)$ can again be temporarily 'frozen' and for the remaining variable X_2 the same Shannon theorem can again be applied. Hence:

$$F(X_1, X_2, ..., X_n) = F(1, 1, X_3, ..., X_n) X_1 X_2 + F(1, 0, X_3, ..., X_n) X_1 \overline{X}_2$$
$$+ F(0, 1, X_3, ..., X_n) \overline{X}_1 X_2 + F(0, 0, X_3, ..., X_n) \overline{X}_1 \overline{X}_2.$$

Repeated application of the Shannon theorem until such time that all the variables have been extracted, leads to the expression:

$$F(X_1, X_2, ..., X_i, ..., X_n)$$
$$= \sum_{e=(00...0...0)}^{e=(11...1...1)} F(e_1, e_2, ... e_i, ..., e_n) X_1^{e_1} X_2^{e_2} ... X_i^{e_i} ... X_n^{e_n} \quad (6.36)$$

with the conventions:

$$X_i^{e_i} = \begin{cases} X_i \quad \text{when } e_i = 1 \\ \overline{X}_i \quad \text{when } e_i = 0 \end{cases}$$

and
$$e = (e_1, e_2, ..., e_n) \quad \text{(binary vector with } n \text{ dimensions)}.$$

In an abridged manner, it can be agreed to regard e as the binary representation of a number.

Hence the abridged notation:

$$F(X_1, X_2, \ldots, X_i, \ldots, X_n) = \sum_{e=0}^{e=2^n-1} F(e) X_1^{e_1} X_2^{e_2} \ldots X_i^{e_i} \ldots X_n^{e_n} \quad (6.37)$$

The algebraic form of the second member of (6.36) takes the name of the *disjunctive canonical form*. With this terminology, it is the sum of products. There are 2^n possible coefficients $F(e)$ and as many monomials of the form:

$$X_1^{e_1} X_2^{e_2} \ldots X_i^{e_i} \ldots X_n^{e_n}$$

It can be said that:

— the canonical form is a homogeneous function of degree n relative to the $2n$ letters $X_i, \bar{X}_i$;
— it is a linear combination of products of n letters (whether complemented or not) (2^n parameters). A product of form:

$$X_1^{e_1} X_2^{e_2} \ldots X_i^{e_i} \ldots X_n^{e_n}$$

is known as a *fundamental product*.

6.6.5.3 Dual Expressions

From the application of the principle of duality can be deduced the following formulae:

$$F(X) = \left[F(1) + \bar{X}\right]\left[F(0) + X\right] \quad (6.38)$$

(Shannon dual theorem). Whence:

$$F(X_1, X_2, \ldots, X_i, \ldots, X_n)$$

$$= \prod_{e=(00\ldots0\ldots0)}^{e=(11\ldots1\ldots1)} [F(\bar{e}_1, \bar{e}_2, \ldots \bar{e}_i, \ldots \bar{e}_n) + X_1^{e_1} + X_2^{e_2} + \cdots + X_i^{e_i} + \cdots X_n^{e_n}]$$

(6.39)

(conjunctive canonical form)

This is a product of sums, the sums being of the form:

$$X_1^{e_1} + X_2^{e_2} + \cdots + X_i^{e_i} + \cdots + X_n^{e_n}$$

and known as *fundamental sums*.

In an abridged form this can again be expressed as:

$$F(X_1, X_2, \ldots, X_i, \ldots, X_n)$$

$$= \prod_{e=0}^{e=2^n-1} [F(2^n - 1 - e) + X_1^{e_1} + X_2^{e_2} + \cdots + X_i^{e_i} \cdots + X_n^{e_n}] \;\;[1] \quad (6.40)$$

[1] Remember that if the numerical value of $e_1, e_2, \ldots, e_n$ is e, that of $\bar{e}_1, \bar{e}_2, \ldots \bar{e}_n$ is the restricted complement $= 2^n - 1 - e$ (cf. Chapters 2,9).

6.6 A Practical Method for the Determination of the Complement

In practice, the determination of the complement of an expression such as $E(X_1, \ldots, X_n)$ does not require the application of the De Morgan theorems; as can easily be seen, it suffices to replace $E(X_1, \ldots, X_n)$ by an expression $\overline{E(X_1, \ldots, X_n)}$, obtained by the correspondence in the following table (Table 6.7):

Table 6.7. *Rules for complementation*

$+$	$\cdot$
$\cdot$	$+$
0	1
1	0
X	$\overline{X}$
$\overline{X}$	X

It will be noted that this table differs from Table 6.1 (Rules for dualisation) in the last two lines.

Example:

$$F(A, B, C) = A\overline{B} + \overline{C}\overline{D} + \overline{A}B + AB + BC \qquad (6.41)$$

Giving:

$$\overline{F}(A, B, C) = \overline{A\overline{B} + \overline{C}\overline{D} + \overline{A}B + AB + BC}. \qquad (6.42)$$

It should be possible to stop at this point. However, if it is desired to avoid a 'pile-up' of the bars of negation, the theorems of De Morgan can be applied. Both methods are examined (step-by-step, or by Table 6.7).

First method:

$$\overline{F} = \overline{A\overline{B} + \overline{C}\overline{D} + \overline{A}B} \cdot \overline{AB + BC}$$
$$= \left(\overline{A\overline{B}} + \overline{\overline{C}\overline{D}} + \overline{\overline{A}B}\right) \cdot \left(\overline{A} + \overline{B}\right)\left(\overline{B} + \overline{C}\right). \qquad (6.43)$$

Second method (Table 6.7):

$$\overline{F} = \left(A\overline{B} + \overline{C}\overline{D} + \overline{A}B\right) \cdot \left(\overline{A} + \overline{B}\right)\left(\overline{B} + \overline{C}\right) \qquad (6.44)$$

or, again:

$$F = \overline{(\overline{A} + B)(C + D)(A + \overline{B})} \cdot (\overline{A} + \overline{B})(\overline{B} + \overline{C}), \text{ etc.} \quad (6.45)$$

6.6.7 Reduction to one Level of the Complementations Contained in an Expression

For a given expression comprising several superimposed complementations such as:

$$F = \overline{\overline{A} + B + \overline{\overline{C} + D + E}}, \quad (6.46)$$

it is always possible to reduce the expression to a form where the complementations appear at one level only (at most). That this is always possible results from the existence of canonical forms. However, by repeated application of the De Morgan and involution theorem it can also be seen that this is so, without the necessity of reverting to the canonical form. For the above example, this would give:

$$F = \overline{(\overline{A} + B)} \, \overline{(C + D + E)} \quad (6.47)$$
$$= (A \cdot \overline{B})(C + D)\overline{E} \quad (6.48)$$

which can also be developed as a sum of products:

$$F = A\overline{B}C\overline{E} + A\overline{B}D\overline{E}. \quad (6.49)$$

6.6.8 Relationship between a Dual Expression and its Complement

For a given expression of a function $F(X_1, ..., X_n)$, Tables 6.1 and 6.7 permit us to deduce 2 expressions associated with the functions:

$$\widetilde{F}(X_1, ..., X_n) \text{ (dual)} \quad \text{and} \quad \overline{F}(X_1, ..., X_n) \text{ (complement)}.$$

Hence the relation:

$$\widetilde{F}(X_1, ..., X_i, ..., X_n) = \overline{F}(\overline{X}_1, ..., \overline{X}_i, ..., \overline{X}_n). \quad (6.50)$$

Effectively, the three procedures for the construction of the expressions F, $\widetilde{F}$, $\overline{F}$ can be summarized in the following table (Table 6.8):

6.6 Boolean Variables and Expressions

Table 6.8

F	$\tilde{F}$	$\bar{F}$
0	1	1
1	0	0
A	$A = \overline{(\bar{A})}$	$\bar{A} = \overline{(\bar{A})}$
$\bar{A}$	$\bar{A} = \overline{(\bar{\bar{A}})}$	$A = \overline{(\bar{A})}$
$A+B$	$A \cdot B = \overline{\bar{A} + \bar{B}}$	$\bar{A} \cdot \bar{B} = \overline{A+B}$
$A \cdot B$	$A + B = \overline{\bar{A}\ \bar{B}}$	$\bar{A} + \bar{B} = \overline{AB}$

nce F is constructed in a finite number of operations $+, \cdot,$ and $-$ based
the variables and the constants 0 and 1, it will be seen from the step-
-step application of the correspondences of Table 6.8, that the relation
50 is indeed true.

Chapter 7

Boolean Functions

7.1 Binary Variables

A variable for which the domain of values is 2 symbols is known as binary variable. The set of values will usually be denoted by $\{0, 1\}$.

7.2 Boolean Functions

7.2.1 Definition

A Boolean function (or switching function) is a mapping of the s $\{0, 1\}^n$ onto the set $\{0, 1\}$.

$\{0, 1\}^n$ designates the cartesian product of $\{0, 1\}$ with itself n time

$$\{0, 1\}^n = \underbrace{\{0, 1\} \times \{0, 1\} \ldots \{0, 1\}}_{n}$$

In other words: a Boolean function having n variables is a functic which, to every n-digit combination $a = (a_1, a_2, \ldots a_n)$, assigns a val 0 or a value 1. In practice such a function can take the form of a tab known as a 'truth table', similar to the tables of the same name used formal logic for the calculus of propositions.

7.2.2 Truth Table for a Boolean Function of n Variables

This is a table which, for each of the 2^n binary combinations of va ables, indicates the corresponding vaue of the function. In practice th can be presented in several forms. The most common is a linear table 2 vertical columns: that on the left listing the 2^n combinations of bina variables, arranged one below the other, that on the right indicating t value (0 or 1) for each of the combinations. In practice it is advantageo to adopt a fixed order for the combinations of variables; that mo frequently used being in order of binary numbers increasing from 0 $2^n - 1$. In this instance, since the order is assumed to be knov reference to the right-hand column (0 or 1) suffices to define the functi

7.2 Boolean Functions

Example: Table 7.1 a illustrates the definition procedure, in the case a 3-variable function.

Table 7.1

$x_1\,x_2\,x_3$	$f(x_1, x_2, x_3)$
0 0 0	1
0 0 1	0
0 1 0	1
0 1 1	0
1 0 0	0
1 0 1	1
1 1 0	1
1 1 1	0

(a)

	$(x_1\,x_2\,x_3)$
	0 1 0 1 0 1 0 1
	0 0 1 1 0 0 1 1
	0 0 0 0 1 1 1 1
00	1 0 0 1 0 1 0 1
$(x_4\,x_5)$ 01	1 0 1 0 1 0 1 0
10	0 1 1 1 1 0 0 1
11	0 0 0 0 1 1 1 0

$f(x_1, x_2, x_3, x_4, x_5)$

(b)

The truth table can also be presented in rectangular (square if necessary) form. The variables $x_1, x_2, \ldots x_n$ are partitioned into two disjoint sets $\{x_{i_1}, x_{i_2}, \ldots x_{i_s}\}$ and $\{x_{i_{s+1}}, x_{i_{s+2}}, \ldots, x_{i_n}\}$. Each of the 2 sets resulting from this separation is allocated to one of 2 rectangular axes on which are shown the sets of corresponding binary combinations (there are thus 2^s combinations on one axis and 2^{n-s} on the other). With each combination of n variables is thus associated, at the intersection of a line and a column, a cell in which is inscribed the value of the function. Table 7.1 b gives an example of this type of table.

Actually, it is in this form that the truth tables will most often be used. It will be seen that a table of this type (Karnaugh map) becomes a tool for the simplification of algebraic expressions. Additionally, it is useful when, for example, it is desired to emphasise the rôle of certain variables: they can be allocated to one side of the table.

Note. The name *"truth table"* used to describe the set of combinations of variables and associated values for a Boolean function is, as will have been noted, borrowed from the language of logic: it has already been used in chapter 5 for the definition of certain elementary logic functions. Here, it will be used in a general way to describe Boolean functions. It can be seen that such a table is indeed a 'truth' table if a suitable viewpoint be adopted. In logic, a truth table constitutes a list of logic values for a compound proposition as a function of those of the component propositions. It will therefore be convenient to adopt the following convention:

$$\begin{cases} \text{let } P_i \text{ be the proposition } (x_i = 1), (i = 1, 2, \ldots, n) \\ \text{and } F \text{ be the proposition } [f(x_1, x_2, \ldots, x_n) = 1] \end{cases}$$

whence the following correspondence:

$$\begin{array}{c|c} P_i = 1 & x_i = 1 \\ P_i = 0 & x_i = 0 \\ F = 1 & f(x_1, x_2, \ldots, x_n) = 1 \\ F = 0 & f(x_1, x_2, \ldots, x_n) = 0. \end{array} \quad (7.$$

A one-to-one correspondence has thus been established between th values 0 and 1 of the variables and of the function on the one hand, an the logic values of the associated propositions on the other. In additio by reason of the conditions established: 0 corresponds with 0 and with 1. A truth table such as that of Fig. 7.1 is thus a table of the sam type as that used in the determination of propositions in order to defir a logic function.

In particular, it is easy to extract an expression for F. Table 7.1 give for example:

$$F = (\overline{P}_1 \wedge \overline{P}_2 \wedge \overline{P}_3) \vee (\overline{P}_1 \wedge P_2 \wedge \overline{P}_3) \vee (P_1 \wedge \overline{P}_2 \wedge P_3) \vee (P_1 \wedge P_2 \wedge \overline{P}_3)$$
(7.2

The expression (7.2) is a logic sum of logic products.

For a given correspondence between the logic values of the abov propositions and the values of the variables, it can be said that:

$$f(x_1, x_2, x_3) = \overline{x}_1 \cdot \overline{x}_2 \cdot \overline{x}_3 + \overline{x}_1 \cdot x_2 \cdot \overline{x}_3 + x_1 \cdot \overline{x}_2 \cdot x_3 + x_1 \cdot x_2 \cdot \overline{x}_3 \quad (7.$$

where $+$ and $\cdot$ are the operations defined in Chapter 6 (§ 6.5.1).

It can thus be seen that:

1. A truth table (general sense) can, with a suitable convention, b considered as the truth table for a logic function.

2. On this basis it is possible to express the function as a sum c products. A return to this algebraic representation of a function (canon cal disjunctive form) will be made later.

7.2.3 Other Means of Defining a Boolean Function

7.2.3.1 Characteristic Vector

The right hand column of the truth table is a binary 'column' vecto If it be assumed that the binary combinations appearing at the left hand side are always arranged in increasing order of binary numbers, knowledge of this vector suffices to define the Boolean function. This wi be known as the characteristic vector for the function f.

7.2 Boolean Functions

Thus is given:

$$V_f = (v_0, v_1, v_2, \ldots, v_k, \ldots, v_{2^n-1}) \tag{7.4}$$

where v_k is the value of the function associated with the binary combination of the variables representing the integer k.

We have for example, in the case of the function defined by Table 1:

$$V_f = (1\ 0\ 1\ 0\ 0\ 1\ 1\ 0) \tag{7.5}$$

2.3.2 Decimal Representations

— *First representation*

The vector V_f constitutes the binary representation of a whole number between 0 and $2^{2^n} - 1$. Thus the function can be defined by making use of the decimal equivalent of this number (which, on average, is 3.32 times shorter).

$$N_{10}^f = \sum_{i=0}^{2^n-1} v_i 2^{2^n-1-i} \tag{7.6}$$

Example:

$$\begin{cases} V_f = 1\ 0\ 1\ 0\ 0\ 1\ 1\ 0 \\ N_{10}^f = 2^7 + 2^5 + 2^2 + 2^1 = 166 \end{cases} \tag{7.7}$$

— *Second representation*

This is the most frequently used representation, although less compact than the first.

Thus is obtained the sequence of decimal representations of the binary combinations for which the value of the function is 1 (n assumed to be known).

In the case of Table 7.1, it can thus be said:

$$f = \sum (0, 2, 5, 6) \tag{7.8}$$

2.4 Number of Boolean Functions Having n Variables

With n binary variables can be formed

$$\underbrace{2 \times 2 \times 2 \times \cdots \times 2}_{n \text{ factors}} = 2^n$$

distinct combinations associated with the 2^n n-digit binary numbers (numbers from 0 to $2^n - 1$ inclusive). In order to define a function, each of the $N = 2^n$ lines of the truth table can be associated with t values 0 or 1, which by the same reasoning as before, shows that the are 2^N characteristic vectors, or distinct Boolean functions, i.e. $2^{(2^n)} =$ (this last symbol having no ambiguity) distinct Boolean functions havi n variables.

Putting $B(n) = 2^{2^n}$, it will be seen that $B(n+1) = [B(n)]$ each time that n increases by one unit, the number $B(n)$ of the functio with n variables is increased by the square. This suggests, if this necessary, that $B(n)$ increases very rapidly with n, which Table 7 confirms.

Table 7.2. *Some values for* $B(n) = 2$

n	$B(n) = 2^{2^n}$
1	4
2	16
3	256
4	65 536
5	4 294 967 296

7.3 Operations on Boolean Functions

By use of the truth tables, which serve to define the Boolea functions, it is now possible to define the operations of sum, product a complementation on the functions.

7.3.1 Sum of two Functions f + g

Given two Boolean functions f and g with n variables, and wi characteristic vectors V_f and V_g, the function "sum of f and g" denot by $f + g$ (read f plus g) is that for which the characteristic vector $V_f + V_g$ (cf. Chap. 6 § 6.5.4).

By definition this gives:
$$V_{f+g} = V_f + V_g. \tag{7.}$$

Example:

Table 7.3

$x_1 x_2$	f	g	$f + g$
0 0	0	0	0
0 1	0	1	1
1 0	1	1	1
1 1	1	0	1

7.3 Operations on Boolean Functions

7.3.2 Product of two Functions f · g

Similarly the product of f and g, written $f \cdot g$ or fg (read : fg) is the function whose characteristic vector is $V_f \cdot V_g$

$$V_{f \cdot g} = V_f \cdot V_g. \qquad (7.10)$$

Example:

Table 7.4

$x_1 x_2$	f	g	$f \cdot g$
0 0	0	0	0
0 1	0	1	0
1 0	1	1	1
1 1	1	0	0

7.3.3 Complement of a Function f̄

This is the function, denoted $\bar{f}$ (read $\bar{f}$), and defined by $\overline{V}_f$. By definition, therefore:

$$V_{\bar{f}} = \overline{V}_f. \qquad (7.11)$$

Example:

Table 7.5

$x_1 x_2$	f	$\bar{f}$
0 0	0	1
0 1	0	1
1 0	1	0
1 1	0	1

7.3.4 Identically Null Function (0)

This function is always equal to 0: it is denoted as 0 or (0). Its characteristic vector is the null vector (formed only from zeros).

7.3.5 Function Identically Equal to 1 (1)

This function is always equal to 1, : it is denoted as 1 or (1). Its characteristic vector is the vector I (formed uniquely from 1).

7.4 The Algebra of Boolean Functions of n Variables

The set of Boolean functions with n variables possesses 2 distinct elements. With the aid of characteristic vectors it has been possible to define the operations $+$, $\cdot$, and $-$ for this set. Since these vectors make up a Boolean algebra of $2^N = 2^{2^n}$ elements (cf. Chap 6) the immediate outcome is that the set of functions of n variables which, with these operations, also constitutes a Boolean algebra having 2^{2^n} elements.

7.5 Boolean Expressions and Boolean Functions

7.5.1 Boolean Expressions

The definition given in the case of an completely general Boolean algebra will be taken up once more and be applied to the particular 2-element Boolean algebra $\langle \{0, 1\}, +, \cdot, - \rangle$ as defined in § 6.5.1. Thus the variables now have for their field of action the set $\{0, 1\}$: which are the binary variables. Thus the Boolean expressions include:
1. The constants 0 and 1,
2. All variables (direct or complements),
3. Every expression derived from a finite number of applications of the operations $+$, $\cdot$, and $-$.

7.5.2 Function Generated by a Boolean Expression

Consider an expression $F(X_1, X_2, ..., X_n)$ in which the variables are successively replaced by the 2^n combinations of values which the variables can take. For each one of these, the determination of the expression gives one element of the algebra, i.e. 0 or 1. It will thus be seen that the expression $F(X_1, X_2, ..., X_n)$ defines a Boolean function $f(X_1, X_2, ..., X_n)$ of the binary variables $X_1, X_2, ..., X_n$. It can be said that *f is the Boolean function generated by F* or *"represented" by F*. It is important to note that the *expression* $F(X_1, X_2, ..., X_n)$ and the *function* $f(X_1, ..., X_n)$ are two different mathematical entities. Nevertheless, it has just been seen that they are not independent: every expression generates a function and, from this point of view, the expression appears as a means of calculation, a procedure which will allow the storage and subsequent restitution of the function f, and serving as a substitute for the (2^n lines) truth table. The expression $F(X_1, ..., X)_n$ can, therefore, be considered to be a *representation* or a *notation* of the function f, which, moreover, in the majority of cases, will prove to be more compact than the truth table.

7.5 Boolean Expressions and Boolean Functions

For this reason the following convention will be adopted: $f(X_1, \ldots, X_n) = F(X_1, \ldots, X_n)$ will be written using only one symbol to represent both expression and function. This is a procedure frequently used in mathematical analysis; when writing, for example, $f(x, y) = x^2 - y^2$, all that has been done is the replacement by an expression $(x^2 - y^2)$ of a function defined for an infinite number of combinations xy and which, for this reason, cannot be presented in tabular form: thus (in practice at any rate) no distinction is made between the expression and the function generated. (This procedure for the definition of a function is not the only one, on the other hand it cannot always be utilised.) Moreover, as in mathematical analysis, where several formulae can be made to represent the same function (e.g. $f(x, y) = x^2 - y^2 = (x + y)(x - y)$), it is evident that a Boolean function can be represented by more than one formula: the law of idempotence, for example, shows that there is an infinity of them.

(e.g.: $A = A + A = A + A + A =$ etc.).

Later the problem of the determination of an algebraic representation of an expression for a given function defined by a truth table will be examined. (It will be seen that this is always possible.)

Example: Expression: $F(A, B, C) = A + B\overline{C}$
Function: this is given by Table 7.6.

Table 7.6

A B C	$f_{A+B\overline{C}}$
0 0 0	0
0 0 1	0
0 1 0	1
0 1 1	0
1 0 0	1
1 0 1	1
1 1 0	1
1 1 1	1

If $A = 1$, $A + B\overline{C} = 1$;

If $A = 0$, $F = 1$ only when $B = 1$ and $C = 0$.

— Notation: for simplification this can be put:

$$f(A, B, C) = f_{A+B\overline{C}} = A + B\overline{C} \qquad (7.12)$$

Remark: determination of the truth table on the basis of an expression: This calculation can often be made more rapidly than would appear to be possible at first sight and this is the case in the above example. It is often possible to determine the value of the function for whole groups of binary combinations at a single 'bite'. In particular, it always 'pays' to apply the following properties, wherever possible:

— if one term of a sum takes the value 1, the entire sum has the value 1, no matter what the other terms.

— if one factor of a product takes the value 0, the product takes the value 0, no matter what the other factors.

7.5.3 Equality of 2 Expressions

In conformance with the general definition (Chapter 6) two algebraic expressions E_1 and E_2 are said to be equal (or equivalent) if they generate the same function. This can be expressed as $E_1 = E_2$. By definition, two equal expressions take the same value for all the values of the variables.

Example:

$$\begin{cases} E_1(A, B, C) = AB\bar{C} + ABC + \bar{A}BC \\ E_2(A, B, C) = B(A + C) \end{cases} \qquad (7.13)$$

Giving $E_1 = E_2$. In fact, the truth tables for the functions generated are the same, as shown by Table 7.7.

Table 7.7. *Equality of 2 expressions*

A B C	$B(A + C)$	$AB\bar{C} + ABC + \bar{A}BC$
0 0 0	0	0
0 0 1	0	0
0 1 0	0	0
0 1 1	1	1
1 0 0	0	0
1 0 1	0	0
1 1 0	1	1
1 1 1	1	1

Table 7.7 can be obtained in the following manner:

$$E_1 = AB\bar{C} + ABC + \bar{A}BC \qquad (7.14)$$

7.5 Boolean Expressions and Boolean Functions

A check is made to ensure that the only values for $E_1 = 1$ are:

$$110 \ (AB\bar{C} = 1), \quad 111 \ (ABC = 1) \text{ and } 011 \ (\bar{A}BC = 1).$$

Any of these combinations is sufficient for one of the three terms in E_1 to be equal to 1.

$$E_2 = B(A + C)$$

$$\begin{cases} \text{if } B = 0 & E_2 = 0 \\ \text{if } B = 1 & E_2 = 1 \text{ when } A = 1 \text{ or } C = 1 \text{ or } (A = 1 \text{ and } C = 1) \end{cases}$$

(7.15)

This procedure, the comparison of the truth tables associated with the two expressions, is general and will often be used in practice to demonstrate the identities. Other methods will be discussed later: algebraic method, the Veitch-Karnaugh diagram method.

7.5.4 Algebraic Expressions for a Boolean Function

The reverse problem is now to be examined: to find an algebraic representation for a given Boolean function with n variables $X_1, \ldots, X_n$, given by its truth table.

For this it will first be necessary to define several distinguished functions of n variables which will prove useful.

7.5.4.1 Fundamental Product

— Definition.

A *fundamental product* or "*minterm*" with n variables takes the form:

$$P = X_1^{e_1} X_2^{e_2} \cdots X_i^{e_i} \cdots X_n^{e_n} = \prod_{i=1}^{i=n} X_i^{e_i} \qquad (7.16)$$

with:

$$\begin{aligned} X_i^{e_i} &= X_i \quad \text{if} \quad e_i = 1 \quad \text{i.e,} \quad X^1 = X \\ X_i^{e_i} &= \bar{X}_i \quad \text{if} \quad e_i = 0 \qquad\qquad X^0 = \bar{X} \end{aligned} \qquad (7.17)$$

In other words, a fundamental product is a product of n variables each of which can be barred or not, and each letter appearing exactly once.

Examples:

- The products $A\bar{B}\bar{C}D$, $ABC\bar{D}$, $\bar{A}\bar{B}\bar{C}D$ are fundamental products having 4 variables (A, B, C, D);
- ABC is not a fundamental product of 4 variables.

Remark: there are 2^n fundamental products corresponding with the 2^n binary combinations of the variables $e_1, e_2, \ldots, e_n$.

Truth table for a fundamental product:

In order for the product

$$P = X_1^{e_1} X_2^{e_2} \cdots X_i^{e_i} \cdots X_n^{e_n}$$

to be equal to 1, it is necessary and sufficient that its factors be equal to 1, i.e. that we have:

$$X_i^{e_i} = 1 \quad (i = 1, 2, \ldots, n).$$

If $e_i = 1$, $X_i^{e_i} = X_i$ and $X_i^{e_i} = 1$ equivalent to $X_i = 1 = e_i$.

If $e_i = 0$, $X_i^{e_i} = \bar{X}_i$ and $X_i^{e_i} = 1$ equivalent to $X_i = 0 = e_i$.

For $X_i^{e_i} = 1$ it is necessary and sufficient that $X_i = e_i$
Whence, finally: for $P = 1$, it is necessary and sufficient that

$$X_i = e_i (i = 1, 2, \ldots, n)$$

or

$$(X_1 X_2, \ldots X_i, \ldots, X_n) = (e_1 e_2, \ldots, e_i, \ldots, e_n). \qquad (7.18)$$

The fundamental product P is thus equal for exactly one combination of variables, the combination $(e_1, e_2, \ldots, e_i, \ldots, e_n)$. Consequently, the truth table of the function generated, has the following form: its right-hand column carries a single 1 in respect of the combination $(e_1, e_2, \ldots, e_i, \ldots, e_n)$ and the value 0 in all other respects.

Conversely, for a given truth table having these properties, where the function takes the value 1 for a single combination of variables $(e_1, e_2, \ldots, e_i, \ldots, e_n)$ it is the table for the function generated by the fundamental product

$$P = X_1^{e_1} X_2^{e_2} \cdots X_i^{e_i} \cdots X_n^{e_n}$$

Examples:

(1) $A\bar{B}\bar{C}D = A^1 B^0 C^0 D^1$ is equalt o 1 only for the combination 1001 of the variables.

7.5 Boolean Expressions and Boolean Functions

(2) Consider the function defined by table 7.8. This is the function generated by the product:
$$P = \bar{A}BC \tag{7.19}$$

Table 7.8. *A fundamental product: truth table*

A B C	f
0 0 0	0
0 0 1	0
0 1 0	0
0 1 1	1
1 0 0	0
1 0 1	0
1 1 0	0
1 1 1	0

(3) Now consider a function having three variables. Refer to Table 7.9 for this. It will be seen, that in conformance with the definition of the sum of products:
$$f = f_1 + f_2 + f_3,$$
where f_1, f_2, f_3 are the functions defined in Table 7.10.

Example:

Table 7.9

A B C	f
0 0 0	0
0 0 1	1
0 1 0	1
0 1 1	0
1 0 0	1
1 0 1	0
1 1 0	0
1 1 1	0

But, each of the functions $f_1, f_2,$ and f_3 is generated by a fundamental product which is denoted as m_{abc}, with:
$$m_{abc} = A^a B^b C^c,$$

where abc is the binary combination of ABC for which $m_{abc} = 1$. Thus denoting abc in the decimal code:

$$\begin{cases} m_1 = \bar{A}\bar{B}C \\ m_2 = \bar{A}B\bar{C} \quad \text{and} \\ m_4 = A\bar{B}\bar{C} \end{cases} \begin{cases} f_1 = m_1 \\ f_2 = m_2 \\ f_3 = m_4 \end{cases} \qquad (7.20)$$

Table 7.10

A B C	f_1	f_2	f_3	$f = f_1 + f_2 + f_3$
0 0 0	0	0	0	0
0 0 1	1	0	0	1
0 1 0	0	1	0	1
0 1 1	0	0	0	0
1 0 0	0	0	1	1
1 0 1	0	0	0	0
1 1 0	0	0	0	0
1 1 1	0	0	0	0

Whence

$$f = f_1 + f_2 + f_3 = m_1 + m_2 + m_4 = \bar{A}\bar{B}C + \bar{A}B\bar{C} + A\bar{B}\bar{C} \qquad (7.21)$$

A recapitulation of how to form the expression f is to be found in Table 7.11, where F is the expression sought.

Table 7.11

A B C	f	$F =$
0 0 0	0	
0 0 1	1	$\bar{A} \cdot \bar{B} \cdot C$
0 1 0	1	$+ \bar{A} \cdot B \cdot \bar{C}$
0 1 1	0	
1 0 0	1	$+ A \cdot \bar{B} \cdot \bar{C}$
1 0 1	0	
1 1 0	0	
1 1 1	0	

Remark: as has been agreed the same letter is used to represent both the function f and any algebraic expression for f, although in reality they are two different mathematical entities. Thus in the present instance:

$$f = \bar{A}\bar{B}C + \bar{A}B\bar{C} + A\bar{B}\bar{C}. \qquad (7.22)$$

7.5 Boolean Expressions and Boolean Functions

7.5.4.2 Disjunctive Canonical Form of a Function of n Variables

It is now clear that the reasoning used in the case of 3 variables and in that of a particular function, is quite general and that every function can be expressed as a Boolean sum of fundamental products, by making use of the following rule:

1. Write down the fundamental products corresponding with the combinations of variables for which the function $f(X_1, X_2, \ldots X_n)$ equals 1.
2. Determine the Boolean sum of all these products. The expression thus obtained constitutes an algebraic representation of the function, known as the *first canonical form* or *disjunctive canonical form*.

From the rule as stated is obtained:

$$f(X_1, X_2, \ldots, X_i, \ldots, X_n)$$
$$= \sum_{e_1 e_2 \ldots e_i \ldots e_n = 0}^{e_1 e_2 \ldots e_i \ldots e_n = 2^n - 1} f(e_1, e_2, \ldots, e_i, \ldots, e_n)\, X_1^{e_1} X_2^{e_2} \cdots X_i^{e_i} \cdots X_n^{e_n} \quad (7.23)$$

(first canonical form or disjunctive canonical form).

The first canonical form is unique: this character of uniqueness results directly from the one-to-one correspondence which exists between the fundamental products comprising the canonical form and the binary combinations of the truth table for which the function equals 1.

7.5.4.3 Fundamental Sums

The *fundamental sum* or *maxterm* is the name given to a sum of the form:

$$S = X_1^{e_1} + X_2^{e_2} + \cdots + X_i^{e_i} + \cdots X_n^{e_n} = \sum_{i=0}^{i=n} X_i^{e_i} \quad (7.24)$$

with the same convention as for $X_i^{e_i}$ above.

In other words, a fundamental sum with n variables is a sum of the n variables, whether barred or not, each of the n letters appearing exactly once.

Truth table for a sum of variables.

In order that the sum

$$S = X_1^{e_1} + X_2^{e_2} \cdots + X_i^{e_i} + \cdots X_n^{e_n}$$

be zero it is necessary and sufficient that all its terms be zero, i.e. that this should give:

$$X_i^{e_i} = 0 \quad (i = 1, 2, \ldots, n) \quad (7.25)$$

it will be seen that, in this instance, this necessitates

$$X_i = \bar{e}_i \quad (i = 1, 2, \ldots, n). \tag{7.26}$$

Thus in order that S be zero it is necessary and sufficient that the combination of variables be identical with that of the complements of the e_i.

Summarizing, S is zero for a single combination of the variables,

$$(\bar{e}_1, \bar{e}_2, \ldots, \bar{e}_i, \ldots, \bar{e}_n)$$

and has the value 1 for all others. Conversely, every function having a table truth of this type, i.e. having the value 0 for a single combination $(a_1, a_2, \ldots, a_n)$ of variables can be represented by a fundamental sum:

$$f = X_1^{\bar{a}_1} + X_2^{\bar{a}_2} + \cdots + X_i^{\bar{a}_i} + \cdots + X_n^{\bar{a}_n}.$$

Examples:

(1) The fundamental sum

$$A + \bar{B} + \bar{C} + D = A^1 + B^0 + B^0 + D^1$$

is zero for the combination 0110 and for this combination only.

(2) Consider the function defined by Table 7.12.

Table 7.12. *A fundamental sum — truth table*

A B C	f
0 0 0	1
0 0 1	1
0 1 0	1
0 1 1	0
1 0 0	1
1 0 1	1
1 1 0	1
1 1 1	1

This may be written: $f = A + \bar{B} + \bar{C}$

(3) Consider the function defined by Table 7.13
It will be seen that it is possible to write: $f = f_1 \cdot f_2 \cdot f_3$, where f_1, f_2, f_3 are defined in Table 7.14. Thus it suffices to take the fundamental sums associated with the zeros of the column f.

7.5 Boolean Expressions and Boolean Functions

Table 7.13

A B C	f
0 0 0	1
0 0 1	0
0 1 0	0
0 1 1	1
1 0 0	0
1 0 1	1
1 1 0	1
1 1 1	1

Table 7.14

A B C	f_1	f_2	f_3	$f = f_1 \cdot f_2 \cdot f_3$
0 0 0	1	1	1	1
0 0 1	0	1	1	0
0 1 0	1	0	1	0
0 1 1	1	1	1	1
1 0 0	1	1	0	0
1 0 1	1	1	1	1
1 1 0	1	1	1	1
1 1 1	1	1	1	1

Moreover:

$$\begin{cases} f_1 = A + B + \overline{C} \\ f_2 = A + \overline{B} + C \\ f_3 = \overline{A} + B + C \end{cases} \qquad (7.27)$$

Whence:

$$f = (A + B + \overline{C})(A + \overline{B} + C)(\overline{A} + B + C). \qquad (7.28)$$

5.4.4 Conjunctive Canonical Form of a Function of n Variables

Reasoning as in the disjunctive form, we see that every function can be expressed as a product of fundamental sums, by use of the following rule:

- Write down all the fundamental sums corresponding with the combinations of variables for which the function $f(X_1, X_2, \ldots, X_n)$ has the value 0.
- Write down the product of these sums.

The expression thus obtained constitutes an algebraic representation of the function known as the "second canonical form" or "conjunctive canonical form".

Thus giving:

$$f(X_1, X_2, \ldots, X_i, \ldots, X_n)$$
$$= \sum_{e_1 e_2 \ldots e_i \ldots e_n = 0}^{e_1 e_2 \ldots e_i \ldots e_n = 2^n - 1} \left[f(\bar{e}_1, \bar{e}_2, \ldots, \bar{e}_i, \ldots, \bar{e}_n) + X_1^{e_1} + X_2^{e_2} + \cdots + X_i^{e_i} + \cdots X_n^{e_n} \right]$$

(7.29)

(second canonical form, conjunctive canonical form).

Remarks:

1. The second canonical form is unique. As for the first uniqueness is the immediate result of the manner in which it is obtained.
2. In the disjunctive canonical form, the terms whose coefficient $f(e_1, e_2, \ldots, e_n)$ disappear when f is zero. In the conjunctive canonical form, the factors disappear from the product when $f(\bar{e}_1, \bar{e}_2, \ldots, \bar{e}_n)$ have the value 1, since the factor is then of form $1 + A = 1$.
3. It is important to note that in the first canonical form, the value $f(e_1, e_2, \ldots, e_n)$ is associated with

$$X_1^{e_1} \cdot X_2^{e_2} \cdot \ldots \cdot X_n^{e_n}.$$

In the second, it is the value $f(\bar{e}_1, \bar{e}_2, \ldots, \bar{e}_n)$ which is associated with

$$X_1^{e_1} + X_2^{e_2} + \cdots + X_n^{e_n}.$$

7.5.4.5 Some Abridged Notations

— *Fundamental products.*

By putting

$$m_i = X_1^{e_1} \cdot X_2^{e_2} \cdot \ldots \cdot X_n^{e_n} \quad (7.30)$$

where $i = e_1 e_2 \cdots e_n$ (i is an index, for which $e_1, e_2, \ldots e_n$ is the binary representation; $0 \leq i \leq 2^n - 1$) is obtained:

$$m_i \cdot m_j = 0 \text{ for } i \neq j. \quad (7.31)$$

— *Fundamental sums*

By putting

$$M_i = X_1^{e_1} + X_2^{e_2} + \cdots + X_n^{e_n} \quad (7.32)$$

with $i = e_1, e_2 \cdots e_n$ as above, is obtained:

$$M_i + M_j = 1 \text{ for } i \neq j. \quad (7.33)$$

— *Disjunctive canonical form*

$$f = \sum_{i=0}^{i=2^{n}-1} f(i) \cdot m_i \quad (7.34)$$

7.5 Boolean Expressions and Boolean Functions

— *Conjunctive canonical form*

$$f = \prod_{i=0}^{i=2^{n-1}} \left[f(2^n - 1 - i) + M_i \right] \qquad (7.35)$$

since $\bar{e}_1, \bar{e}_2, \ldots, \bar{e}_n = 2^n - 1 - i$ when putting $i = e_1 e_2 \cdots e_n$).

5.4.6 Partial Expansions of a Boolean Function

The procedure by which the canonical forms of a *Boolean* function of n variables above was obtained shows that it is possible to define partial expansions with respect to s variables $(1 \leq s \leq n)$, the case $s = n$ corresponding with the canonical forms with n variables. The sums of products can thus be written:

$$f(X_1, X_2, \ldots, X_n)$$
$$= \sum_{j=0}^{j=2^s-1} f(j_1, j_2, \ldots, j_s, X_{s+1}, X_{s+2}, \ldots, X_n) \cdot X_1^{j_1} \cdot X_2^{j_2} \cdot \ldots X_s^{j_s} \quad (j = j_1 \, j_2 \cdots j_s)$$

or, more briefly:

$$f(X_1, X_2, \ldots, X_n) = \sum_{j=0}^{j=2^s-1} f_j(X_{s+1}, X_{s+2}, \ldots, X_n) m_j(X_1, X_2, \ldots, X_s)$$

$$\left(f_j(X_{s+1}, X_{s+2}, \ldots X_n) = f(j_1, j_2, \ldots, j_s, X_{s+1}, X_{s+2}, \ldots, X_n), \right.$$
$$\left. m_j(X_1, X_2, \ldots, X_s) = X_1^{j_1} \cdot X_2^{j_2} \ldots X_s^{j_s} \right)$$

Table 7.15

$$\overrightarrow{j(X_1, X_2, X_3)}$$

		0	1	0	1	0	1	0	1
		0	0	1	1	0	0	1	1
		0	0	0	0	1	1	1	1
	00	1	0	0	1	0	1	0	1
$i(X_4 X_5)$	01	1	0	1	0	1	0	1	0
(f_i)	10	0	1	1	1	1	0	0	1
	11	0	0	0	0	1	1	1	0

(f_j)

$$f = \sum_{j=0}^{j=7} f_j(X_4, X_5) \, m_j(X_1, X_2, X_3)$$

$$f = \sum_{i=0}^{i=0} f_i(X_1, X_2, X_3) \, m_i(X_4, X_5)$$

With the analogous abridged notation, we can also write:

$$f(X_1, X_2, ..., X_n) = \sum_{i=0}^{i=2^{n-s}-1} f(X_1, X_2, ... X_s) \, m_i(X_{s+1}, X_{s+2}, ..., X_n)$$

If, with the aid of a rectangular truth table $f(X_1, X_2, ..., X_n)$ b represented by the 2^s combinations of j the first variables $(j = j_1, j_2 ... j$ inscribed on the horizontal axis, and the 2^{n-s} combinations of $i = (i_1 i$ on the other, the function f_i will have as its characteristic vector th line i, while the function f_j will have as its characteristic vector th column j (Table 7.15).

7.6 Dual Functions

For a given function $f(X_1, X_2, ... X_n)$ the dual function, denote $\tilde{f}(X_1, X_2, ..., X_n)$, is given by the relation:

$$\tilde{f}(X_1, X_2, ..., X_n) = f\overline{\left(\overline{X_1}, \overline{X_2}, ..., \overline{X_n}\right)}. \tag{7.3}$$

Properties:

7.6.1 Dual of a Dual Function

$$\tilde{\tilde{f}} = f \tag{7.3}$$

In fact:

$$\tilde{\tilde{f}} = \widetilde{(\tilde{f})} = \overline{\left[f(\overline{\tilde{X}_1}, \overline{\tilde{X}_2}, ..., \overline{\tilde{X}_n})\right]} = f(X_1, X_2, ..., X_n) \tag{7.3}$$

7.6.2 Dual of a Sum

$$\widetilde{\sum_i X_i} = \overline{\sum_i \overline{X}_i} = \prod_i X_i \tag{7.3}$$

Example: dual of a fundamental sum:

$$\tilde{M}_k = m_k. \tag{7.4}$$

7.6.3 Dual of a Product

$$\widetilde{\prod_i X_i} = \overline{\prod_i \overline{X}_i} = \sum_i X_i \tag{7.4}$$

Example: dual of a fundamental product

$$\tilde{m}_k = M_k. \tag{7.4}$$

6.4 Dual of a Complement

$$\left(\tilde{\bar{A}}\right) = \left(\bar{\tilde{A}}\right) = \bar{A}. \tag{7.43}$$

6.5 Dual of a Constant Function

$$f(X_1, X_2, \ldots, X_n) \equiv 0, \tag{7.44}$$

then

$$f(\bar{X}_1, \bar{X}_2, \ldots, \bar{X}_n) \equiv 0 \tag{7.45}$$

and

$$\tilde{f} = \bar{f}(\bar{X}_1, \bar{X}_2, \ldots, \bar{X}_n) \equiv 1. \tag{7.46}$$

Similarly, if

$$f(X_1, X_2, \ldots, X_n) \equiv 1 \tag{7.47}$$

$$\tilde{f} \equiv 0. \tag{7.48}$$

It will thus be seen that for a given expression representing a function, the expression for the dual function is obtained by replacing $+$ by $\cdot$, $\cdot$ by $+$, leaving the variables (barred or not) untouched and by replacing 0 by 1, 1 by 0. These are the rules given in a more general context in Chapter 6.

7 Some Distinguished Functions

Some distinguished functions, which will be of use later in this work, are now to be defined. Certain of these can be defined for any number n of variables and have already been encountered for $n = 2$ in the 2-variable functions tables of Chapter 5.

7.1 n-Variables AND Function

This is the function whose value is:

$$\begin{cases} 1 \text{ when } X_1 = X_2 = \cdots = X_k = \cdots = X_n = 1 \\ 0 \text{ in all other cases} \end{cases} \tag{7.49}$$

This may be written:

$$f = X_1 \cdot X_2 \cdot X_3 \cdot \cdots \cdot X_k \cdots \cdot X_n \tag{7.50}$$

$$X_1 X_2 X_3 \cdots X_k \cdots X_n.$$

It can also be said that f is the function which takes as its value the minimum value of the inputs (on the understanding that $0 < 1$).

7.7.2 n-Variables OR Function

This is the function whose value is:

$$\begin{cases} 0 \text{ when } X_1 = X_2 = \cdots = X_k = \cdots X_n = 0 \\ 1 \text{ in all other cases} \end{cases} \quad (7.51)$$

we write:

$$f = X_1 + X_2 + \cdots + X_k + \cdots + X_n. \quad (7.52)$$

It can also be said that the function OR is the function which takes as its value the maximum value of the inputs.

Remark: The same symbols have been used for the functions with n variables.

$$\cdots, \text{ or } + + + \cdots +.$$

This *convention* could have been obtained in another way: given that

$$X_1 + (X_2 + X_3) = (X_1 + X_2) + X_3,$$

we may define a quantity, denoted by $X_1 + X_2 + X_3$, and which the common value of the expressions above. This will then be defined step-by-step:

$$X_1 + X_2 + \cdots + X_n = (X_1 + X_2 + \cdots + X_{n-1}) + X_n.$$

It is easily verified that this definition coincides with the direct definition for the operations AND and OR with n variables. It must be remembered that the symbols $+$ and $\cdot$ as defined in Chapter 6 were intended to represent operations of functions and that, consequently, their use for functions of n variables necessitates a complementary definition.

7.7.3 Peirce Function of n-Variables (OR-INV-NOR)

This is the function whose value is:

$$\begin{cases} 1 \text{ when } X_1 = X_2 = \cdots = X_k = \cdots = X_n = 0 \\ 0 \text{ in all other cases} \end{cases} \quad (7.5)$$

It is written:

$$f = X_1 \downarrow X_2 \downarrow \cdots \downarrow X_k \downarrow \cdots \downarrow X_n = \downarrow_{i=1}^{i=n} X_i \quad (7.5)$$

— *case of one variable:*

By definition $\downarrow X = X\downarrow$ is the function whose value is

$$\begin{cases} 1 \text{ when } X = 0 \\ 0 \text{ when } X = 1 \end{cases} \qquad (7.55)$$

Thus
$$\downarrow X = X\downarrow = \overline{X}. \qquad (7.56)$$

— *case of two variables:*

the function $X \downarrow Y$ has already been encountered in chapter 5.

It must be noted that the symbol $\downarrow$, when repeated, has a different signification from that of $\downarrow$ used in the case of two variables: the use of $\downarrow$ repeated $n-1$ times serves to indicate the n-ary operation defined earlier, but here it does not signify that the result of the operation will be obtained by repeated application of the binary $\downarrow$ (cf. 7.7.5.6). In fact:

$$X_1 \downarrow X_2 \cdots \downarrow X_k \cdots \downarrow X = \overline{X}_1 \cdot \overline{X}_2 \cdots \overline{X}_k \cdots \overline{X}_n$$
$$= \overline{X_1 + X_2 + \cdots + X_k \cdots + X_n} \qquad (7.57)$$

It can thus be checked, for example, for $n = 3$ that:

$$X_1 \downarrow (X_2 \downarrow X_3) \neq (X_1 \downarrow X_2) \downarrow X_3. \qquad (7.58)$$

In other words, the binary operation $\downarrow$ is not associative.

The notation $\downarrow$ in n variables is not, therefore, so easily associated with that for 2 variables as in the case of AND and OR and, consequently, will be less convenient in use. Its properties will be discussed in greater detail at a later stage.

7.7.4 Sheffer Function in n-Variables (AND-INV; NAND)

This is the function whose value is:

$$\begin{cases} 0 \text{ when } X_1 = X_2 = \cdots = X_k = \cdots = X_n = 1 \\ 1 \text{ in all other cases} \end{cases} \qquad (7.59)$$

It can be expressed:

$$f = X_1/X_2/X_3/\cdots/X_k/\cdots/X_n = \Big/_{i=1}^{i=n} X_i \qquad (7.60)$$

— *case with one variable:*

It can again be written:
$$|X = X| \tag{7.61}$$
$$|X = X| = \overline{X} \tag{7.62}$$

— *case with 2 variables:* once again the operation X/Y as defined in Chapter 5 is encountered.

We have:
$$X_1/X_2/ \cdots /X_k/ \cdots / X_n = \overline{X}_1 + \overline{X}_2 + \cdots + \overline{X}_k + \cdots + \overline{X}_n$$
$$= \overline{X_1 X_2 \cdots X_k \cdots X_n} \tag{7.63}$$

It can easily be verified that the operation / is not associative in general, by considering, for example, the binary operation:
$$X_1 / (X_2 / X_3) \neq (X_1 / X_2) / X_3 \tag{7.64}$$

7.7.5 Some Properties of the Operations $\downarrow$ and /. Sheffer Algebrae

7.7.5.1 Expressions by Means of the Operations $(+)$, $(\cdot)$ and $(-)$

In accordance with the definition given it can immediately be stated that:
$$\downarrow_i X_i = X_1 \downarrow X_2 \downarrow \ldots \downarrow X_n$$
$$= \overline{X}_1 \cdot \overline{X}_2 \cdot \cdots \cdot \overline{X}_n \quad \text{or} \quad \downarrow_i X_i = \prod_i \overline{X}_i$$
$$= \overline{X_1 + X_2 + \cdots + X_n} \qquad = \overline{\sum_i X_i} \tag{7.65}$$

$$/_i X_i = X_1 / X_2 / \cdots / X_n$$
$$= \overline{X_1 X_2 \cdots X_n} \quad \text{or} \quad /_i X_i = \sum_i \overline{X}_i$$
$$= \overline{X}_1 + \overline{X}_2 + \cdots + \overline{X}_n \qquad = \overline{\prod_i X_i} \tag{7.66}$$

7.7.5.2 Properties of Duality for $\downarrow$ and /

Consider the Peirce function:
$$X_1 \downarrow X_2 \downarrow \cdots \downarrow X_n \text{ in } n \text{ variables}$$

7.7 Some Distinguished Functions

We can write:

$$\overline{\overline{X}_1 \downarrow \overline{X}_2 \downarrow \cdots \downarrow \overline{X}_n} = \overline{\prod_i \overline{\overline{X}}_i} = \overline{\prod_i X_i} = \sum_i \overline{X}_i \qquad (7.67)$$

$$= X_1 \mid X_2 \mid \ldots \mid X_n \qquad (7.68)$$

Thus:

The dual of the Peirce function is the Sheffer function ($/$). Consider now the Sheffer function.

Here:

$$\overline{\overline{X}_1 \mid \overline{X}_2 \mid \ldots \mid \overline{X}_n} = \overline{\sum_i \overline{\overline{X}}_i} \qquad (7.69)$$

$$= \overline{\sum_i X_i} = \prod_i \overline{X}_i = X_i \downarrow X_x \downarrow \cdots \downarrow X_n. \qquad (7.70)$$

Therefore:

The dual of the Sheffer function ($/$) is the Peirce function ($\downarrow$). This second result could be demonstrated in another manner, noting that:

$$(\overline{\overline{f}}) = f.$$

Thus we have the relations

$$\begin{cases} \overline{\displaystyle\downarrow_i X_i} = /_i \overline{X}_i \\ \overline{\displaystyle/_i X} = \downarrow_i \overline{X}_i \end{cases} \qquad (7.71)$$

which, in a certain sense, generalise the De Morgan formulae:

$$\begin{cases} \overline{\displaystyle\sum_i X_i} = \prod_i \overline{X}_i \\ \overline{\displaystyle\prod_i X_i} = \sum_i \overline{X}_i \end{cases} \qquad (7.72)$$

7.7.5.3 Sheffer Algebrae

A Boolean algebra has, in an axiomatic manner, been defined above as a set equipped with certain internal laws of composition ($+$, $\cdot$, $-$) which satisfies certain axioms. The algebraic structure thus defined was, by reason of this axiomatic character, capable of application to very diverse switching networks. In a similar manner (Hammer-Ivănescu) a new algebraic structure can be defined, that of a Sheffer algebra.

The name *Sheffer algebra* (Shefferian algebra) is given to the algebraic structure consisting of a set S having at least 2 elements, equipped with a system of functions with n variables, denoted by

$$F_n = \downarrow_{i=1}^{i=n} X = X_1 \downarrow X_2 \downarrow \cdots \downarrow X_n \quad \text{for} \quad n > 1 \text{ and } \overline{X}_1 \text{ for } n = 1$$

and satisfying the following axioms:

Axioms for a Sheffer algebra:

$$\left.\begin{aligned} A &= A \\ A \downarrow A &= \overline{A} \\ A \downarrow B &= B \downarrow A \\ A \downarrow (A \downarrow B) &= A \downarrow \overline{B} \\ (A \downarrow B) \downarrow (A \downarrow C) &= \overline{A \downarrow (\overline{B} \downarrow \overline{C})} \\ A_1 \downarrow \cdots \downarrow A_n &= \overline{A_1 \downarrow \cdots \downarrow A_{n-1}} \downarrow A_n \quad (n>2) \\ A \downarrow \overline{A} &= B \downarrow \overline{B} \quad \text{for every pair } A, B \end{aligned}\right\} \quad (7.73)$$

7.7.5.4 *Expressions for the Operations* $\cdot$, $+$ *and* $-$ *with the aid of th Operations* $\downarrow$ *and* $/$ (n *Variables*)

— *Operation* ($\cdot$) (product)

$$X_1 \downarrow X_2 \downarrow \cdots \downarrow X_i \downarrow \cdots \downarrow X_n = \overline{X}_1 \cdot \overline{X}_2 \cdot \cdots \cdot \overline{X}_i \cdot \cdots \cdot \overline{X}_n$$

Putting $X_i = \overline{A}_i$ we obtain the relation:

$$\prod_i A_i = A_1 . A_2 . \cdots . A_i . \cdots . A_n$$
$$= \overline{A}_1 \downarrow \overline{A}_2 \downarrow \cdots \downarrow \overline{A}_i \downarrow \cdots \downarrow \overline{A}_n = \downarrow \overline{A}_i \quad (7.74)$$

— *Operation* ($+$) (sum)

In a similar manner:

$$X_1 / X_2 / \cdots / X_i / \cdots X_n = \overline{X}_1 + \overline{X}_2 + \cdots + \overline{X}_i + \cdots + \overline{X}_n$$

In posing $X_i = \overline{A}_i$, this time is obtained:

$$\sum_i A_i = A_1 + A_2 + \ldots + A_i \ldots + A_n$$
$$= \overline{A}_1 / \overline{A}_2 / \ldots / \overline{A}_i / \ldots / \overline{A}_n = /_i \overline{A}_i \quad (7.75)$$

— *Complement* ($-$)

It has been seen that:

$$\overline{A} = \overline{A + A + \cdots + A} = A \downarrow A \downarrow \cdots \downarrow A \quad (7.76)$$

7.7 Some Distinguished Functions

In a similar manner:

$$\bar{A} = \overline{A \cdot A \cdot \cdots \cdot A} = A \mid A \mid \cdots \mid A.$$

7.5.5 Canonical Forms

From the two normal canonical forms, it is easy to obtain expressions by use of the symbols ↓ and /, which can be called the Sheffer canonical forms.

Consider the following function defined by its truth table (Table 7.16). For the first canonical form of F we have:

$$F = \bar{A}BC + \bar{A}\bar{B}C + AB\bar{C}. \tag{7.77}$$

a) This can be written as:

$$F = \overline{\overline{(\bar{A}BC)}\,\overline{(\bar{A}\bar{B}C)}\,\overline{(AB\bar{C})}} \quad \text{(De Morgan theorem and double negation)}$$

Whence, by definition of $\mid$:

$$F = (\bar{A} \mid B \mid C) \mid (\bar{A} \mid \bar{B} \mid C) \mid (A \mid B \mid \bar{C}). \tag{7.78}$$

Hence: having knowledge of the first canonical form, an expression for the function is obtained with the aid of the operator / only by replacing all the operation signs + and · by / and placing in parentheses the groups of symbols corresponding with the fundamental products.

b) Next consider the second canonical form; which is written (cf. truth table 7.16) as:

$$F = (A + B + C)(A + \bar{B} + C)(\bar{A} + B + C)(\bar{A} + B + \bar{C})$$

$$(\bar{A} + \bar{B} + \bar{C}) \tag{7.79}$$

Table 7.16

$A\ B\ C$	F
0 0 0	0
0 0 1	1
0 1 0	0
0 1 1	1
1 0 0	0
1 0 1	0
1 1 0	1
1 1 1	0

Thus can be written

$$F = \overline{(\overline{A}\,\overline{B}\overline{C})}\,\overline{(\overline{A}B\overline{C})}\,\overline{(A\overline{B}\overline{C})}\,\overline{(A\overline{B}C)}\,\overline{(ABC)} \qquad (7.80)$$

$$= (\overline{A}\,\overline{B}\overline{C}) \downarrow (\overline{A}B\overline{C}) \downarrow (A\overline{B}\overline{C}) \downarrow (A\overline{B}C) \downarrow (ABC) \qquad (7.81)$$

$$F = (A \downarrow B \downarrow C) \downarrow (A \downarrow \overline{B} \downarrow C) \downarrow (\overline{A} \downarrow B \downarrow C) \downarrow (\overline{A} \downarrow B \downarrow \overline{C})$$

$$\downarrow (\overline{A} \downarrow \overline{B} \downarrow \overline{C}) \qquad (7.82)$$

(NOR / NOR Form)

Hence the rule:

An expression for the function is obtained from the second canonical form with the aid of the operator $\downarrow$ only, by replacing $+$, $\cdot$ by $\downarrow$ throughout and in placing in parentheses the groups corresponding with the fundamental sums in the second canonical form.

N. B. Mnemotechnically, it will be noted that it is the first operative symbol encountered in a term not reduced to a single variable, which gives the desired Sheffer symbol:

First canonical form:

$$\text{AND} \cdot \rightarrow / \text{ (AND-INV)}$$

Second canonical form:

$$\text{OR} + \rightarrow \downarrow \text{ (OR-INV)}$$

Other forms

Other forms can still be obtained, by mixing the functions AND, OR and AND-INV, OR-INV.

Reverting to the first canonical form, we can write:

$$F = \overline{(A + \overline{B} + \overline{C})(A + B + \overline{C})(\overline{A} + \overline{B} + C)} \qquad (7.83)$$

$$F = (A + \overline{B} + \overline{C}) / (A + B + \overline{C}) / (\overline{A} + \overline{B} + C) \qquad (7.84)$$

(Form OR / NAND)

But this could equally be written:

$$F = \overline{(A + \overline{B} + \overline{C})} + \overline{(A + B + \overline{C})} + \overline{(\overline{A} + \overline{B} + C)} \qquad (7.85)$$

$$F = (A \downarrow \overline{B} \downarrow \overline{C}) + (A \downarrow B \downarrow \overline{C}) + (\overline{A} \downarrow \overline{B} \downarrow C) \qquad (7.86)$$

(Form OR-NOR / OR)

7.7 Some Distinguished Functions

Reverting now to the second canonical form, gives:

$$F = \overline{(\overline{A}\,\overline{B}\overline{C})}\,\overline{(\overline{A}\,B\overline{C})}\,\overline{(A\,\overline{B}\overline{C})}\,\overline{(A\,\overline{B}C)}\,\overline{(ABC)}. \qquad (7.87)$$

Hence:

$$F = (\overline{A}\mid \overline{B}\mid \overline{C})(\overline{A}\mid B\mid \overline{C})(A\mid \overline{B}\mid \overline{C})(A\mid \overline{B}\mid C)(A\mid B\mid C) \qquad (7.88)$$

or again (7.81) reproduced below:

$$F = (\overline{A}\,\overline{B}\overline{C})\downarrow(\overline{A}\,B\overline{C})\downarrow(A\,\overline{B}\overline{C})\downarrow(A\,\overline{B}C)\downarrow(ABC). \qquad (7.89)$$

The different possible forms are summarised below:

$$\begin{cases} 1)\ F = \overline{A}BC + \overline{A}\,\overline{B}C + AB\overline{C} \\ 2)\ F = (\overline{A}\mid B\mid C)/(\overline{A}\mid \overline{B}\mid C)/(A\mid B\mid \overline{C}) \\ 3)\ F = (A+\overline{B}+\overline{C})/(A+B+\overline{C})/(\overline{A}+\overline{B}+C) \\ 4)\ F = (A\downarrow \overline{B}\downarrow \overline{C}) + (A\downarrow B\downarrow \overline{C}) + (\overline{A}\downarrow \overline{B}\downarrow C) \end{cases} \qquad (7.90)$$

$$\begin{cases} 1)\ F = (A+B+C)(A+\overline{B}+C)(\overline{A}+B+C)(\overline{A}+B+\overline{C}) \\ \qquad \cdot (\overline{A}+\overline{B}+\overline{C}) \\ 2)\ F = (A\downarrow B\downarrow C)\downarrow(A\downarrow \overline{B}\downarrow C)\downarrow(\overline{A}\downarrow B\downarrow C)\downarrow(\overline{A}\downarrow B\downarrow \overline{C})\downarrow \\ \qquad (\overline{A}\downarrow \overline{B}\downarrow \overline{C}) \\ 3)\ F = (\overline{A}\,\overline{B}\overline{C})\downarrow(\overline{A}B\overline{C})\downarrow(\overline{A}\,\overline{B}C)\downarrow(A\overline{B}C)\downarrow(ABC) \\ 4)\ F = (\overline{A}\mid \overline{B}\mid \overline{C})(\overline{A}\mid B\mid \overline{C})(A\mid \overline{B}\mid \overline{C})(A\mid \overline{B}\mid C)(A\mid B\mid C) \end{cases} \qquad (7.91)$$

Remarks:

1. In formulae 1 and 2 of each group, the literal symbols are identical with those of the usual canonical form for the type with 1 operator.
2. In the formulae 3 and 4 of each group, the literal symbols are the complements of those of the associated canonical forms for the type with 2 operators.

7.7.5.6 Some General Properties of the Sheffer and Peirce Operators

— *Pseudo-associativity*

The Sheffer and Peirce operators are not associative.

Consider the expressions:

$$F = A \downarrow (B \downarrow C) \text{ and } F' = (A \downarrow B) \downarrow C$$

Then

$$F \neq F'$$

Indeed:

$$F = \overline{A + (B \downarrow C)} = \overline{A} \; \overline{B \downarrow C} = \overline{A} \; \overline{\overline{B + C}}$$

$$= \overline{A}(B + C) = \overline{A}BC + \overline{A}B\overline{C} + \overline{A}\overline{B}C \qquad (7.92)$$

$$F' = \overline{(A \downarrow B) + C} = \overline{A \downarrow B} \cdot \overline{C}$$

$$= \overline{\overline{A + B}} \cdot \overline{C} = (A + B)\overline{C} = AB\overline{C} + A\overline{B}\overline{C} + \overline{A}B\overline{C}$$

Thus, for a given system of values ABC, the two members are not necessarily equal. In other words, *the operation $\downarrow$ is not associative*. In a similar manner it will be seen that *the operation $/$ is not associative*. There is, however, the following interesting property (with $\downarrow$):

$$A \downarrow \overline{B \downarrow C} = \overline{A \downarrow B} \downarrow C \qquad (7.93)$$

which can be called *pseudo-associativity*.

This can be seen as follows:

$$A \downarrow \overline{B \downarrow C} = A \downarrow (B + C) = \overline{A} \; \overline{B + C}$$

$$= \overline{A}\overline{B}\overline{C} = A \downarrow B \downarrow C$$

$$\overline{A \downarrow B} \downarrow C = (A + B) \downarrow C = \overline{A + B\overline{C}}$$

$$= \overline{A}\overline{B}\overline{C}$$

hence the relation (7.93) above.

This property can be generalised to the case of n variables as follows:

$$\overline{X_1 \downarrow X_2 \downarrow \cdots \downarrow X_k} \downarrow X_{k+1} \downarrow \cdots \downarrow X_n = X_1 \downarrow X_2 \downarrow \cdots \downarrow X_k \downarrow X_{k+1} \downarrow \cdots \downarrow X_n \qquad (7.94)$$

In the same manner:

$$\overline{X_1 / X_2 / \cdots / X_k} / X_{k+1} / \cdots / X_n = X_1 / X_2 / \cdots / X_k / X_{k+1} / \cdots / X_n. \qquad (7.95)$$

7.7 Some Distinguished Functions

Other useful relations:

The following relations result immediately from the definition:

$$\begin{cases} X_1 \downarrow X_2 \downarrow \cdots \downarrow X_n \downarrow 0 = X_1 \downarrow X_2 \downarrow \cdots \downarrow X_n & (n > 1) \\ X_1 \downarrow X_2 \downarrow \cdots \downarrow X_n \downarrow 1 = 0 \end{cases} \quad (7.96)$$

$$\begin{cases} X_1 / X_2 / \cdots / X_n / 1 = X_1 / X_2 / \cdots / X_n & (n > 1) \\ X_1 / X_2 / \cdots / X_n / 0 = 1 \end{cases}$$

$$\begin{cases} 0 \downarrow X = X \downarrow 0 = \overline{X} \\ 1/X = X/1 = \overline{X} \end{cases} \quad (n = 1)$$

is provides a second means for obtaining the complement:

$$\begin{cases} \text{or} & \overline{X} = 0 \downarrow X = X \downarrow 0 \\ & \overline{X} = 1 / X = X / 1, \end{cases} \quad (7.97)$$

e first consisting of the use of the formulae determined above:

$$\begin{cases} \overline{X} = X \downarrow X \downarrow \cdots \downarrow X \; (\downarrow \text{ of } n \text{ variables}, & n > 1) \\ \overline{X} = X / X \cdots / X \; (/ \text{ of } n \text{ variables}, & n > 1) \end{cases} \quad (7.98)$$

so:

$$X \downarrow \overline{X} = X / X = 0.$$

7.6 Operation $\oplus$

This operation has been given the following names: exclusive OR, equivalence, "addition modulo 2".

For the case with 2 variables this function is defined by the following ble:

X Y	X $\oplus$ Y
0 0	0
0 1	1
1 0	1
1 1	0

he first two names have been borrowed from formal logic. The origin the appelation "addition modulo 2" is as follows: if 0 and 1 are considered as integers, and if the sum of X and Y modulo 2 (denoted $|X + Y|_2$) be determined, this will give:

$$\begin{cases} 0 + 0 \equiv 0 \ (\text{mod. } 2) \\ 0 + 1 \equiv 1 \ (\text{mod. } 2) \\ 1 + 0 \equiv 1 \ (\text{mod. } 2) \\ 1 + 1 \equiv 2 \ (\text{mod. } 2) = 0 \ (\text{mod. } 2). \end{cases} \qquad (7.99)$$

It can also be said that $X \oplus Y$ equals 1, if the combination (XY) contains 1 an odd number of times. It will be seen that the operation $|X + Y|_2$ is thus formally identical with the operation $X \oplus Y$.

The last name (sum modulo 2) is the one which will be used most often.

It is easy to check that:

$$x \oplus y = x\bar{y} + \bar{x}y \qquad (7.100)$$

It will be easy to check the following properties (this may be done either with the aid of a definition table for $\oplus$, or by the use of the canonic expression (7.100) above which is its equivalent):

1. $x \oplus y = y \oplus x$ (commutativity)
2. $x \oplus (y \oplus z) = (x \oplus y) \oplus z$ (associativity)
3. $x(y \oplus z) = xy \oplus xz$ (distributivity of $\cdot$ with respect to $\oplus$)
4. $\begin{cases} x \oplus 0 = x \\ x \oplus 1 = \bar{x} \end{cases}$
5. $\begin{cases} x \oplus x = 0 \\ x \oplus \bar{x} = 1 \end{cases}$
6. Function $x_1 \oplus x_2 \oplus \ldots \oplus x_n$ (sum modulo 2 of $x_1, x_2, \ldots, x_n$).

The property of associativity permits $x_1 \oplus x_2 \oplus x_3$ to be defined as being $(x_1 \oplus x_2) \oplus x_3$; hence

$$x_1 \oplus x_2 \oplus x_3 \oplus x_4 = (x_1 \oplus x_2 \oplus x_3) \oplus x_4, \ \text{etc.}$$

The expression in n terms denoted by $x_1 \oplus x_2 \oplus \cdots \oplus x_n$ can be obtained from the repeated application of the two-term operation $\oplus$ with arbitrary grouping of the variables. It will be seen that this is a function whose value is 1 when there is an *odd* number of 1 among the values of the variables and thus 0 when there is an *even* number of 1. This property can thus serve directly to define the operation $x_1 \oplus \cdots \oplus x_n$ with n variables.

7. $x \oplus y = \bar{x} \oplus y = x \oplus \bar{y}$ \hfill (7.101)

7.7 Some Distinguished Functions

and, more generally:

$$\overline{x_1 \oplus x_2 \oplus \ldots \oplus x_n} = \overline{x}_1 \oplus x_2 \oplus \cdots \oplus x_n$$
$$= x_1 \oplus \overline{x}_2 \oplus \cdots \oplus x_n$$
$$= x_1 \oplus x_2 \oplus \cdots \oplus \overline{x}_n. \quad (7.103)$$

For complementing a sum modulo 2, it suffices to complement any *one* term, and only *one* (and, more generally; $2k+1$ terms).

3. Algebraic representation of a switching function by means of the operations $\cdot$ and $\oplus$.

a) *Function given by its truth table*

Consider for example the 3-variable functions $f(X, Y, Z)$. The first canonical form of $f(X, Y, Z)$ is written:

$$f(X, Y, Z) = \sum_{i=0}^{i=2^3-1} f(i) m_i$$

$$\begin{cases} m_i = X^{i_1} Y^{i_2} Z^{i_3} \\ i = (i_1 i_2 i_3) \end{cases} \quad (7.104)$$

we have:
$$m_i \cdot m_j = 0 \quad \text{for} \quad i \neq j. \quad (7.105)$$

Consequently:
$$f(i) m_i \cdot f(j) m_j = 0 \quad \text{for} \quad i \neq j. \quad (7.106)$$

This results in:
$$\sum_{i=0}^{i=7} f(i) m_i = [f(0) m_0] + \left[\sum_{i=1}^{i=7} f(i) m_i\right]$$
$$= [f(0) m_0] \oplus \left[\sum_{i=1}^{i=7} f(i) m_i\right] \quad (7.107)$$

Similarly:
$$\sum_{i=1}^{i=7} f(i) m_i = f(1) m_1 + \sum_{i=2}^{i=7} f(i) m_i = f(1) m_1 \oplus \left[\sum_{i=2}^{i=7} f(i) m_i\right] \quad (7.108)$$

It will be seen that by a step-by-step operation this will give:

$$f(0) m_0 + f(1) m_1 + f(2) m_2 + f(3) m_3 + f(4) m_4 + f(5) m_5$$
$$+ f(6) m_6 + f(7) m_7 \quad (7.109)$$
$$= f(0) m_0 \oplus f(1) m_1 \oplus f(2) m_2 \oplus f(3) m_3 \oplus f(4) m_4$$
$$\oplus f(5) m_5 \oplus f(6) m_6 \oplus f(7) m_7. \quad (7.110)$$

It will thus be seen that by replacing $+$ by $\oplus$ in the first canonical form an equivalent expression can be obtained. It will be seen that this is true for any number of variables.

Next consider a fundamental product containing one or more barred letters. If $\bar{A}$ be one of these, it can be expressed as $\bar{A}B$

$$\bar{A}B = (A \oplus 1) B = AB \oplus B. \tag{7.111}$$

In this second expression the letter A appears only in its unbarred form. Since it does not enter into B, it is clear that the new expression $AB \oplus B$ no longer contains A in its barred form.

It will be seen that this can be obtained from $\bar{A}B$ by a process known as 'splitting' ("with respect to A") and which consists in the following operations:

Splitting of $\bar{A}B$
- *Remove the bar* from the letter A (thus obtaining a first term AB)
- *Remove the barred letter* (thus obtaining a second term B)
- *Link the two terms with* $\oplus$.

Each fundamental product appearing in the first canonical form may ultimately contain barred variables. The operation of splitting can be applied repeatedly until such time as all the signs $-$ have disappeared.

This then gives rise to the following rule:

1. Write down the disjunctive canonical form of the function.
2. Express each fundamental product with the aid of the operations $\oplus$ and $\cdot$ by applying the splitting operation until all signs $-$ have been removed.
3. Link the expressions thus obtained for each fundamental product by the signs $\oplus$.
4. Simplify the expressions by deleting the monomials appearing an even number of times and write in once only those which appear an odd number of times.

Examples:

1. $x\bar{y} + \bar{x}y = (xy \oplus x) \oplus (xy \oplus y) = x \oplus y.$ \hfill (7.112)

2. Sum and remainder of the binary addition. We obtain expressions of the form (cf. § 9.4.2.1):

$$\begin{cases} S = \bar{x}\bar{y}r + \bar{x}y\bar{r} + x\bar{y}\bar{r} + xyr \\ r' = \bar{x}yr + x\bar{y}r + xy\bar{r} + xyr \end{cases} \tag{7.113}$$

1.1 Determination of S

With:
$$\bar{x}\bar{y}r = x\bar{y}r \oplus \bar{y}r = xyr \oplus xr \oplus yr \oplus r$$
$$\bar{x}y\bar{r} = xy\bar{r} \oplus y\bar{r} = xyr \oplus xy \oplus yr \oplus y$$
$$x\bar{y}\bar{r} = xy\bar{r} \oplus x\bar{r} = xyr \oplus xy \oplus xr \oplus x$$
$$xyr = xyr.$$
(7.114)

Whence:
$$S = \bar{x}\bar{y}r \oplus \bar{x}y\bar{r} \oplus x\bar{y}\bar{r} \oplus xyr$$
$$= (xyr \oplus xr \oplus yr \oplus r) \oplus (xyr \oplus xy \oplus yr \oplus y)$$
$$\oplus (xyr \oplus xy \oplus xr \oplus x) \oplus xyr$$
(7.115)

whence:
$$S = x \oplus y \oplus r$$
(7.116)

1.2 Determination of r'

$$\bar{x}yr = xyr \oplus yr$$
$$x\bar{y}r = xyr \oplus xr$$
$$xy\bar{r} = xyr \oplus xy$$
(7.117)

$$r' = (xyr \oplus y\, r) \oplus (xyr \oplus xr) \oplus (xyr \oplus xy) \oplus xyr \quad (7.118)$$

whence:
$$r' = xy \oplus yr \oplus rx.$$
(7.119)

b) Function given by an algebraic expression

If this expression takes one of the two canonical forms, it is easy to refer back to the preceding case. On the other hand, for any expression it is possible to proceed in two manners:
- establish the truth table and make use of the method above;
- proceed by algebraic manipulations, making use of the identities:

$$\begin{cases} A + B = A \oplus B \oplus AB \\ \bar{A} = A \oplus 1 \end{cases}$$
(7.120)

The first method can always be used. It will be seen that it is merely a particular case of the second. The second method is indicated for the case of an expression containing a large number of variables, but of simple form.

7.8 Threshold Functions

7.8.1 Definition

For a given number n of real numbers (of any sign) $m_1, \ldots, m_n$ known as *weight* Δ and an $(n+1)$th S known as *threshold*, a threshold function corresponds, by definition, to the following operation:

$$f(X_1, \ldots, X_i, \ldots X_n) = 1 \quad \text{if} \quad \sum_{i=1}^{i=n} m_i X_i \geq S$$
$$f(X_1, \ldots, X_i, \ldots, X_n) = 0 \quad \text{if} \quad \sum_{i=1}^{i=n} m_i X_i < S \quad (7.121)$$

The variables $X_1, \ldots, X_n$ have two meanings in the above expressions. In the expression $f(X_1, \ldots, X_n)$ they have the normal sense of *Boolean variables* (switching variables) represented by the symbols 0 and 1. On the other hand, for the determination of arithmetic sums

$$\sum_{i=1}^{i=n} m_i X_i$$

these same symbols are considered as those for the integers 0 and 1. Moreover, the n weights m_i can take on any real values in magnitude and sign and not necessarily integral ones.

In practice, a given threshold function $f(X_1, \ldots, X_n)$ is not necessarily given in the form (7.121) but may be generated by an expression (e.g. canonical), by a truth table, etc. To demonstrate its character as a threshold function it is necessary to show that it is possible to find $n+$ numbers $m_1, \ldots, m_i, \ldots, m_n$ and S such that the relations (7.121) are satisfied.

7.8.2 Geometrical Interpretation

Within a space of n dimensions, an equation of the type

$$\sum_{i=1}^{i=n} m_i X_i = S$$

represents (by definition) a hyperplane H. Additionally, it can be shown that the hyperplane thus defined divides the space into two zones aracterised by the inequalities:

$$\begin{cases} \sum_{i=1}^{i=n} m_i X_i \geq S \\ \sum_{i=1}^{i=n} m_i X_i < S \end{cases} \quad (7.122)$$

e "geometrical" interpretation of (7.121) is thus as follows: the aracteristic set of the function is to be found on the same side of the perplane H, or again on H, whereas that of the function $\bar{f}$ is to be ind on the side and outside the hyperplane.

.3. Particular Case: Majority Function

A function which obeys the following relations is known as a *majority iction* or *quorum function*:

$$\begin{cases} f(X_1, \ldots, X_n) = 1 \quad \text{when} \quad \sum_{i=1}^{i=n} X_i \geq m \\ \qquad\qquad (m \text{ integer}, \ 0 \leq m \leq n) \quad (7.123) \\ f(X_1, \ldots, X_n) = 0 \quad \text{when} \quad \sum_{i=1}^{i=n} X_i < m \end{cases}$$

will be seen that this is a particular type of threshold function. The lue of the function is 1 when m or more variables equal 1, and is 0 in other cases.

$$\left.\begin{array}{ll} \text{If } m=0 & f(X_1, \ldots, X_n) \equiv 1 \\ \text{If } m=1 & f(X_1, \ldots, X_n) = X_1 + X_2 + \cdots + X_n \\ \text{If } m=n & f(X_1, \ldots, X_n) = X_1 \cdot X_2 \cdot \ldots \cdot X_n \end{array}\right\} \quad (7.124)$$

$m > n$ were allowed then this would give $f(X_1, \ldots, X_n) \equiv 0$).
If, in addition n is an odd number, hence of form $n = 2p - 1$, then at for $m = p = (n+1)/2$ is known as the *simple majority function*.

Example: simple majority function with 3 variables:

is corresponds with the truth table 7.17:

Table 7.17. *Simple majority function (3 variables)*

X Y Z	$f(X, Y, Z)$
0 0 0	0
0 0 1	0
0 1 0	0
0 1 1	1
1 0 0	0
1 0 1	1
1 1 0	1
1 1 1	1

Its value is 1 if there are 2 or 3 symbols 1 among the three variables, i[.e.] if there is a majority of 1's. It is generally denoted:

$$\text{Maj}(X, Y, Z) \quad \text{or} \quad X \# Y \# Z. \tag{7.12}$$

From table 7.17:

$$X \# Y \# Z = \overline{X}YZ + X\overline{Y}Z + XY\overline{Z} + XYZ$$
$$= \overline{X}YZ + X\overline{Y}Z + XY\overline{Z} + XYZ + XYZ + XYZ$$
$$= (\overline{X}YZ + XYZ) + (X\overline{Y}Z + XYZ) + (XY\overline{Z} + XYZ)$$
$$= YZ + XZ + XY \tag{7.12}$$

thus:

$$X \# Y \# Z = XY + YZ + ZX. \tag{7.12}$$

It is also easily verified that:

$$X \# Y \# Z = (X + Y)(X + Z)(Z + X). \tag{7.12}$$

It will be noted that these two expressions are symmetric, i.e. by p[er]mutation of the variables an equivalent expression is obtained. T[his] is a particular case of *symmetric function*.

It will also be noted that this function has the peculiarity of bei[ng] equal to its own dual function, as is shown by the expressions (7.127) a[nd] (7.128).

7.9 Functionally Complete Set of Operators

A set of operators (functions) in 2 or more variables is said to [be] *functionally complete* if, with these variables, it is possible to express t[he] three operations $+$, $\cdot$ and $^-$.

7.9 Functionally Complete Set of Operators

Some examples of functionally complete sets will now follow.

a) *Peirce function* $A \downarrow B$ (Function "OR-NOT" "OR-INV", Function NOR")

By definition:
$$A \downarrow B = \overline{A + B}. \tag{7.129}$$

The set of functions which contains the unique function $A \downarrow B$ is functionally complete. In fact:

$$\begin{cases} 1.\ A \downarrow A = \overline{A + A} = \overline{A} \\ \quad \text{whence: } \overline{A} = A \downarrow A \\ 2.\ A + B = \overline{\overline{A + B}} = \overline{(\overline{A + B})} = \overline{A \downarrow B} \\ \quad = (A \downarrow B) \downarrow (A \downarrow B) \\ 3.\ A \cdot B = \overline{\overline{A} + \overline{B}} = \overline{A} \downarrow \overline{B} \quad \overline{A} = A \downarrow A \\ \quad\quad\quad\quad\quad\quad\quad\quad\quad\quad\quad \overline{B} = B \downarrow B \end{cases} \tag{7.130}$$

Whence: $A \cdot B = (A \downarrow A) \downarrow (B \downarrow B)$

In other words the three operations $+, \cdot, ^-$ can be expressed with the help of the single operator $\downarrow$, thanks to the formulae:

$$\begin{cases} \overline{A} = A \downarrow A \\ A + B = (A \downarrow B) \downarrow (A \downarrow B) \\ A \cdot B = (A \downarrow A) \downarrow (B \downarrow B) \end{cases} \tag{7.131}$$

Consequently, any given algebraic expression formed with the aid of the operations $(+, \cdot, ^-)$ can be re-written in making use of the operator $\downarrow$, uniquely. This property is extremely useful when dealing with transistor logic. From a knowledge of the functions which must be provided by a network, the establishment of an algebraic formula is facilitated, subsequently leading to the development of a schematic circuit diagram making use of the operator $\downarrow$ only. This can immediately be generalised to the case with n variables.

b) *Sheffer function* $A \mid B$ (Function AND-NOT, "AND-INV", NAND "Incompatibility").

We proceed in an analogous manner and express the operations $+$, $^-$ by means of the operation $(/)$.

By definition:
$$A \mid B = \overline{A \cdot B}. \tag{7.132}$$
Whence the properties:

$$\begin{cases} A \mid A = \overline{A \cdot A} = \overline{A} \\ A = A \mid A \\ A + B = \overline{\overline{A} \cdot \overline{B}} \\ = \overline{A} \mid \overline{B} \\ = (A \mid A)(B \mid B) \\ A + B = (A \mid A) \mid (B \mid B) \\ A \cdot B = \overline{\overline{A \cdot B}} = \overline{A \mid B} \\ = (A \mid B) \mid (A \mid B) \\ A \cdot B = (A \mid B) \mid (A \mid B) \end{cases} \tag{7.133}$$

Summarizing, this gives the following formulae (generalised to n variables):
$$A = A \mid A$$
$$A + B = (A \mid A) \mid (B \mid B) \tag{7.134}$$
$$A \cdot B = (A \mid B) \mid (A \mid B).$$

Int follows then, that the Sheffer function itself also constitutes a functionally complete set, and it is possible to express every function (Boolean) in an algebraic form making use of this operator only.

c) *Set* $(A \# B \# C, ^-, 0, 1)$

The value of the function $A \# B \# C$ is 1 when 2 or 3 of the variables have the value 1, i.e. when there is a majority of 1. This is a ternary operation.
$$A \# B \# C = AB + BC + CA \tag{7.135}$$
giving:
$$A \cdot B = A \# 0 \# B$$
$$A + B = A \# 1 \# B \tag{7.136}$$
$$\overline{A} = \overline{A}$$

Thus $A \cdot B$, $A + B$, $\overline{A}$ can be expressed by means of majority functions, negation, 0 and 1 which constitute a complete set.

Remark: The above relationships have been given merely to show that the set $(A \# B \# C, \bar{}, 0, 1)$ is functionally complete. Here too, there exist other methods of expressing $(+, \cdot, \bar{})$ with the aid of these operators. For example, it can be said that:

$$A + B = \overline{\bar{A} \# \bar{B} \# 0}. \tag{7.137}$$

d) *Set* $(+, -)$

Here:
$$\begin{cases} A + B = A + B \\ A \cdot B = \overline{\bar{A} + \bar{B}} \\ \bar{A} = \bar{A} \end{cases} \tag{7.138}$$

e) *Set* $(\cdot, -)$.

Here again:
$$\begin{aligned} A \cdot B &= A \cdot B \\ A + B &= \overline{\bar{A} \cdot \bar{B}} \\ \bar{A} &= \bar{A} \end{aligned} \tag{7.139}$$

f) *Set* $(\oplus, \cdot, 1)$.

Here
$$\begin{aligned} A + B &= A \oplus B \oplus AB \\ A \cdot B &= A \cdot B \\ \bar{A} &= A \oplus 1 \end{aligned} \tag{7.140}$$

The set $(\oplus, \cdot, 1)$ is complete.

7.10 Determination of Canonical Forms

1) *Function given in the form of a truth table*

The two canonical forms are obtained directly from the truth table, by means of the formulae:

$$\begin{aligned} f &= \sum_{i=0}^{i=2^n-1} f(i) m_i \\ f &= \prod_{i=0}^{i=2^n-1} [f(i) + M_i] \end{aligned} \quad \begin{cases} i = (i_1 i_2 \cdots i_n) \\ \bar{i} = (\bar{i}_1 \bar{i}_2 \cdots \bar{i}_n) \\ m_i = X_1^{i_1} \cdot X_2^{i_2} \cdot \ldots \cdot X_n^{i_n} \\ M_i = X_1^{i_1} + X_2^{i_2} + \cdots + X_n^{i_n} \end{cases} \tag{7.141}$$

The truth table 7.18 represents a function as well as the determination of of the two canonical forms:

Table 7.18. *A function and its two canonical forms*

A B C	f	f =	f =
0 0 0	0		$(A + B + C)$
0 0 1	1	$\bar{A}\bar{B}C$	
0 1 0	1	$+ (\bar{A}B\bar{C})$	
0 1 1	0		$\cdot (A + \bar{B} + \bar{C})$
1 0 0	0		$\cdot (\bar{A} + B + C)$
1 0 1	1	$+ (A\bar{B}C)$	
1 1 0	0		$\cdot (\bar{A} + \bar{B} + C)$
1 1 1	0		$\cdot (\bar{A} + \bar{B} + \bar{C})$

whence:
$$f = \bar{A}\bar{B}C + \bar{A}B\bar{C} + A\bar{B}C \tag{7.142}$$

$$f = (A + B + C)(A + \bar{B} + \bar{C})(\bar{A} + B + C)(\bar{A} + \bar{B} + C) \\ (\bar{A} + \bar{B} + \bar{C}). \tag{7.143}$$

2) *Function given by one of its algebraic expressions*

First method: Write down the truth table and proceed as above.
Second method:
Disjunctive canonical form:

1) Reduce the expression to a sum of products.
2) For a given monomial of the expression from which the letter $X_{i_1}, X_{i_2} \cdots X_{i_k}$ are missing, multiply by the term
$(X_{i_1} + \bar{X}_{i_1})(X_{i_2} + \bar{X}_{i_2}) \cdots (X_{i_k} + \bar{X}_{i_k})$ (which equals 1)

Carry out this operation for the monomials. Develop and make any reductions necessary to arrive at a sum of products. The form thus obtained is the canonical form sought.

The reason for this is easily seen: the result is a sum of the products of all the variables, whether barred or not, i.e. of fundamental products. Deletion of the terms appearing several times, in accordance with the law of idempotence, will produce the disjunctive canonical form.

Example:
$$f(A, B, C, D) = A + B(\bar{A}D + AC\bar{D}) + \bar{B}C \tag{7.144}$$

gives:
$$f(A, B, C, D) = A + \bar{B}C + \bar{A}BD + ABC\bar{D}. \tag{7.145}$$

7.10 Determination of Canonical Forms

We write:

$$\begin{aligned}
f(A, B, C, D) &= A(B + \bar{B})(C + \bar{C})(D + \bar{D}) \\
&\quad + (\bar{A} + A)\bar{B}C(D + \bar{D}) \\
&\quad + \bar{A}B(C + \bar{C})D + AB C\bar{D} \\
&= A(BCD + BC\bar{D} + B\bar{C}D + B\bar{C}\bar{D} \\
&\quad + \bar{B}CD + \bar{B}C\bar{D} + \bar{B}\bar{C}D + \bar{B}\bar{C}\bar{D}) \\
&\quad + (AD + A\bar{D} + \bar{A}D + \bar{A}\bar{D})\bar{B}C \\
&\quad + \bar{A}BCD + \bar{A}B\bar{C}D + ABC\bar{D} \\
&= ABCD + ABC\bar{D} + AB\bar{C}D + AB\bar{C}\bar{D} \\
&\quad + A\bar{B}CD + A\bar{B}C\bar{D} + A\bar{B}\bar{C}D + A\bar{B}\bar{C}\bar{D} \\
&\quad + A\bar{B}CD + A\bar{B}C\bar{D} + \bar{A}\bar{B}CD + \bar{A}\bar{B}C\bar{D} \\
&\quad + \bar{A}BCD + \bar{A}B\bar{C}D + ABC\bar{D}
\end{aligned} \quad (7.146)$$

The terms $ABC\bar{D}$, $A\bar{B}CD$, $A\bar{B}C\bar{D}$ each appear twice in this sum. Thus, from the law of idempotence:

$$\begin{aligned}
f(A, B, C, D) &= ABCD + ABC\bar{D} + AB\bar{C}D + AB\bar{C}\bar{D} \\
&\quad + A\bar{B}CD + A\bar{B}C\bar{D} + A\bar{B}\bar{C}D + A\bar{B}\bar{C}\bar{D} \\
&\quad + \bar{A}\bar{B}CD + \bar{A}\bar{B}C\bar{D} \\
&\quad + \bar{A}BCD + \bar{A}B\bar{C}D
\end{aligned} \quad (7.147)$$

Conjunctive canonical form

1) Reduce the expression to a product of the sums of variables, whether barred or not.
2) In each sum, for each missing letter X, add in the term $X\bar{X}$. Develop, to get back to a product of sums. After deletion of the repeated terms the required canonical form will be obtained.

Example:

$$f(A, B, C, D) = (AB + CD)(A\bar{C} + D) \quad (7.148)$$

giving:

$$\begin{aligned}
f(A, B, C, D) &= (AB + C)(AB + D)(A + D)(\bar{C} + D) \\
&= (A + C)(B + C)(A + D)(B + D)(A + D)(\bar{C} + D) \\
&= (A + C)(B + C)(A + D)(B + D)(\bar{C} + D) \\
A + C &= A + C + B\bar{B} + D\bar{D} \\
&= (A + C + B\bar{B} + D)(A + C + B\bar{B} + \bar{D}) \\
&= (A + B + C + D)(A + \bar{B} + C + D) \\
&\quad \cdot (A + B + C + \bar{D})(A + \bar{B} + C + \bar{D}) \quad \text{etc.}
\end{aligned} \quad (7.149)$$

Whence:

$$\begin{aligned}f(A,B,C,D) = &(A+B+C+D)(A+\bar{B}+C+D)(A+B+C+\bar{D})\\&\cdot(A+\bar{B}+C+\bar{D})(A+B+C+\bar{D})(A+B+C+\bar{D})\\&\cdot(\bar{A}+B+C+D)(\bar{A}+B+C+\bar{D})(A+B+C+D)\\&\cdot(A+B+\bar{C}+D)(A+\bar{B}+C+D)(A+\bar{B}+\bar{C}+D)\\&\cdot(A+B+C+D)(A+B+\bar{C}+D)(\bar{A}+B+C+D)\\&\cdot(\bar{A}+B+\bar{C}+D)(A+B+\bar{C}+D)(A+\bar{B}+\bar{C}+D)\\&\cdot(\bar{A}+B+\bar{C}+D)(\bar{A}+\bar{B}+\bar{C}+D)\end{aligned} \qquad (7.150)$$

By reducing the terms appearing several times, we have:

$$\begin{aligned}f(A,B,C,D) = &(A+B+C+D)(A+\bar{B}+C+D)(A+B+C+\bar{D})\\&\cdot(A+\bar{B}+C+\bar{D})(\bar{A}+B+C+D)(\bar{A}+B+C+\bar{D})\\&\cdot(A+B+\bar{C}+D)(A+\bar{B}+\bar{C}+D)(\bar{A}+B+\bar{C}+D)\\&\cdot(\bar{A}+\bar{B}+\bar{C}+D).\end{aligned}$$

3) *Function given in decimal form*

Notation: (Disjunctive form)

$$f = \sum i_k$$

we have, by definition

$$f = \sum m_{i_k} \qquad (7.151)$$

Use is then made of the truth table to obtain the actual algebraic canonical form.

7.11 Some Operations on the Canonical Forms

7.11.1 Passage from one Canonical Form to Another

First method
Proceed by way of the truth table. It is easy to establish from one or other form).

Example:

$$f(A,B,C) = \bar{A}\bar{B}C + \bar{A}B\bar{C} + A\bar{B}\bar{C}. \qquad (7.152)$$

Giving the following truth table (Table 7.19)

7.11 Some Operations on the Canonical Forms

Table 7.19

A B C	$f(A, B, C)$
0 0 0	0
0 0 1	1
0 1 0	1
0 1 1	0
1 0 0	1
1 0 1	0
1 1 0	0
1 1 1	0

hence:

$$f(A, B, C) = (A + B + C)(A + \bar{B} + \bar{C})(\bar{A} + B + \bar{C})(\bar{A} + \bar{B} + C) \\ (\bar{A} + \bar{B} + \bar{C}) \quad (7.153)$$

The inverse transformation, from the conjunctive canonical form to the disjunctive canonical form is made in a similar manner.

Second method

By putting $i = (i_1 i_2 \ldots i_n)$ and

$$\begin{cases} m_i = X_1^{i_1} X_2^{i_2} \ldots X_n^{i_n} & \text{(fundamental product)} \\ M = X_1^{i_1} + X_2^{i_2} + \ldots + X_n^{i_n} & \text{(fundamental sum)} \end{cases}$$

one has:

$$f = \sum_{i=0}^{i=2^n-1} f_i m_i = \prod_{i=0}^{i=2^n-1} [f(\bar{i}_1, \bar{i}_2, \ldots, \bar{i}_n) + M_i] \quad (7.154)$$

From consideration of the numbers $\bar{i} = \bar{i}_n$ and $i = i_1 i_2 \ldots i_n$:

$$\bar{i} + i = 1\,1\,1\,1\cdots 1 = 2^n - 1$$

when performing this addition in binary, the sum digit value is always equal to 1 and there is no carry from one rank bit position to the next-most significant one).
Thus:

$$f = \prod_{i=0}^{i=2^n-1} [f(2^n - 1 - i) + M_i] = \prod_{i=0}^{i=2^n-1} [f(i) + M_i] \quad (7.155)$$

whence the method of conversion from the disjunctive canonical form to the conjunctive canonical form:

1. Write, in the decimal code, the indices i_k of the fundamental products comprising the given canonical form.

2. Write, in the decimal code, the numbers $\bar{i}_k = 2^n - 1 - i_k$.
3. Eliminate the fundamental sums $M_{\bar{i}_k}$. Those which remain comprise the factors of the disjunctive canonical form.

Example: Reverting to the above example, gives:

$$f(A, B, C) = m_1 + m_2 + m_4 \quad (n = 3, \ 2^n - 1 = 7);$$

1) indices i : 1, 2, 4;
2) indices $\bar{i}$: 6, 5, 3;
3) fundamental sums of product:

$$M_0, \ M_1, \ M_2, \ M_4, \ M_7.$$

$$\begin{aligned}
f &= M_0 \cdot M_1 \cdot M_2 \cdot M_4 \cdot M_7 \\
&= (A + B + C)(A + \bar{B} + \bar{C})(\bar{A} + B + \bar{C})(\bar{A} + \bar{B} + C) \\
&\quad \cdot (\bar{A} + \bar{B} + \bar{C}).
\end{aligned} \tag{7.156}$$

The rule for the inverse transformation (conjunctive canonical form to disjunctive canonical form) is similar.

1. Write down the indices i_k of the terms M_{i_k} for the given conjunctive cannonical form.
2. Write down the indices $\bar{i}_k = 2^n - 1 - i_k$.
3. Eliminate the fundamental products $m_{\bar{i}_k}$. The required canonical form is the sum of those which remain.

Example: Reverting once again to the same function:

1. Indices $i_k = 0, 1, 2, 4, 7$.
2. Indices $\bar{i}_k = 7, 6, 5, 3, 0$.
3. Indices $f = m_1 + m_2 + m_4$.

7.11.2 Complement of a Function f Given in Canonical Form

If f is in disjunctive canonical form the disjunctive canonical form of $\bar{f}$ can be obtained by summing the fundamental products which do not intervene in the expression for f.

If f is in conjunctive canonical form the conjunctive canonical form of $\bar{f}$ will be obtained from the product of the fundamental sums which do not intervene in the expression for f.

Examples

$$\begin{cases} f(A,B,C) = \bar{A}\bar{B}\bar{C} + A\bar{B}C + AB\bar{C} \\ \bar{f}(A,B,C) = AB\bar{C} + \bar{A}BC + \bar{A}B\bar{C} + \bar{A}\bar{B}C + ABC \end{cases} \quad (7.157)$$

and

$$\begin{cases} f(A,B,C) = (\bar{A}+\bar{B}+C)(A+\bar{B}+\bar{C})(A+\bar{B}+C)(A+B+\bar{C}) \\ \qquad\qquad\qquad (\bar{A}+\bar{B}+\bar{C}) \\ \bar{f}(A,B,C) = (A+B+C)(\bar{A}+B+\bar{C})(\bar{A}+B+C) \end{cases} \quad (7.158)$$

Remark: if it is not required to find the complement with the same type of canonical form, then it is simple to pass from 7.157 to 7.158, in one direction or the other by repeated applications of the De Morgan theorems.

7.12 Incompletely Specified Functions

Frequently it happens that the value of a function is not specified for certain combinations of variables. It is possible, in fact, that certain combinations of input variables do not occur or, also, that they do occur but that the effect on the logic network in question is indifferent. In the first case, for example, it is possible to imagine the case where the inputs are provided from another system which may not be able to supply certain combinations. In the second case, for example, it could occur in the case of a computer for which the output is only sampled at a certain instant. In practice the function can be given the value 0 or 1 in a perfectly arbitrary manner. However, as will be seen, an attempt will be made to utilise this circumstance in simplifying to the maximum the associated logic networks. For the time being, only the notations used to describe an incompletely specified function will be defined. It is as well to note that such a function is, in a certain sense, a family of functions of which one will be chosen in terms of particular criteria; if there are k non-specified combinations, then 2^k functions can be completely specified which will fulfil the required task.

Notations:

1. *Truth table*

With respect to the combinations for which the function has not been specified, a dash — or the sign $\emptyset$ will be written in (interpreted as 0 or 1)

Example:

Table 7.20. *An incompletely specified function*

A B C	$f(A, B, C)$	
0 0 0	0	0
0 0 1	1	1
0 1 0	—	∅
0 1 1	1 or	1
1 0 0	—	∅
1 0 1	—	∅
1 1 0	0	0
1 1 1	1	1

2. *Algebraic expressions*

This can be written in the canonical form

$$f(X_1, X_2, \ldots, X_n) = \sum_{i=0}^{i=2^n-1} a_i m_i$$

by replacing a_i by 0 or 1, when the function is specified, and by leaving a_i in the form of an algebraic variable in the contrary case.

Thus the above function can be written:

$$f(A, B, C) = m_1 + a_2 m_2 + m_3 + a_4 m_4 + a_5 m_5 + m_7 \quad (7.159)$$
$$= (m_1 + m_3 + m_7) + (a_2 m_2 + a_4 m_4 + a_5 m_5). \quad (7.160)$$

7.13 Characteristic Function for a Set

For a given set A, belonging to a universal set I the characteristic function associated with A is the function f_A defined in I, and taking the value 1 for the elements of A, and 0 for the others.

7.14 Characteristic Set for a Function

For a given completely specified Boolean function f, the combination of n variables for which the function is equal to 1 is known as the set E_f. Thus a function f appears as the characteristic function of its characteristic set E_f.

Chapter 8

Geometric Representations of Boolean Functions

8.1 The n-Dimensional Cube

8.1.1 Example: 2-Variable Functions

For a given completely specified Boolean function having 2 variables and defined by a truth table indicating the values (0 or 1) for each of the binary combinations XY of the function $f(X, Y)$, each (ordered) combination XY can be considered as being the coordinates of a point in a plane. With rectangular axes and with an identical scale in both axes, the 4 corresponding points are located at the vertices (summits) of a square. Consider for example the function defined by Table 8.1:

Table 8.1

$X\ Y$	$f(X, Y)$
0 0	1
0 1	1
1 0	1
1 1	0

The 4 combinations 00, 01, 11 and 10 are represented geometrically by the 4 points A, B, C and D respectively (Fig. 8.2a), the vertices of a square in the plane OXY. To specify a function $f(X, Y)$ is to associate with

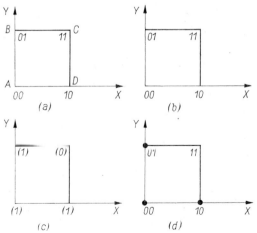

Fig. 8.2

certain of these vertices the value 1 and the value 0 with others, or equivalently, to choose a certain subset of the set of vertices for which the function $f(X, Y)$ takes the value 1. This subset is simply the characteristic set $f(X, Y)$ and its characteristic function is $f(X, Y)$. For a completely specified function, rather than indicating the value 0 or 1 alongside each vertex (as would be the case for a truth table) it is more convenient to indicate the characteristic set $f(X, Y)$ only. For this purpose it suffices to mark in a 'heavy' point at each vertex which forms the characteristic set (Fig. 8.2 b, c, d).

8.1.2 3-Variables Functions

In a similar manner, the $2^3 = 8$ combinations of 3 variables XYZ can be associated with the vertices of a cube, represented in perspective in Fig. 8.3. A 3-variable function will be defined by a set of vertices made to 'stand out' by the use of the 'heavy' points as above. Thus the function defined by Table 8.4:

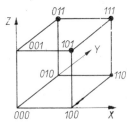

Fig. 8.3. Representation of a function on a cube

Table 8.4

X Y Z	$f(X, Y, Z)$
0 0 0	0
0 0 1	0
0 1 0	0
0 1 1	1
1 0 0	0
1 0 1	1
1 1 0	1
1 1 1	1

is represented in this way in Figure 8.3.

8.1.3 Functions of n Variables (any n)

For $n = 2$ or $n = 3$, the set (E_1) of the combinations of n binary variables can be *represented geometrically* by the set (E_2) of the vertices of a square and of a cube respectively. In the case of the cube, a planar (2-dimensional) projection is used in practice. Conversely the set (E_1) can be considered as the *analytical representation* of (E_2). What happens should there be a need to study the functions having more than 3 variables in this same manner?

8.1.3.1 Sets (E_1) and (E_2)

— (E_1) (algebraic representation): can always be defined as the set of binary combinations of n variables, a set with 2^n elements.

8.1 The n-Dimensional Cube

— (E_2) (geometric representation): in the case of n dimensions, the geometric entity, known as an *n-dimensional hypercube* (or, more briefly, an *n-dimensional cube*) which generalizes the squares and cubes of ordinary geometry (with representations limited to 2 or 3 dimensions) can only be defined indirectly:

— with the aid of 2 or 3-dimensional geometric representations, or
— through the intermediary of an analytical representation.

If the only interest lay in the set of summits of this n-dimensional cube, points in a plane would suffice for a geometrical representation. The analytical definition is simply the set (E_1) of binary combinations with coordinates. In fact, it will later be shown that the simplification of the functions takes account of the important role of a *relation* (cf. Chap 1) between vertices and which must be added to the definition of the cube so as to endow it with two forms, analytic and geometric.

8.1.3.2 Relation of Adjacency Between 2 Vertices

— *Analytically*

Two vertices are said to be *adjacent* if the corresponding binary relations differ in one co-ordinate, and one only. In other words, if their Hamming distance (see § 8.1.4 below) equals 1.

— *Geometrically*

Two vertices A and B are said to be adjacent if they are situated at the 2 extremities of a single edge.

In both cases (analytic and geometric) the n-dimensional cube will be defined as the set of 2^n vertices equipped with the adjacency relation. Hence a conventional geometric representation.

8.1.3.3 Geometric Representation

This representation takes account of neither the size nor the exact shape of the cube, but indicates only:

— the vertices,
— the absence or presence of the direct link (edge) between 2 vertices.

In this sense, it merely constitutes a planar representation of the *adjacency relation* (8.1.3.2) between vertices. Under these conditions, the planar representation of an n-dimensional cube can be constructed as follows:

1. Draw a set of parallel lines numbered from 0 to n.

2. For every m, on line m, mark off C_n^m points, associated with the combinations of weight m (i.e. possessing m symbols of value and $n-m$ of value 0), opposite the corresponding binary combination.
3. For every pair of analytically adjacent points, join the 2 points of the pair with a straight line segment. Two points in such a pair are only to be found on two consecutive lines, thus for every m and for every point on line m it suffices to join these points to those on line $m+1$ which are effectively at a distance 1.

In addition it will be noted that:

— n lines extend from each vertex,
— by suitable spreading of the points (vertices) of the different lines can be made to appear (in perspective) the squares and cubes contained in the n-dimensional cube, and which will be known in the text which follows as "sub-cubes" of 2 and 3 dimensions respectively (Figs. 8.2, 8.3, 8.5a).
— this could equally well have been done by determination of the cube of order n based on that of order $n-1$: for which it suffices to "double" the drawing of the cube and to join the homologous vertices by new edges. This is an example of the way in which a new cube may be formed by the association of 2 symmetrical squares with respect to a plane (passing through the centre of the cube) (Fig. 8.5a).

8.1.4 Other Representations and Definitions

8.1.4.1 The n-Dimensional Cube

An n-dimensional cube can be represented by the notation:

$$K = (X_1, X_2, \ldots X_n) \tag{8.1}$$

or
$$K = (-, -, \ldots, -). \tag{8.2}$$

The letters X_i ($i=1, \ldots, n$) or the dashes (—) representing the variables: generation of all possible combinations of the constants 0 and 1 provides the set (E_1) of the 2^n binary combinations with n positions. The notations (8.1) and (8.2) thus constitute compact formulae which will be frequently employed for (E_1) in the text and, with the implicit assumption of the addition of the adjacency relation, for the cube itself.

8.1.4.2 The k-Dimensional Sub-Cube of an n-Dimensional Cube ($0 \leq k \leq n$)

If $n-k$ variables of the n variables of the formulae (8.1) or (8.2) are fixed (by assigning them the values 0 or 1) then, by definition, a k-dimen

8.1 The n-Dimensional Cube

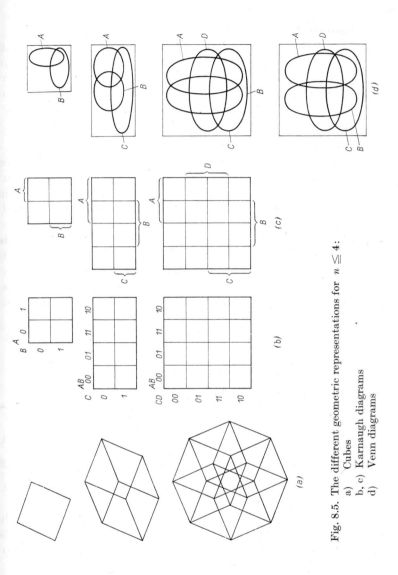

Fig. 8.5. The different geometric representations for $n \leqq 4$:
a) Cubes
b), c) Karnaugh diagrams
d) Venn diagrams

sional sub-cube is obtained. For $k = 0$ the different vertices are obtained for $k = n$, is obtained the n-dimensional cube itself.

Example:

$$(0 \ X_2 \ X_3 \ X_4 \ 1 \ X_6)$$

or

$$(0 \ - \ - \ - \ 1 \ -)$$

(8.3)

constitute two notations for a 4 (6 − 2)-dimensional sub-cube of the 6-dimensional cube $(X_1 X_2 X_3 X_4 X_5 X_6)$ or $(- - - - - -)$.

8.1.4.3 Distance of 2 Vertices of an n-Dimensional Cube

Consider for example the vertex 000 of a 3-dimensional cube. From this vertex 3 lines extend at the ends of which are to be found 3 other vertices (Fig. 8.3). For each of these the associated binary combination can be derived from 000 by complementation of one coordinate, and one only, corresponding with the axis of the coordinate followed in passing from 000 to the vertex in question. In addition, it is convenient to assume that the length of the edge plays the part of a "quantum" of distance, the unit length, for the various paths that can be formed on the cube with the edges. The distance between 2 geometrically adjacent vertices (separated by a single edge) and measured in this unit is, therefore, 1. This leads to the following definition:

Hamming distance:

For 2 given vertices of the n-dimensional cube

$$\begin{cases} X = (x_n x_{n-1} \cdots x_i \cdots x_1) \\ Y = (y_n y_{n-1} \cdots y_i \cdots y_1) \end{cases}$$

(8.4)

By definition, the "Hamming distance" (or the distance) of X and Y is the quantity

$$D(X, Y) = \sum_{i=1}^{i=n} |x_i + y_i|_2$$

(8.5)

($\sum$ being the summation in the ordinary arithmetic sense, $|x_i + y_i|_2$ the residue modulo 2 of the sum $x_i + y_i$, and $|x_i + y_i|_2$ formally identical to $x_i \oplus y_i$).

It will be seen that:

— $D(X, Y)$ is the number of positions by which the words X and Y differ.
— it is also the weight (number of 1's) of the vector $X \oplus Y$:

$$D(X, Y) = P(X \oplus Y).$$

The Hamming distance is very useful in the theory of codes: if X is the word transmitted and Y the word received in its stead, then $X \oplus Y$ represents the error, rank by rank, committed on X, and $D(X, Y)$ is the number of positions which do not tally ("weight" of the error).

Examples:

1. $\begin{cases} X = 0\ 1\ 0 \\ Y = 1\ 1\ 0 \end{cases}$ (8.7)

   $X \oplus Y = 1\ 0\ 0$ (8.8)

   $D(X, Y) = 1 + 0 + 0 = 1$ (8.9)

2. $\begin{cases} X = 1\ 0\ 1\ 1\ 0\ 1 \\ Y = 0\ 1\ 0\ 1\ 0\ 0 \end{cases}$ (8.10)

   $X \oplus Y = 1\ 1\ 1\ 0\ 0\ 1$ (8.11)

   $D(X, Y) = 1 + 1 + 1 + 0 + 0 + 1 = 4$ (8.12)

Consequently, if the distances on the cube are measured by taking the edge as the unit, it will be seen that:

1. For $n = 2, 3$, two geometrically adjacent vertices are also adjacent analytically.
2. For $n > 3$, this same property remains true by reason of the very process of construction of the geometric representation. Crossing a geometric edge is thus equivalent analytically to complementing a variable.

More generally, if the Hamming distance between 2 vertices is K, this signifies that the shortest path between these 2 vertices consists of K edges.

Summarizing, although in practice it is only possible to work on an n-dimensional cube through the intermediary of 2-dimensional representations (3-dimensional representations are never used), this geometric interpretation of a set 2^n of n-dimensional binary vectors and of the adjacency relation, remains of value even in the n-dimensional case: thanks to the representation described and, in fact, to diverse variants, to be treated later (Veitch and Karnaugh diagrams), problems of functions of 5 and 6 variables can be treated *in practice*. Additionally, in all theoretical considerations, it provides a convenient geometric language as well as a 'bolster' for intuition.[1]

---

[1] A type of intuition comparable with that of the "intuition for the pure number" of which *Poincaré* spoke.

4 Chinal, Design Methods

8.2 Venn Diagrams

As indicated in Chapter 1, a graphical representation of the relation between sets and, in particular, of the operations for union, intersection and complementation is known as a Venn diagram. It was shown (Chapter 6) that every algebra of sets with the above operations is a Boolean algebra. Thus it is not surprising that Venn diagrams may be used to represent the operations on switching functions which, themselves, constitute a Boolean algebra. For this it suffices to use the following correspondence between the sets depicted in the Venn diagrams and the Boolean functions: with every *set* E_i is associated its *characteristic function* X_i. This then produces the following correspondence between operations on sets and characteristic functions (Table 8.6).

Table 8.6. *Operations on sets and on characteristic functions*

$E_i \cap E_j$	$X_i \cdot X_j$
$E_i \cup E_j$	$X_i + X_j$
$\overline{E}_i$	$\overline{X}_i$

Functions

For n given sets $E_1, \ldots, E_n$ on the Venn diagram, further (2^n) subsets can be defined by forming all possible combinations of these sets or of their complements, just as the 2^n minterms of a switching function would be formed. Every set appearing in the Venn diagram is, therefore, defined by a particular selection of these subsets. Neither the form nor the exact dimensions of these subsets play any part from the point of view of the algebra of classes, and they may be considered as "quanta", as elements or points from which all other sets are defined. (In particular, it is unnecessary to take a 'finer' definition of these points) (Figs. 8.7 and 8.8). (A defect of the Venn diagram is that the parts of each subset are not necessarily adjacent) (cf. Fig. 8.5d).

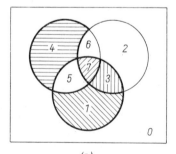 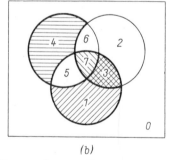

(a)   (b)

Fig. 8.7. Two Venn diagrams for a set E

8.2 Venn Diagrams

Figure 8.7 illustrates the above correspondence for the cases of 3 and 4 variables. Here the set E has been represented as a shaded area. The set X is a function of sets: it can be defined by a binary vector with 2^n

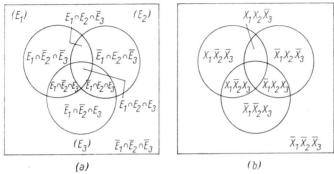

Fig. 8.8. Sets and functions.

components indicating the selection of elements (among the possible 2^n) which compose it.
The characteristic function X of the set E (Table 8.9) is thus represented in this sense by the set E of the Venn diagram.

Table 8.9

$X_1 X_2 X_3$	X
0 0 0	0
0 0 1	1
0 1 0	0
0 1 1	1
1 0 0	1
1 0 1	0
1 1 0	0
1 1 1	1

Algebraic expressions

The set E can also be defined by an expression (in the algebra of sets)

$$E = \left(\bar{E}_1 \cap \bar{E}_2 \cap E_3\right) \cup \left(\bar{E}_1 \cap E_2 \cap E_3\right) \cup \left(E_1 \cap \bar{E}_2 \cap \bar{E}_3\right) \cup (E_1 \cap E_2 \cap E_3) \tag{8.13}$$

for which, by application of the rules of Table 8.6, it can immediately be deduced that:

$$X = \overline{X}_1\overline{X}_2X_3 + \overline{X}_1X_2X_3 + X_1\overline{X}_2\overline{X}_3 + X_1X_2X_3 \tag{8.14}$$

More generally, by application of these same rules a Boolean expression in terms of $\cdot$, $+$, and $-$ can be made to correspond with an expression in the operations $\cap$, $\cup$, $^-$, and vice versa. Thus these expressions, except for the difference in notation, are identical.

Example of application: simplification of expressions.

The expressions for E or X can be simplified. The determination of E and X is the same except for the notation it will nevertheless be reproduced as an illustration of the above (but detailing only the more important operations):

Table 8.10

$$E = \left(\overline{E}_1 \cap \overline{E}_2 \cap E_3\right) \cup \left(\overline{E}_1 \cap E_2 \cap E_3\right) \cup \left(E_1 \cap \overline{E}_2 \cap \overline{E}_3\right) \cup \left(E_1 \cap E_2 \cap E_3\right)$$
$$= (E_1 \cap E_2 \cap E_3) \cup \left(\overline{E}_1 \cap E_2 \cap E_3\right) \cup \left(\overline{E}_1 \cap \overline{E}_2 \cap E_3\right) \cup \left(\overline{E}_1 \cap E_2 \cap E_3\right)$$
$$\cup \left(E_1 \cap \overline{E}_2 \cap \overline{E}_3\right)$$
$$= \left[(E_1 \cup \overline{E}_1) \cap E_2 \cap E_3\right] \cup \left[\overline{E}_1 \cap \left(E_2 \cup \overline{E}_2\right) \cap E_3\right] \cup \left(E_1 \cap \overline{E}_2 \cap \overline{E}_3\right)$$
$$= (E_2 \cap E_3) \cup \left(\overline{E}_1 \cap E_3\right) \cup \left(E_1 \cap \overline{E}_2 \cap \overline{E}_3\right)$$

$$X = \overline{X}_1\overline{X}_2X_3 + \overline{X}_1X_2X_3 + X_1\overline{X}_2\overline{X}_3 + X_1X_2X_3$$
$$= \left(X_1X_2X_3 + \overline{X}_1X_2X_3\right) + \left(\overline{X}_1\overline{X}_2X_3 + \overline{X}_1X_2X_3\right) + X_1\overline{X}_2\overline{X}_3$$
$$= \left[(X_1 + \overline{X}_1)X_2X_3\right] + \left[\overline{X}_1(X_2 + \overline{X}_2)X_3\right] + X_1\overline{X}_2\overline{X}_3$$
$$= X_2X_3 + \overline{X}_1X_3 + X_1\overline{X}_2\overline{X}_3$$

The final expressions contain fewer symbols than the expressions (disjunctive canonical form) which were used at the start.

So far the Venn diagram has supplied only the canonical form of E. It can also be used, and it is here where the real interest of this representation lies, to perform the transformations of the upper half of Table 8.10 and, even better, to 'short-circuit' some of the intermediary steps. For the associated algebraic expression X, advantage will automatically be taken of this economy in calculation.

Example: In Fig. 8.7 it will be seen that the set E can be obtained in several ways, and in particular:

1. As the union of 4 disjoint sets (Fig. 8.7a).
2. As the union of 3 sets (Fig. 8.7b).

The following equation is obtained directly by simple inspection of Fig. 8.7a:

$$E = \left(E_1 \cap \bar{E}_2 \cap \bar{E}_3\right) \cup \left(E_2 \cap E_3\right) \cup \left(\bar{E}_1 \cap E_3\right) \qquad (8.15)$$

and its counterpart:

$$X = X_1 \bar{X}_2 \bar{X}_3 + X_2 X_3 + \bar{X}_1 X_3$$

obtained earlier from the calculation of Table 8.10.

Summarizing; the Venn diagram enables the algebraic operations to be replaced by an examination of the goemetric configurations, and the method can be used, in particular, for the simplification of expressions. However, as it stands, it presents certain minor disadvantages in application; if the form of the sets is free certain of the subsets are not necessarily adjacent so that, in practice, the use of a modified version is preferred, in which the sets take *standardised* forms: i.e. the Karnaugh diagram.

8.3 The Karnaugh Diagram

8.3.1 The Case of 4 Variables (or less)

Consider a Venn diagram for 2 sets A and B. In its conventional form it is composed of 2 circles or ovals whose internal areas have one part in common. It is clear that the exact shape of the 2 zones plays no part at this time and that as a result they can be deformed progressively into two rectangles in the manner shown in Fig. 8.5c, d. An analogous operation can be performed in the case of 3 and 4 variables (Fig. 8.5c, d). In each of the figures a 'square' represents the intersection of n ($n = 2, 3$ or 4) sets of rectangular shape taken from among A, B, C, D or their complements. With the aid of Table 8.6 each 'square' is thus associated with a fundamental product. Two 'squares' which have a common side, or which are situated at the extremities of the same line or of the same column, are said to be adjacent. A check is made to ensure that in passing from one 'square' to an adjacent 'square' the variable leaves only one of the sets $A, B, C, D, \bar{A}, \bar{B}, \bar{C}, \bar{D}$ but remains in the others, which is tantamount to saying that one single variable changes (is complemented): it is not complemented in one of the fundamental products but is complemented in the other. The union of these two adjacent squares gives a 3-variables expression, the union of two sets of this type, in its turn, gives a 4-square set represented by a 2-variable expression, etc.

Examples (Fig. 8.11):

$$ABC\overline{D} : (1 \text{ 'square'})$$
$$ABC \phantom{\overline{D}} : (2 \text{ 'squares'})$$
$$AB \phantom{C\overline{D}} : (4 \text{ 'squares'})$$
$$B \phantom{ABC\overline{D}} : (8 \text{ 'squares'})$$

(8.17)

More generally, for 2 given zones represented by monomials (with variables or less) which differ in one variable only, their union is represented by a monomial containing one fewer variable (that which changes in passing from one zone to the other). Summarizing, with the aid of Venn or Karnaugh diagrams, it has been shown how a set E may be associated

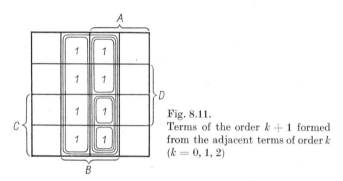

Fig. 8.11.
Terms of the order $k + 1$ formed from the adjacent terms of order k ($k = 0, 1, 2$)

with every *function f*. This set may be represented by diverse expressions in the algebra of sets with which (Table 8.6) can be associated just as many expressions for the function f.

The Karnaugh diagram may also be considered as a planar representation of 2, 3 or 4-dimensional cubes. Consider the 'square' of Fig. 8.5a. It can be associated with the diagram of Fig. 8.5b: each vertex is represented by a 'square', two adjacent vertices on the diagram correspond to adjacent 'squares' (as defined above). This correspondence can similarly be established for the case of 3 or 4 variables.

The 3-variable diagram may be considered to be composed of 2 diagrams associated with two opposite faces ('squares') of the cube. The 4-variables diagram is formed in a similar manner.

Fig. 8.5 indicates how this binary numbering can be assigned directly to the 'squares' of the diagram. The disposition of the sets A, B, C, D corresponds exactly with the numbering adopted for the rows and columns (Cyclic codes; cf. Chapter 3), which is that of Karnaugh.[1]

---

[1] The Veitch diagram which came earlier, used the numbering 00, 01, 10, 11.

There are 2^n 'squares' in the n-variables diagram, as many as there are vertices in the n-dimensional cube.

Figure 8.5 illustrates the correspondence between the diverse representations decribed above.

8.3.2 Karnaugh Diagrams in 5 and 6 Variables

8.3.2.1 The 5-Variable Case

For 5 given variables A, B, C, D, E a 5-variable diagram can be formed with the help of 2 ordinary 4-variable diagrams. One of these will be assigned the value $E = 0$ and the other will be assigned the value $E = 1$. (Fig. 8.12a). In these circumstances, 2 'squares' situated in the same location in both diagrams, correspond to binary combinations which differ by only one variable, i.e. at 2 adjacent vertices. It is sometimes

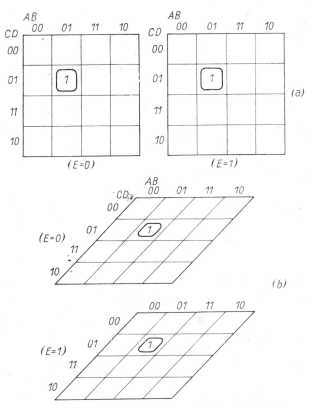

Fig. 8.12. Karnaugh diagrams in 5 variables

convenient to represent the 2 partial diagrams as situated one below the other (Fig. 8.12b). Thus the 2 'squares' represented in the diagrams (8.12a) and (8.12b) are adjacent. In practice, the first representation will be used, by drawing the two 4-variable diagrams side by side.

8.3.2.2 The Case of 6 Variables

This time 4 diagrams with 4 variables themselves grouped into 2 5-variable diagrams are used; in practice they are laid out in grid form (Fig. 8.13). Each diagram corresponds to one of the 4 possible combina

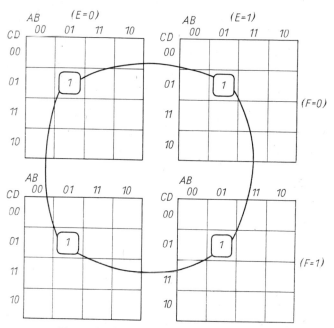

Fig. 8.13. Karnaugh diagram in 6 variables

tions (EF) and the internal 'squares' represent the diverse binary combinations of the variables A, B, C, D. The 4 grids shown by way of an example in Fig. 8.13 are adjacent 2 by 2. In a general way, in addition to the ordinary adjacencies of 4-variables diagrams, every 'square' is adjacent to the two corresponding 'squares' situated in the adjacent diagrams.

Note: it is possible to continue beyond $n = 6$, but the practical interest in the diagrams thus obtained is much reduced and they will not be described.

8.3.3 Decimal Notation

The decimal notation for the functions (Chap. 7) can be transposed directly onto the Karnaugh diagrams (which are truth tables). The numbers carried in these 'squares' are the decimal representations of the corresponding binary combinations (Fig. 8.14).

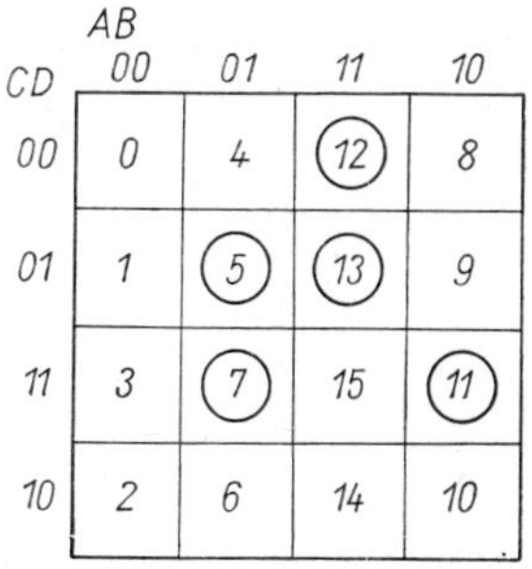

```
       AB
  CD   00  01  11  10
  00 |  0 | 4 |(12)| 8 |
  01 |  1 |(5)|(13)| 9 |
  11 |  3 |(7)| 15 |(11)|
  10 |  2 | 6 | 14 | 10 |
```

Fig. 8.14.
Decimal notation of the 'squares'

8.3.4 The Karnaugh Diagram and the Representation of Functions in Practice

8.3.4.1 Notations

A function f is represented by its characteristic set (set of 'squares') as follows:

— *Completely specified functions*

The value 1 is inscribed in every 'square' associated with a binary combination for which $f = 1$. The other 'squares' are left empty, which suffices to indicate the value 0 assigned to them.

— *Incompletely specified functions*

For the specified values the procedure is as above. For the others the symbol — or $\varnothing$ (which can be read as 0 or 1) is inscribed in the corresponding 'squares'.

8.3.4.2 Derivation of a Karnaugh Diagram

In practice, the method to be adopted depends on the manner in which the function is specified. A brief description of the way in which the set of 1's for a completely specified function is derived is as follows:

1. Function given by an algebraic expression

There is an interest in making use of the form 8.5c which enable important subsets of the characteristic set to be written in 'en bloc' without the necessity of detailing all the 'squares' which make it up

Example:
$$f(A, B, C) = BC + BD + \overline{C}D + AC\overline{D} + \overline{A}\,\overline{B}\,\overline{C} \qquad (8.18)$$

As can be seen (Fig. 8.15) the 5 blocks, surrounded by a heavy line associated with the 5 functions for which the expressions are BC, BD, $\overline{C}D$ etc. can be completed directly.

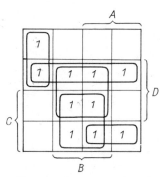

Fig. 8.15. Representation of a function

Table 8.16

A B C D	f
0 0 0 0	1
0 0 0 1	1
0 0 1 0	0
0 0 1 1	0
0 1 0 0	0
0 1 0 1	1
0 1 1 0	1
0 1 1 1	1
1 0 0 0	0
1 0 0 1	1
1 0 1 0	1
1 0 1 1	0
1 1 0 0	0
1 1 0 1	1
1 1 1 0	1
1 1 1 1	1

CD \ AB	00	01	11	10
00	1			
01	1	1	1	1
11		1	1	
10		1	1	1

Fig. 8.17

2. Function given by a truth table

In a certain sense this a particular case of the preceding one. The form 8.5b or 8.5c may be used without distinction. The diagram (Fig. 8.17) is obtained from Table 8.16; it is as before.

8.3 The Karnaugh Diagram

3. *Function given in decimal form*

The diagram is ready for use with the numbers (decimal) already inscribed in the 'squares'. The vertices for which $f = 1$ are indicated by encircling the corresponding numbers in the 'squares' (Fig. 8.14). Summarizing:

a) With each *'square'* (element) of the Karnaugh diagram can be associated a *fundamental product*, or *binary combination* which renders this product equal to 1.

Example:
$$A \cap \bar{B} \cap C \cap \bar{D} / A\bar{B}C\bar{D} / 1010 \qquad (8.19)$$

Every set of one or of several 'squares' may thus be regarded as the characteristic set of a function.

b) If a function is given by one of its algebraic expressions X, Table 8.6 enables a set E, which is the characteristic set of the function created by X, to be associated with this expression.

c) If a function is given in the form of a truth table, its characteristic set is the union of the 'squares' associated with the binary combinations for which the function equals 1.

d) Finally, the Karnaugh diagram may be regarded as:

— a 2-dimensional *truth table*, intended to show the adjacencies between vertices.

— a *standardized Venn diagram*.

— a *planar representation of the n-dimensional cube*.

It should be noted that the truth table normally used (Chapter 7) is, itself, in a sense, a planar representation of the n-dimensional cube, but by reason of its linear form, it does not show the adjacent vertices (at unit distance). In the Karnaugh diagram, however, the 2 adjacent vertices stand out, partly from their *geometric* adjacency and partly with the aid of a *supplementary convention*, that which consists of considering as adjacent any 2 'squares' situated at the 2 extremities of a line or of a column.

It will finally be noted that the Karnaugh diagram, like the other representations, enables operations on the *functions* to be performed geometrically. It is only indirectly, through the choice of particular representations for the functions in question, that it enables *algebraic expressions* to be manipulated.

8.4 The Simplification of Algebraic Expressions by the Karnaugh Diagram Method

It was shown above that the Karnaugh diagram provides a plana geometric representation of Boolean functions. Now to be defined ar the diverse operations that may be performed with this type of diagram which, as will be seen, can be substituted for the algebraic operations o the Boolean *functions*.

8.4.1 Karnaugh Diagram Interpretations for the Sum, Product and Complementation Operations

8.4.1.1 Sum

If E_1 and E_2 are the respective characteristic sets for the completely specified functions f_1 and f_2, the set $E_1 \cup E_2$ (the 'union' of the sets o 1's) is the characteristic set of the sum $f_1 + f_2$ (the Karnaugh diagram in particular). Fig. 8.18 illustrates this property.

8.4.1.2 Product

In a similar manner, the set $E_1 \cap E_2$ corresponds to the produc $f_1 \cdot f_2$ ("intersection of the sets of 1's" in the Karnaugh diagram) Fig. 8.19 gives an example of this operation.

8.4.1.3 Complement

Finally, if E corresponds to f, then $\bar{E}$ is associated with $\bar{f}$ ("the complement of the sets of 1's" in the Karnaugh diagram) (Fig. 8.20).

8.4.2 Representation of Certain Remarkable Functions

8.4.2.1 Variables

On a diagram having n variables (2^n 'squares'), each variable i represented by 2^{n-1} elementary 'squares' (example: Fig. 8.5).

8.4.2.2 Product of k Variables $(k > 1)$

A product of k variables is represented by a square or rectangle having 2^{n-k} 'squares' (Figs. 8.21—8.23).

8.4 The Simplification of Algebraic Expressions   203

Note: it will be seen that the fundamental products are represented by the smallest number of non-empty sets possible ('squares'). This is one of the origins of the name minterm (= minimum term).

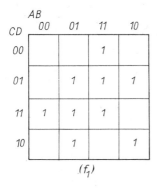

Fig. 8.18. Sum of 2 functions   Fig. 8.19. Product of 2 functions

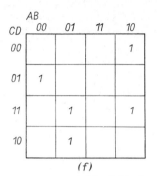

Fig. 8.20. Complement of a function

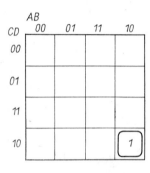

Fig. 8.21.
A fundamental product

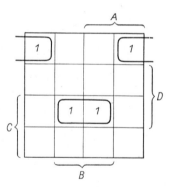

Fig. 8.22. A 3-variable product

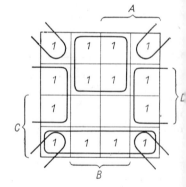

Fig. 8.23. Some 2-variable products

8.4.2.3 Fundamental Sums

A fundamental sum takes the value zero for a single combination only. In other words, the representation of such a sum in the Karnaugh diagram includes one single empty 'square' only (of value 1). (Fig. 8.24)

8.4 The Simplification of Algebraic Expressions

ives the representation of the sum $\bar{A} + B + \bar{C} + D$ (fundamental for he variables A, B, C, D).

Note: The expression (Fig. 8.24):

$$\bar{A} + B + \bar{C} + D = \overline{A\bar{B}C\bar{D}} \tag{8.20}$$

ndicates the relation between the two types of function in this particular case.

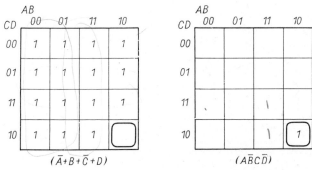

Fig. 8.24. A fundamental product and the complementary fundamental sum

Note 2: The fundamental sums are represented by the largest sets which are not identical with the universal set. Hence the name 'maxterm' by which they are also known.

8.4.3 The Use of the Karnaugh Diagram for the Proof of Algebraic Identities

As was shown above, algebraic operations can be replaced by operations on sets. This process can be used notably to prove algebraic identities.

Examples:

a) $$ABC\bar{D} + ABCD = ABC \tag{8.21}$$

Algebraically this would give:

$$ABC\bar{D} + ABCD = ABC(D + \bar{D}) = AB\dot{C} \tag{8.22}$$

geometrically: Fig. 8.11 reveals the set ABC, formed by the union of 2 adjacent 'squares', corresponding to $ABCD$ and $ABC\tilde{D}$.

b) $$AB + BC + CA = AB \oplus BC \oplus CA. \tag{8.23}$$

For example, in writing:
$$AB \oplus BC \oplus CA = (AB \oplus BC) \oplus CA,$$
the required relation is obtained (Fig. 8.25)

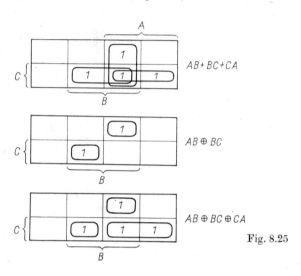

Fig. 8.25

c) $\qquad A + \bar{A}B = A \oplus \bar{A}B = A + B.$ \hfill (8.24)

These relations are evident in Fig. 8.26

d) $\qquad (A + B)(A \oplus B) = (A \oplus B).$ \hfill (8.25)

By taking account of the intersection of the terms $A + B$ and $A \oplus B$ this equality is obtained (Fig. 8.27)

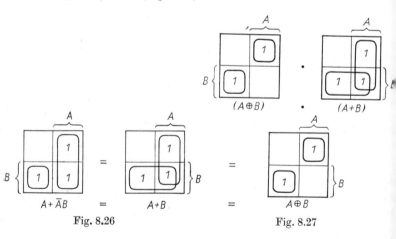

Fig. 8.26 \qquad\qquad Fig. 8.27

8.4 The Simplification of Algebraic Expressions

e)
$$\begin{cases} A\bar{B} + \bar{A}B = (A+B)(\bar{A}+\bar{B}) \\ AB + \bar{A}\bar{B} = (A+\bar{B})(\bar{A}+B) \end{cases} \qquad (8.26)$$

cf. Fig. 8.28).

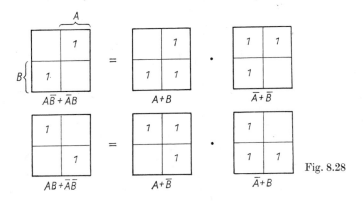

Fig. 8.28

f)
$$AB + \bar{A}C + BC = AB + \bar{A}C. \qquad (8.27)$$

It can be seen in Fig. 8.29 that of the three terms AB, $\bar{A}C$ and BC, the terms AB and $\bar{A}C$ suffice to cover the total set, whence the required identity.

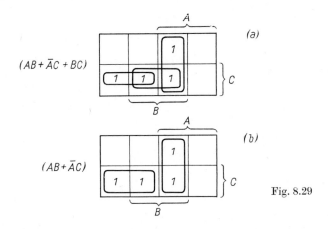

Fig. 8.29

By manipulation of the sets in the Karnaugh diagram different equivalent expressions for the same function can be derived. In particular, expressions of a simple character (with respect to some criteria) can be found.

15 Chinal, Design Methods

g) Consider the expression

$$f(A, B, C) = AB\bar{C} + \bar{A}B\bar{C} + A\bar{B}C + \bar{A}\bar{B}C \qquad (8.28)$$

(disjunctive canonical form).

From this the covering of Fig. 8.30 is easily deduced. Hence:

$$f(A, B, C) = B\bar{C} + \bar{B}C \quad (= B \oplus C).$$

h) Consider the expression:

$$f(A, B, C) = AB\bar{C} + ABC + \bar{A}B\bar{C} + \bar{A}BC + A\bar{B}C. \qquad (8.29)$$

The characteristic set f can be covered by the square and rectangle of Fig. 8.31, from which the following expression can be deduced:

$$f(A, B, C) = B + AC. \qquad (8.30)$$

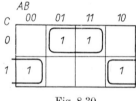

Fig. 8.30

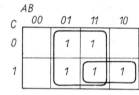

Fig. 8.31

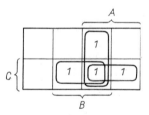

Fig. 8.32

i) Consider the function:

$$f(A, B, C) = AB\bar{C} + A\bar{B}C + ABC + \bar{A}BC. \qquad (8.31)$$

The 3-inputs majority function as defined in Chapter 7 can be recognised in the canonical form. Its characteristic set, represented in Fig. 8.32, can be covered by 3 rectangles, as shown in the diagram, thus providing the simplified expression:

$$f(A, B, C) = AB + BC + CA \qquad (8.32)$$

which was established in Chapter 7 by purely algebraic operations.

8.4 The Simplification of Algebraic Expressions

j) Consider the following (non-canonical) function, given in the form of sum of products:

$$f(A, B, C) = AB\bar{C}\bar{D} + B\bar{C}D + BCD + \bar{B}C. \tag{8.33}$$

With each of these four monomials corresponds a rectangle as shown in Fig. 8.33 a. Their union forms the characteristic set f, which can be covered as in Fig. 8.33 b. This new covering corresponds to the (simplified) expression

$$f(A, B, C, D) = AB\bar{C} + BD + \bar{B}C. \tag{8.34}$$

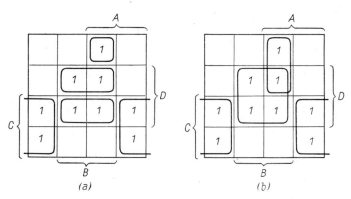

Fig. 8.33

k) Consider the function

$$f(A, B, C, D)$$
$$= \bar{A}\bar{B}\bar{C}D + \bar{A}\bar{B}CD + \bar{A}BCD + \bar{A}BC\bar{D} + ABCD + ABC\bar{D}$$
$$+ A\bar{B}\bar{C}\bar{D} + A\bar{B}C\bar{D} \tag{8.35}$$

– *Sums of products*

The following stages are to be followed:
1) representation of the function by its characteristic set E (Fig. 8.34),
2) determine the largest rectangles (or squares) contained in E (it is useless to retain the rectangles or squares contained in these latter),
3) determine one minimal covering: here it will be found to be unique, giving:

$$f(A, B, C, D) = \bar{A}\bar{B}\bar{C}D + A\bar{B}\bar{D} + BC + C\bar{D}. \tag{8.36}$$

5*

— *Products of sums*

Problem: to find a minimal form for the product of sums for the above function f. We have (Fig. 8.35):

$$\bar{f}(A, B, C, D) = B\bar{C} + A\bar{C}D + \bar{A}\bar{C}\bar{D} + \bar{B}CD. \qquad (8.37)$$

the form obtained by use of the above method for the sums of products and applied to $\bar{f}$ (the set of 0's can be considered, without the need for rewriting the diagram, as the characteristic set $E_{\bar{f}} = \bar{E}_f$.

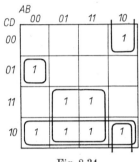

Fig. 8.34

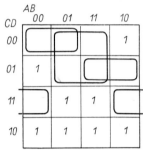

Fig. 8.35

Whence:

$$(A, B, C, D) = (\bar{B} + C)(\bar{A} + C + \bar{D})(A + C + D)(B + \bar{C} + \bar{D}). \qquad (8.38)$$

Notes:

— It will be checked that in this last case the operation of complementation does not change the "cost" (the number of symbols) of the expression.
— That there may be several minimal coverings (e.g. Fig. 8.35) will be seen in further examples to be given in Chapter 10.
— It is possible that a function may not be capable of simplification if the search be restricted only to expressions such as a product of sums or of a sum of products. Consider for example the function:

$$f(A, B, C, D) = A \oplus B \oplus C \oplus D.$$

It will be seen (Fig. 8.36) that no pair of adjacent elementary 'squares' is present which explains the fact that any 2 of 4 fundamental products differ in 2 positions. The function in question cannot, therefore, be simplified in this manner.

8.4 The Simplification of Algebraic Expressions

— For the products of sums, a direct study could be made of the representative sets of sums for which the intersection would be the representative set of the function. In practice this method is less convenient (cf. Fig. 8.24) and use will be made of the above method only (use of the complement).

	AB			
CD	00	01	11	10
00		1		1
01	1		1	
11		1		1
10	1		1	

Fig. 8.36

— *Simplification of the 5 or 6-variable function*

In principle the method is exactly the same as that for $n \leq 4$. It differs from the $n \leq 4$ case only with respect to its application, which, in this case, no longer necessitates a single diagram, but 2 or 4 dependent on whether $n = 5$ or $n = 6$.

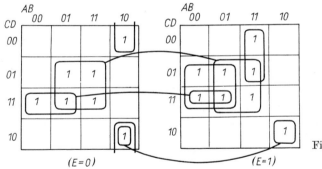

Fig. 8.37

Examples:
The reader will be able to simplify the following functions:

) $f(A, B, C, D, E) = \bar{A}\bar{B}CD\bar{E} + \bar{A}B\bar{C}D\bar{E} + \bar{A}BCD\bar{E} + AB\bar{C}D\bar{E}$
  $+ ABCD\bar{E} + A\bar{B}\bar{C}\bar{D}\bar{E} + A\bar{B}C\bar{D}\bar{E} + \bar{A}\bar{B}\bar{C}DE + \bar{A}\bar{B}CDE$
  $+ \bar{A}B\bar{C}DE + \bar{A}BCDE + AB\bar{C}\bar{D}E + AB\bar{C}DE + ABCDE$
  $+ A\bar{B}C\bar{D}E$

(15 fundamental products, 75 symbols).

Simplified form (Fig. 8.37):
$$f(A, B, C, D, E) = BD + \bar{A}CD + \bar{A}DE + AB\bar{C}E$$
$$+ A\bar{B}C\bar{D} + A\bar{B}D\bar{E}$$
(6 terms, 20 symbols).

m) $f(A, B, C, D, E, F) = \bar{A}\bar{B}\bar{C}D\bar{E}\bar{F} + \bar{A}\bar{B}CD\bar{E}\bar{F} + \bar{A}B\bar{C}D\bar{E}\bar{F}$
$+ \bar{A}BCD\bar{E}\bar{F} + AB\bar{C}D\bar{E}\bar{F} + ABCD\bar{E}\bar{F} + \bar{A}\bar{B}\bar{C}DE\bar{F} + \bar{A}\bar{B}CDE\bar{F}$
$+ \bar{A}B\bar{C}DE\bar{F} + \bar{A}BCDE\bar{F} + ABCDE\bar{F} + A\bar{B}\bar{C}\bar{D}E\bar{F} + A\bar{B}\bar{C}DE\bar{F}$
$+ \bar{A}\bar{B}\bar{C}D\bar{E}F + \bar{A}\bar{B}CD\bar{E}F + \bar{A}B\bar{C}D\bar{E}F + \bar{A}BCD\bar{E}F + \bar{A}BC\bar{D}EF$
$+ AB\bar{C}\bar{D}\bar{E}F + A\bar{B}\bar{C}D\bar{E}F + A\bar{B}C\bar{D}\bar{E}F + \bar{A}\bar{B}\bar{C}DEF + \bar{A}\bar{B}CDEF$
$+ \bar{A}B\bar{C}DEF + \bar{A}BCDEF + AB\bar{C}\bar{D}EF + A\bar{B}\bar{C}\bar{D}EF + A\bar{B}\bar{C}DEF$

(28 fundamental products, 168 symbols).

we obtain the following cover (Fig. 8.38):
$$f(A, B, C, D, E, F) = \bar{A}D + BCD\bar{F} + A\bar{B}\bar{C}E + A\bar{C}\bar{D}F + \bar{A}BC\bar{E}F$$
$$+ BD\bar{E}\bar{F} + A\bar{B}\bar{D}\bar{E}F$$
(7 terms, 28 symbols).

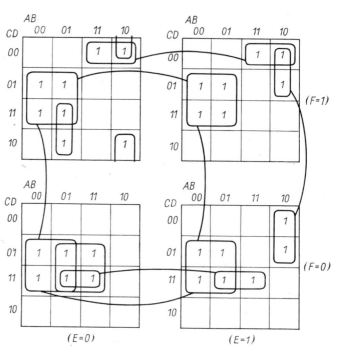

Fig. 8.38

Chapter 9

Applications and Examples

9.1 Switching Networks. Switching Elements

The contact circuits introduced in Chapter 4 provide examples of binary switching circuits where the basic building element is the contact relay. In general, consideration will be given to a circuit having n binary inputs and m binary outputs and known as a binary switching circuit[1]. It will be seen that, generally, such a circuit is produced with the aid of one or more relatively simple basic functional elements, each used in one or more examples, to be known as *switching elements*. A circuit of this type ($[m, n]$ network) will be represented symbolically by a box having n inputs and m outputs $(n, m \geq 1)$ (Fig. 9.1).

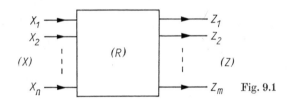

Fig. 9.1

Dependent on the way in which the outputs are conditioned by the inputs, it is necessary to distinguish between 2 types of networks:

1. *combinational* networks: whose outputs at the instant t depend uniquely on the inputs at this same instant t;
2. *sequential* networks: where the outputs depend not only on the input in the present state but also on the inputs at later instants, in other words, on the 'history' of the inputs.

9.2 Electronic Logic Circuits. Gates

9.2.1 Gates

Contact relays permit the realisation of the AND, OR and complementation functions. In a similar manner, by means of the different arrangements of the electronic elements such as diodes, transistors,

[1] or logic circuit.

capacitors, elementary circuits can be constructed to be used as basic functional blocks for the realisation of more complex networks. Such elementary circuits, often called *'gates'* or *decision circuits* perform functions such as AND, OR, —, NOR, exclusive-OR, majority etc., and make up a complete set (Chapter 7) of logic operations. They may be composed of discrete elements assembled on a base and soldered, or in the form of integrated circuits (semi-conductors, thin films, hybrids). Nevertheless, in both cases, the designer of the logic circuit is obliged, in general, to consider the functional blocks at his disposal to be indivisible, which in practice often arises from the physical presentation of the circuit: plate, module (where the elements may be sealed in an epoxy resin) or a 'potted' semi-conductor integrated circuit. In fact, this indivisible character of the functional blocks will remain the rule in every case, irrespective of the implementation of the circuits, by reason of the strict manner in which the electrical performance characteristics are determined for the circuit as a whole.

9.2.2 Logics

That which, in practice, is known as a *'logic'* is, in the language of the circuit manufacturer, none other than the electronic realisation of the logic operation. At the same time, this logic is characterised by a specific choice of the semi-conductor elements employed (diodes, transistors, resistors) and by the method in which they are combined in the performance of the logic operation. Thus the terms diode-resistor logic, diode-transistor logic etc., will be encountered. It should be noted, before proceeding to a description of the more common among them, that these different logic circuits can often be defined in terms of conventional circuits utilising discrete elements as well as in terms of integrated circuits. However, the practical value of a logic circuit, linked to diverse criteria such as cost, reliability, speed, insensitivity to parasitic noise, closely determines the technology which will effectively be adopted. Thus, a priori, a logic which makes use of many transistors is better suited to the technology of integrated circuits. This is why in practice certain logics are seldom encountered except in integrated circuits. In short, the wealth of equipment supplied by the circuit manufacturers for a given logic will include several (3 to 6 for example) gates to which are often added one or more regenerative circuits and one or more types of flip-flop, the gates sufficing to produce a combinational network.

A number of logic circuits will be described later as also will the symbols most commonly used to represent the gates. It must be said that the types described below, do not exhaust the large variety of solutions existing at the present time and for which the reader should consult the specialised literature.

9.2.3 Logic Conventions

For the contact networks the symbols 0 and 1 have been used to code the two states "open" and "closed". Similarly, for the electronic networks there are two operations:

1. The selection of reference states: this is more often made in terms of voltages, sometimes in terms of currents (e.g. high voltage/low voltage, high current flow/negligible current flow).
2. The logic convention.

After having chosen 2 states for the significant parameters (voltages for example) known as 'high' and 'low' and denoted as $+$ and $-$ respectively, the two symbols 0 and 1 can be assigned to these states in two ways (Table 9.2)

Table 9.2. *The 2 logic conventions*

	(P.L)	(N.L)
$+$	1	0
$-$	0	1

The first convention is known as "positive logic" (P. L.), the second as "negative logic" (N. L.). It is important to note that, for a given logic circuit, the logic function produced depends on the convention employed. Tables 9.3 a, b, c illustrate this phenomenon.

Table 9.3. *Two logic operators produced by a single circuit*

X Y	F		X Y	F		X Y	F
$-$ $-$	$-$		0 0	0		1 1	1
$-$ $+$	$-$		0 1	0		1 0	1
$+$ $-$	$-$		1 0	0		0 1	1
$+$ $+$	$+$		1 1	1		0 0	0
(a)			(b)			(c)	

b) $F = X \cdot Y$
c) $F = X + Y$

More generally, if with one convention there exists a Boolean function $f(x_1, x_2, \ldots, x_n)$, then, the opposite convention will give the function

$$\bar{f}(\bar{x}_1, \bar{x}_2, \ldots, \bar{x}_n) = \tilde{f}(x_1, x_2, \ldots, x_n),$$

i.e. the dual function of the first (cf. Chapter 7).

In the text which follows, except where specific mention is made to the contrary, the positive logic convention will be used.

9.2.4 Diode Logic

Figure 9.4 represents a diode-resistor gate. Assuming, as a first approximation, the diodes to be perfect, it will be seen that if one or more of the input voltages are "low", i.e. less than the voltage $+$, the output will automatically follow this low value. Conversely, if all the inputs are high and, for example, equal to $+$, the output will be high. It will be noted that with imperfect diodes (finite resistance in the non-conducting

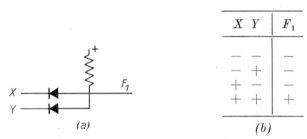

Fig. 9.4. Diode-resistance type AND gate.

mode, non-null in the conducting mode) this performance can be preserved. With a positive logic the table of Fig. 9.4 b corresponds to an AND gate. The same layout can be transposed for the case where there are more than 2 variables. In a similar manner with a positive logic the circuit of Fig. 9.5 produces the function OR. It should be noted that these

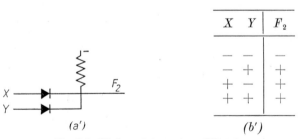

Fig. 9.5. Diode-resistance type OR gate.

two gates do not include any form of signal amplification (they are passive circuits). Furthermore, from the point of view of logic, the production of networks necessitates the addition of inverters (if the complements of the variables are not otherwise available) which are produced with the aid of common-emitter mounted transistors (cf. Fig. 9.6). The regeneration of the signals can be achieved with the aid of emitter-follower type circuits.

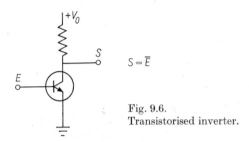

Fig. 9.6.
Transistorised inverter.

9.2.5 Direct Coupled Transistor Logic (DCTL)

The figures represent a NAND and a NOR. In the first case (a), the voltage in F is low (near zero) only when the 3 transistors are conducting (saturated) i.e. if the 3 voltages X, Y and Z are sufficiently high to unblock the corresponding transistors. In the second case (b),

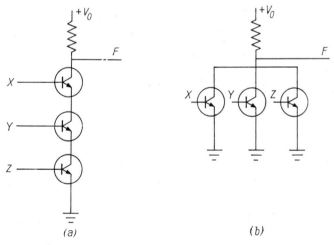

Fig. 9.7. (a) NAND gate — transistor logic $\left(F = \overline{X \cdot Y \cdot Z}\right)$.
(b) NOR gate — transistor logic $\left(F = \overline{X + Y + Z}\right)$.

the output F becomes low when one or more of the transistors are saturated, it is high only when all are blocked. It will be noted, however that in the first case the residual saturation voltages of the transistor are additive which, in general, renders this type of gate more delicate than in the second case, if the given residual voltages cannot be reduced. In particular, when two stages are to be interconnected, it is necessary that the upstream stage be able to block the downstream stage. This type of logic, fairly widespread in integrated circuits, by reason of it simplicity of production, presents however a greater sensitivity to parasitic phenomena (noise) than other types of logic. It will be noted that the inverter is only one particular case of this circuit (Fig. 9.6).

9.2.6 Diode-Transistor Logic (DTL)

Essentially, this logic combines the diode-resistor (passive circuit) gates with a transistorised inverter, the transistor playing the rôle of the regenerative (active) circuit at the same time. The functional block thus obtained is, dependent on the case, one or other of the two function NOR or NAND (Fig. 9.8). This logic has been widely used both in con

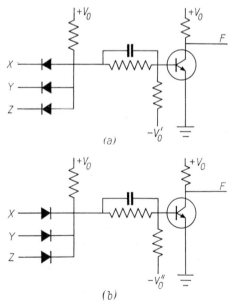

Fig. 9.8. Diode-transistor logic gates.
a) NAND gate. — b) NOR gate.

ventional circuits with discrete elements and in integrated circuits. It remains relatively economical to produce and has a good degree of immunity to noise. In operation it frequently requires two supply voltages V_0 and V'_0).

Fig. 9.9 gives the symbols for the more common logic operators (with 3 inputs).

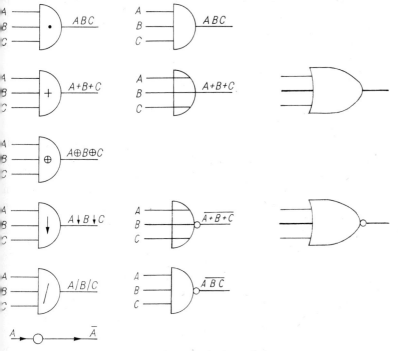

Fig. 9.9. Operator logic symbols

9.3 Combinational Networks. Function and Performance. Expressions and Structure

A combinational circuit is a circuit for which the outputs are Boolean functions of the inputs. The m outputs may, for example, be specified in the form of truth tables. From the viewpoint of logic, they constitute the 'performance', the 'behaviour' of the system. Moreover, with every realisation of the circuit with the aid of gates can be associated an expression (per output). Conversely, from a given expression causing different logic operators to intervene, it is possible to deduce a network which can be produced if the corresponding functional blocks are available. In this sense, the algebraic expressions can be considered as notations for the circuits, for which they describe the structure. For given output functions, there exists an infinity of circuits of different structure which are able to provide the input-output correspondence in question (cf. Fig. 9.10).

The problem of network analysis consists in determination of input-output correspondence produced from a layout. This can, for example, take the form of a truth table. For this purpose it is often convenient

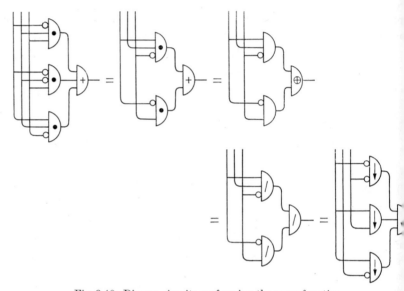

Fig. 9.10. Diverse circuits performing the same function.

to act through the intermediary of one or more algebraic expressions (per output) derived from the circuit. By suitable algebraic manipulation it is always possible to arrive at a form particularly suited to the formulation of the performance of the circuit (canonical forms, for example).

Synthesis is the reverse problem: to design a circuit capable of achieving the required correspondence at minimum cost on the basis of the specifications defining the m outputs as functions of the n inputs. Assuming the specifications are given in the form of truth tables, the following steps are necessary:

1. The determination of algebraic expressions for the outputs.
2. The minimisation of the circuit, the final algebraic expressions giving rise to a given set of logic operators.
3. The establishment of the logic layout based on these expressions.

9.3 Combinational Networks. Function and Performance

It should be noted that, in practice, phases 2 and 3 may need to be repeated in order to determine any modification necessary to the initial layout. The corresponding circuits will be entirely equivalent from the point of view of logic, but may be distinguishable by other characteristics electrical performance, steepness of impulse fronts, etc.).

The number of logic levels

Consider the circuit of Fig. 9.11. The input information passes through 2 types of gate, each of which contributes to the weakening and delaying of the signal. It is convenient to describe this as being a "2-level (or layer) logic circuit" layout. That of Fig. 9.12 (with 2 input gates) is

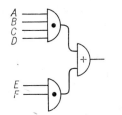

Fig. 9.11. A 2-level circuit.

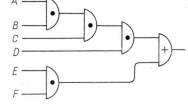

Fig. 9.12. A 4-level circuit.

4-level logic circuit. Conventionally speaking, it is possible to maintain this concept of logical levels for any operation: the number of levels of gated circuit being taken as the maximum number of gates likely to be traversed between an input and an output. This is but one definition, others take account of the electronic structure of the gates, giving a somewhat more precise sense to a parameter which, in practice, is intended to give a rough estimate of the rate of propagation of the signal.

Fan-in and fan-out

The fan-in is the number of inputs which a gate will accept. The fan-out is the number of other circuits (of one or more types taken as references) that the output can drive. In practice the first parameter is equal to the number of input paths provided by the gate. The second is usually indicated by the manufacturer. The values indicated for this parameter generally result from calculations on the electrical performances obtained for the different gate configurations. These calculations, often complex, which of necessity take account of the statistical scatter in the parameters (resistors, transistor gain, saturation voltages etc.) and their variations throughout the field of operation (temperature), can be made in various ways: least favourable case, statistical calculation. For these design methods, beyond the scope of this present work, the reader should refer to the excellent book [123] by A. Pressman.

The remainder of this chapter will be devoted to certain applications of Boolean algebra and to exercises: in some cases the aim will be to give the reader an exercise in computation, others are directed more towards a study of analysis and synthesis of logic networks.

9.4 Examples

9.4.1 An Identity

Problem: to prove the identity:

$$AB + BC + CA = AB \oplus BC \oplus CA. \qquad (9.1)$$

a) *Use of the truth table:*

Table 9.13 demonstrates that the above identity is proven.

Table 9.13

ABC	AB	BC	CA	$AB+BC+CA$	$AB \oplus BC \oplus CA$
0 0 0	0	0	0	0	0
0 0 1	0	0	0	0	0
0 1 0	0	0	0	0	0
0 1 1	0	1	0	1	1
1 0 0	0	0	0	0	0
1 0 1	0	0	1	1	1
1 1 0	1	0	0	1	1
1 1 1	1	1	1	1	1

b) *Use of algebraic manipulations*

The two members of the identity are now to be reduced to a common algebraic form.

1. Assume the common form to be the left-hand member of (9.1), then

$$\begin{aligned} AB \oplus BC &= AB\overline{BC} + \overline{AB}BC \\ &= AB(\overline{B} + \overline{C}) + (\overline{A} + \overline{B})BC \qquad (9.2) \\ &= AB\overline{C} + \overline{A}BC \end{aligned}$$

9.4 Examples

then:
$$(AB \oplus BC) \oplus CA = (AB\bar{C} + \bar{A}BC)\overline{CA} + \overline{(AB\bar{C} + \bar{A}BC)}(CA)$$
$$= (AB\bar{C} + \bar{A}BC)(\bar{C} + \bar{A}) + (\bar{A} + \bar{B} + C) \cdot$$
$$(A + \bar{B} + \bar{C})CA$$
$$= AB\bar{C} + \bar{A}BC + A\bar{B}C + AC$$
$$= AB\bar{C} + \bar{A}BC + AC$$
$$= (AB\bar{C} + AC) + (\bar{A}BC + AC)$$
$$= A(B\bar{C} + C) + C(\bar{A}B + A)$$
$$= A(B + C) + C(B + A)$$
$$= AB + BC + CA \qquad (9.3)$$

2. Assume the common form to be the right-hand member of (9.1).
Utilising the identity $X + Y = Y \oplus Y \oplus XY$, gives:
$$AB + BC + CA = (AB + BC) + CA$$
$$= (AB \oplus BC \oplus ABBC) + CA$$
$$= (AB \oplus BC \oplus ABC) + CA$$
$$= (AB\bar{C} \oplus BC) + CA$$
$$= (AB\bar{C} \oplus BC) \oplus CA \oplus (AB\bar{C} \oplus BC)CA$$
$$= AB\bar{C} \oplus ABC \oplus BC \oplus CA$$
$$= AB \oplus BC \oplus CA. \qquad (9.4)$$

3. Assume the common form to be the disjunctive canonical form.

In a certain sense, the determination of the truth table for each member constitutes a means of obtaining the canonical forms. The detailed algebraic calculations are as follows:

As before
$$(AB \oplus BC) \oplus CA = AB\bar{C} + \bar{A}BC + A\bar{B}C + AC,$$
and multiplying this last term by $B + \bar{B}$, reduces to:
$$(AB \oplus BC) \oplus CA = AB\bar{C} + \bar{A}BC + A\bar{B}C + ABC. \qquad (9.5)$$

Moreover, it can be said that:
$$AB + BC + CA = AB(C + \bar{C}) + (A + \bar{A})BC + C(B + \bar{B})A$$
$$= ABC + AB\bar{C} + ABC + \bar{A}BC + ABC + A\bar{B}C$$
$$= AB\bar{C} + \bar{A}BC + A\bar{B}C + ABC \qquad (9.6)$$

The two expressions (9.1) are equal to the same canonical form: they are, therefore, equal.

Note: the comparison of the 2 truth tables could have been shortened in the following manner: the first number equals 1 when 2 or 3 of the variables equal 1. If exactly 2 variables equal 1, one of the products AB, BC, CA equals 1 and the two members are equal. If the 3 variables equal 1 the three products also equal 1 and the second member therefore equals $1 \oplus 1 \oplus 1 = 1$, as did the first.

Another method, the most simple in fact, consists in using the Karnaugh diagram, as was done in Chapter 8.

9.4.2 Binary Adder and Subtractor

Chapter 2 gave tables defining the correspondence between the variables (arithmetic) x_i, y_i, r_i, s_i and r_{i+1}, defined in respect of the addition or subtraction of 2 binary numbers. By representation of the arithmetic 1 by the (Boolean) logic 1, the arithmetic 0 by the Boolean 0 and using, in a general manner, the same symbols for the 2 types of variables, the truth tables defining the addition and subtraction processes can be extracted, digit by digit, from the tables of Chapter 2.

9.4.2.1 Addition

— *Truth tables*

Table 9.14 gives the 2 truth tables for s_i and r_{i+1}.

Table 9.14

$x_i y_i r_i$	s_i	r_{i+1}
0 0 0	0	0
0 0 1	1	0
0 1 0	1	0
0 1 1	0	1
1 0 0	1	0
1 0 1	0	1
1 1 0	0	1
1 1 1	1	1

— *Boolean equations for an adder. Some expressions*

Some algebraic expressions representative of the functions s_i and r_{i+1} are now to be given. For simplification the suffix i will be "factored ut" in the following manner:

$$x_i y_i = (xy)_i \quad x_i + y_i = (x+y)_i, \quad \text{etc.}$$

) Canonical forms:

$$\begin{cases} s_i = (\bar{x}\bar{y}r + \bar{x}y\bar{r} + x\bar{y}\bar{r} + xyr)_i \\ r_{i+1} = (\bar{x}yr + x\bar{y}r + xy\bar{r} + xyr)_i \end{cases} \qquad (9.7)$$

) Other expressions:

$_i$ can be expressed:

$$\begin{aligned} s_i &= (\bar{x}\bar{y} + xy)_i r_i + (\bar{x}y + x\bar{y})_i \bar{r}_i \\ &= \overline{(x_i \oplus y_i)} r_i + (x_i \oplus y_i)\bar{r}_i \\ &= \overline{(x_i \oplus y_i)} \oplus r_i \\ &= x_i \oplus y_i \oplus r_i. \end{aligned} \qquad (9.8)$$

Moreover, it can be checked that:

$$r_{i+1} = x_i y_i + y_i r_i + r_i x_i = x_i \# y_i \# r_i. \qquad (9.9)$$

The expressions (9.8) and (9.9) could have been derived from those of 2.4.2. Taking into account the agreed conventions, the formulae (2.33) and (2.37):

$$\begin{cases} s_i = |x_i + y_i + r_i|_2 & (9.10) \\ r_{i+1} = |x_1 y_i + y_i r_i + r_i x_i|_2 & (9.11) \end{cases}$$

become:

$$\begin{cases} s_i = x_i \oplus y_i \oplus r_i \\ r_{i+1} = x_i y_i \oplus y_i r_i \oplus r_i x_i \end{cases} \qquad (9.12)$$

The first formula is that as determined above. The second reduces to the formula (9.9) thanks to the identity (9.1).

– *Half-adders*

The equations for s_i and r_{i+1} can be rewritten in the form:

$$s_i = (x_i \oplus y_i) \oplus r_i \qquad (9.13)$$
$$r_{i+1} = x_i y_i (r_i + \bar{r}_i) + r_i (x_i \bar{y}_i + x_i y_i) \qquad (9.14)$$
$$r_{i+1} = x_i y_i + r_i (x_i \oplus y_i). \qquad (9.15)$$

In consequence, s_i and r_{i+1} can be obtained with the aid of 2 circuits giving the sum $\oplus$ and the $\cdot$, the first giving $x_i \oplus y_i$, $x_i y_i$ and the second giving $(x_i \oplus y_i) \oplus r_i$, $(x_i \oplus y_i) \cdot r_i$. Each of these circuits is known as a half-adder. When connected as in Fig. 9.17 together with a supplementary OR gate to give r_{i+1}, they provide a "full-adder" with the 3 inputs x_i, y_i, r_i.

Other algebraic transformations provide other layouts.

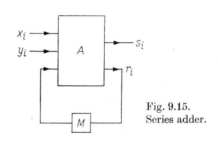

Fig. 9.15. Series adder.

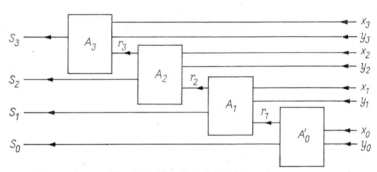

Fig. 9.16. Parallel adder (4-stage).

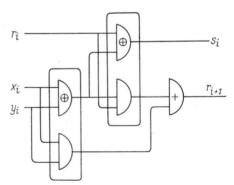

Fig. 9.17. Elementary adder obtained by composing 2 half-adders.

9.4 Examples

— *Explicit expression for the 'carry' r_{i+1}*

The above expressions for s_i and for the carry r_{i+1} are of the recurring type, in the sense that the quantities relating to the rank i are given as functions of those of rank $i-1$. In turn the latter are functions of those of rank $i-2$ etc. In a general manner:

$$\begin{cases} s_i = x_i \oplus y_i \oplus r_i \\ r_i = f(x_{i-1}, y_{i-1}, x_{i-2}, y_{i-2}, \ldots, x_{i-k}, y_{i-k}, \ldots, x_1, y_1, x_0, y_0) \end{cases} \quad (9.16)$$

By the step-by-step application of the carry equation the expression for r_i can be determined.

We have:

$$\begin{aligned} r_i &= x_{i-1}y_{i-1} + r_{i-1}(x_{i-1} \oplus y_{i-1}) \\ &= x_{i-1}y_{i-1} + (x_{i-1} \oplus y_{i-1})[x_{i-2}y_{i-2} + (x_{i-2} \oplus y_{i-2})r_{i-2}] \\ &= x_{i-1}y_{i-1} + x_{i-2}y_{i-2}(x_{i-1} \oplus y_{i-1}) + r_{i-2}(x_{i-1} \oplus y_{i-1})(x_{i-2} \oplus y_{i-2}) \end{aligned} \quad (9.17)$$

Descending to rank 0 (least significant weight) we obtain:

$$r_i = x_0 y_0 \prod_{k=1}^{k=i-1} (x_k \oplus y_k) + x_1 y_1 \prod_{k=2}^{k=i-1} (x_k \oplus y_k) + \cdots$$
$$\cdots + x_j y_j \prod_{k=j+1}^{k=i-1} (x_k \oplus y_k) + \cdots + x_{i-1}y_{i-1} \quad (9.18)$$

$$r_i = \sum_{j=0}^{j=i-1} \left[x_j y_j \prod_{k=j+1}^{k=i-1} (x_k \oplus y_k) \right] \quad (9.19)$$

a formula for which the interpretation is obvious. It is easily proved that the terms $(x_k \oplus y_k)$ can be replaced by $(x_k + y_k)$ without affecting the value of r_i.

A circuit performing the functions s_i and r_{i+1} (9.12) is an *elementary adder*. It can be used either in a *serial adder* or in a *parallel adder*.

Serial adder

The digits of the operands X and Y $(x_i y_i)$ are presented at successive instants i. A memory unit M enables the carry r_{i+1}, developed at the instant i, to be stored until the instant $i+1$ (Fig. 9.15).

Parallel adder

This is a circuit at whose inputs are presented (simultaneously) in parallel the digits x_i, y_i $(i = 0, \ldots, n)$ of the operands X and Y. With $n+1$ digits, it possesses $2(n+1)$ inputs (x_i, y_i) and $n+1$ outputs. In the most conventional version (iterative form) it possesses $n+1$

blocks or 'stages'. 1 stage is known as a half-adder (rank 0) and n stage comprise elementary adders of the type described above (ranks $i > 0$) The carry developed at rank $i(r_{i+1})$ is presented at the stage $i +$ (cf. Fig. 9.16, $n = 3$). It will be noted that the maximum length to be traversed by the signals is $n + 1$ stages, i.e. $2(n + 1)$ levels when these are of 2 levels. For certain additions ($X = 2^{n+1} - 1$, $Y = 1$, fo example) the length of "propagation of the carry" is effectively equal to this value.

Conversely, with gates having an unlimited fan-in, formula (9.18) theoretically supplies an adder for which the number of levels to be traversed until an output becomes constant, and equal to 4 if so desired Thus the use of this type of adder, relative to the preceding type, per mits the propagation of the carry to be accelerated.[1] It should, however be noted that, in practice, it is necessary to modify the expression (9.18) to take account of the finite fan-in, but that the speed advantage is usually maintained. Finally, it must be said that the cost of this type of adde is no longer proportional to n, but increases much more rapidly.

9.4.2.2 Subtraction $(X - Y)$

The following expressions are obtained in a similar manner:

$$\begin{cases} s_i = x_i \oplus y_i \oplus r_i & (9.20) \\ r_{i+1} = \bar{x}_i y_i + y_i r_i + \bar{x}_i r_i = \bar{x}_i \# y_i \# r_i & (9.21) \end{cases}$$

only the second equation differs from that obtained for addition: from which it is derived by the replacement of x_i by $\bar{x}_i$. Other methods o effecting the subtraction (the utilisation of the true and restricted complements introduced in Chapter 3) will be examined later.

9.4.3 Restricted Complement of a Number (Complement "to 1")

The restricted complement of a given number $X = x_{n-1}, x_{n-2} \cdots x_k \cdots x_1 x_0$, is the number $C_1(X)$ represented by $\bar{X}$.

$$C_1(X) = \bar{X} = \bar{x}_{n-1}\bar{x}_{n-2} \cdots \bar{x}_k \cdots \bar{x}_1\bar{x}_0 \qquad (9.22)$$

(X is assumed to be an integer)

---

[1] Whence the name often used, adder with "look-ahead carry". Alongside the formu lations (9.18), (9.19) it is possible to envisage two other expressions (AND-OR at 2 levels, for example). In practice this often leads to intermediary formulae between the iterative forms and those of a number of constant level with respect to n, o type (9.18) or others: for example formulae of the type (9.18) can be applied to blocks of length less than n and, themselves, linked step wise. These formulae correspond to various cost-speed compromises; other methods based on differen principles (conditional sums etc.) are available. (cf. [51], for example).

9.4 Examples

Example:

$$X = 0\ 1\ 0\ 1\ 1\ 0$$
$$C_1(X) = 1\ 0\ 1\ 0\ 0\ 1 \qquad (9.23)$$

then:
$$X + C_1(X) = X + \overline{X} = 2^n - 1.$$

1. In effect, performing the sum digit by digit will give:

$$\begin{cases} s_k = x_k \oplus \overline{x}_k \oplus r_k = r_k \oplus 1 = \overline{r}_k \\ r_{k+1} = x_k \overline{x}_k + r_k(x_k + \overline{x}_k) = r_k \end{cases} \qquad (9.24)$$

Since $r_0 = 0$, we have, for every k ($k = 0, 1, \ldots, n-1$)

$$\begin{cases} s_k = 1 \\ r_k = 0 \end{cases} \qquad (9.25)$$

$$S = \overbrace{1\ 1\ 1\ \cdots\ 1\ \cdots\ 1\ 1}^{n \text{ digits}} = 2^n - 1 \qquad (9.26)$$

2. It can also be said that:

$$C_1(X) = \sum_{k=0}^{k=n-1} (1 - x_k)\, 2^k = \sum_{k=0}^{k=n-1} 2^k - \sum_{k=0}^{k=n-1} x_k 2^k = 2^n - 1 - X$$

$$C_1(X) + X = 2^n - 1 \qquad (9.27)$$

9.4.4 True Complement of a Number

Similarly, for a given integer X the true complement (or complement to 2) is the quantity $C_2(X) = 2^n - X$

Thus:
$$C_2(X) = C_1(X) + 1.$$

With:
$$X = (x_{n-1}, x_{n-2} \cdots x_k \cdots x_1 x_0) \qquad (9.28)$$

we have:
$$C_1(X) = (\overline{x}_{n-1}, \overline{x}_{n-2} \cdots \overline{x}_k \cdots \overline{x}_1 \overline{x}_0). \qquad (9.29)$$

Adding 1 to $C_1(X)$ will be seen to give:

$$C_2(X) = (y_{n-1} y_{n-2} \cdots y_k \cdots y_1 y_0) \qquad (9.30)$$

with:
$$y_k = \overline{x}_k \oplus \prod_{i=0}^{i=k-1} \overline{x}_i = x_k \oplus 1 \oplus \prod_{i=0}^{i=k-1} \overline{x}_i = x_k \oplus \overline{\left(\prod_{i=0}^{i=k-1} x_i \right)} \qquad (9.31)$$

$$y_k = x_k \oplus \left(\sum_{i=0}^{i=k-1} x_i \right) \qquad (9.32)$$

Whence the standard rule: to obtain the true complement for the word X, leave unchanged all zeros to the right of the first 1 encountered in proceeding from the least to the most significant digit, leave this 1 also unchanged and complement all other digits to its left.

Example:
$$\begin{cases} X = 0\ 1\ 1\ 0\ 1\ 0\ 0 \\ C_2(X) = 1\ 0\ 0\ 1\ 1\ 0\ 0 \end{cases} \quad (9.33)$$

Giving:
$$X + C_2(X) = 2^n \equiv 0 \bmod 2^n.$$

This indicates that performing the operations modulo 2^n is the same for effecting $X - Y$ and $X + C_2(Y)$.

By way of an exercise this result will be checked by use of Boolean algebra. Put:
$$C_2(Y) = Y' = (y'_{n-1} y'_{n-2} \cdots y'_k \cdots y'_1 y'_0). \quad (9.34)$$

The operations $X - Y$ and $X + C_2(Y) = X + Y'$ are defined in a recurrent manner by the equations:

$$(X - Y) \begin{cases} s_k = x_k \oplus y_k \oplus r_k & (9.35) \\ r_{k+1} = \bar{x}_k y_k \oplus y_k r_k \oplus \bar{x}_k r_k & (r_0 = 0) \quad (9.36) \end{cases}$$

$$(X + Y') \begin{cases} s'_k = x_k \oplus y'_k \oplus r'_k & (9.37) \\ r'_{k+1} = x_k y'_k \oplus y'_k r'_k \oplus x_k r'_k & (r'_0 = 0) \quad (9.38) \end{cases}$$

Assuming that at rank k;
$$s'_k = s_k \quad (9.39)$$
this would then give:
$$x_k \oplus y_k \oplus r_k = x_k \oplus \left[y_k \oplus \left(\sum_{i=0}^{i=k-1} y_i\right)\right] \oplus r'_k \quad (9.40)$$

i.e.
$$r'_k = r_k \oplus \left(\sum_{i=0}^{i=k-1} y_i\right) \quad (9.41)$$

from which it can immediately be deduced that:
$$r'_{k+1} = x_k \left[y_k \oplus \left(\sum_{i=0}^{i=k-1} y_i\right)\right] \oplus \left[y_k \oplus \left(\sum_{i=0}^{i=k-1} y_i\right)\right] \left[r_k \oplus \left(\sum_{i=0}^{i=k-1} y_i\right)\right]$$
$$\oplus x_k \left[r_k \oplus \left(\sum_{i=0}^{i=k-1} y_i\right)\right] \quad (9.42)$$
$$= x_k y_k \oplus y_k r_k \oplus x_k r_k \oplus \left(\sum_{i=0}^{i=k-1} y_i\right)(y_k \oplus r_k \oplus 1) \quad (9.43)$$

Whence:

$$s'_{k+1} = r_{k+1} \oplus y_k \oplus r_k \oplus \left(\sum_{i=0}^{i=k-1} y_i\right)(r_k \oplus y_k \oplus 1)$$

$$= r_{k+1} \oplus \left[y_k \oplus y_k \left(\sum_{i=0}^{i=k-1} y_i\right) \oplus \left(\sum_{i=0}^{i=k-1} y_i\right)\right] \oplus r_k \left(\prod_{i=0}^{i=k-1} \overline{y}_i\right)$$

$$= r_{k+1} \oplus \left(\sum_{i=0}^{i=k} y_i\right) \oplus r_k \left(\prod_{i=0}^{i=k-1} \overline{y}_i\right) \tag{9.44}$$

Consideration of the last term gives:

$$r_k \left(\prod_{i=0}^{i=k-1} \overline{y}_i\right) = (\overline{x}_{k-1} y_{k-1} \oplus y_{k-1} r_{k-1} \oplus \overline{x}_{k-1} r_{k-1}) \left(\prod_{i=0}^{i=k-1} \overline{y}_i\right)$$

$$= \overline{x}_{k-1} \overline{y}_{k-1} r_{k-1} \prod_{i=0}^{i=k-2} (\overline{y}_i) \tag{9.45}$$

hence:

$$r_k \left(\prod_{i=0}^{i=k-1} \overline{y}_i\right) = \left(\prod_{i=0}^{i=k-1} \overline{x}_i \overline{y}_i\right) r_0 = 0, \text{ since } r_0 = 0 \tag{9.46}$$

hence, step-by-step:

$$r'_{k+1} = r_{k+1} \oplus \left(\sum_{i=0}^{i=k} y_i\right) \tag{9.47}$$

which implies:

$$s'_{k+1} = s_{k+1} \tag{9.48}$$

Thus $s_k' = s_k$ implies $s'_{k+1} = s_{k+1}$. But $s'_0 = s_0$ ($= x_0 \oplus y_0$) and consequently $s'_k = s_k$, for every rank k.

9.4.5 Even Parity Check

This is the name given to the binary digit which is sometimes added to a binary word in order to detect errors in transmission. For a given n-position word X:

$$X = x_n x_{n-1} \cdots x_1 \tag{9.49}$$

word X is transmitted through $n + 1$ positions:

$$X' = x_n x_{n-1} \cdots x_1 P \tag{9.50}$$

where P is such that the total number of 1's in the word X' is even.

Example:

$$\begin{cases} X = 1\,1\,0\,1\,0\,1\,1\,0 \\ X' = 1\,1\,0\,1\,0\,1\,1\,0\,1 \end{cases} \quad \begin{aligned} X &= 1\,0\,0\,1\,0\,1\,1\,0 \\ X' &= 1\,0\,0\,1\,0\,1\,1\,0\,0 \end{aligned} \tag{9.51}$$

It is possible to posit another convention and define an odd parity digit such that the total number of 1's be odd. With this new convention there is no word X' uniquely formed of 0's. This combination is disallowed and its appearance can, dependent on the system in question, serve to indicate certain faults (such as a supply failure which would give 0 throughout).

With the correspondence:

$$\begin{cases} P = 1 \\ P = 0 \end{cases} \begin{vmatrix} \sum_{i=1}^{i=n} x_i & \text{is odd} \\ \sum_{i=1}^{i=n} x_i & \text{is even} \end{vmatrix} \qquad (9.52)$$

which may be expressed as:

$$P = \left| \sum_{i=1}^{i=n} x_i \right|_2$$

where P, x_i are arithmetic variables. With the usual convention, where P and x_i are Boolean variables:

$$P = x_1 \oplus x_2 \oplus \cdots x_i \oplus \cdots x_n = \bigoplus_{i=1}^{i=n} x_i \qquad (9.53)$$

It will be noted that the even parity digit enables all transmission errors to be detected with respect to an odd number of positions.

9.4.6 Addition of 2 Numbers Having even Parity Digits

Consider two binary numbers X and Y having n digits:

$$\begin{cases} X = x_{n-1} x_{n-2} \cdots x_k \cdots x_1 x_0 \\ Y = y_{n-1} y_{n-2} \cdots y_k \cdots y_1 y_0 \end{cases} \qquad (9.54)$$

to which are added the parity digits p_X and p_Y respectively

$$\begin{cases} p_X = \bigoplus_{k=0}^{k=n-1} x_k \\ p_Y = \bigoplus_{k=0}^{k=n-1} y_k \end{cases} \qquad (9.55)$$

Let p_s be the parity digit for the sum $X + Y$. This will be determined by use of the equations giving the sum and the carry s_k and r_{k+1}. Put:

$$S = s_{n-1} s_{n-2} \cdots s_k \cdots s_1 s_0 \qquad (9.56)$$

it will be assumed that S has also n digits, i.e. that the sum $S = X + Y$ does not' overspill'). Thus:

$$p_S = s_{n-1} \oplus s_{n-2} \oplus \ldots \oplus s_k \oplus \ldots \oplus s_1 \oplus s_0$$

$$= \bigoplus \sum_{k=0}^{k=n-1} s_k = \bigoplus \sum_{k=0}^{k=n-1} (x_k \oplus y_k \oplus r_k) \qquad (9.57)$$

$$= \bigoplus \sum_{k=0}^{k=n-1} x_k \oplus \left(\bigoplus \sum_{k=0}^{k=n-1} y_k \right) \oplus \left(\bigoplus \sum_{k=0}^{k=n-1} r_k \right)$$

$$p_S = p_X \oplus p_Y \oplus \left(\bigoplus \sum_{k=0}^{k=n-1} r_k \right) \qquad (9.58)$$

The parity digit p_S of $S = X + Y$ is thus the sum modulo 2 of the parity digits p_X and p_Y of X and Y and of the carries r_k ($k = 0, 1, \ldots, n-1$).

Note: When the digit p'_s determined from the sum effectively obtained and the calculated digit p_s are different, an error is present. Given that all arithmetic operations carried out by a computer are, in the last analysis, reduced to additions, this process can, as was shown by H. Garner, in theory be extended for the verification of any arithmetic operation.

9.4.7 Comparison of 2 Binary Numbers

For a given binary word:

$$X = x_{n-1} x_{n-2} \cdots x_k \cdots x_1 x_0 \qquad (9.59)$$

$X(k)$ will denote the truncated word formed from the $k + 1$ digits of least significant weight (ranks 0 to k).

$$X(k) = x_k x_{k-1} \cdots x_1 x_0, \qquad X(n-1) = X \qquad (9.60)$$

Next consider 2 numbers X and Y ($X, Y \geq 0$) and the binary variable defined by the correspondence:

S_k	$X(k) - Y(k)$
0	≤ 0
1	> 0

(9.61)

Under these conditions, it will be seen that if

$$x_k = y_k \tag{9.62}$$

then

$$S_k = S_{k-1}. \tag{9.63}$$

Conversely, if $x_k \neq y_k$, the inequalities between x_k and y_k on the one hand, and $X(k)$ and $Y(k)$ on the other are of the same sense. This situation is summarized in Table 9.18:

Table 9.18

S_{k-1}	x_k	y_k	S_k
0	0	0	0
0	0	1	0
0	1	0	1
0	1	1	0
1	0	0	1
1	0	1	0
1	1	0	1
1	1	1	1

Thus

$$S_k = \bar{S}_{k-1} x_k \bar{y}_k + S_{k-1} (\bar{x}_k \bar{y}_k + x_k \bar{y}_k + x_k y_k)$$
$$= (S_{k-1} + \overline{S_{k-1}}) x_k \bar{y}_k + (x_k y_k + \bar{x}_k \bar{y}_k) S_{k-1} \tag{9.64}$$
$$= x_k \bar{y}_k + \overline{(x_k \oplus y_k)} S_{k-1}. \tag{9.65}$$

and again:

$$S_k = x_k \bar{y}_k + x_k y_k S_{k-1} + \bar{x}_k \bar{y}_k S_{k-1}$$
$$= (x_k \bar{y}_k + x_k y_k S_{k-1}) + (x_k \bar{y}_k + \bar{x}_k \bar{y}_k S_{k-1})$$
$$= x_k \bar{y}_k + x_k S_{k-1} + x_k \bar{y}_k + \bar{y}_k S_{k-1} \tag{9.66}$$
$$S_k = x_k \bar{y}_k + x_k S_{k-1} + \bar{y}_k S_{k-1}. \tag{9.67}$$

The expression for the carry of the operation $Y - X$ will be recognised in (9.67). It will also be seen that S_{n-1} will equal 1 when the operation $Y - X$ produces a carry of rank $n - 1$, i.e. as might have been expected when $X > Y$; it will also be noted that in order to compare 2 numbers it is not necessary to determine the difference: the last carry of this difference (S_{n-1}) suffices. Next consider the variable E_k defined by:

E_k	$X(k) = Y(k)$
0	No
1	Yes

(9.68)

Evidently:
$$E_k = \overline{(x_k \oplus y_k)} \, E_{k-1} = (x_k \leftrightarrow y_k) \, E_{k-1} \tag{9.69}$$

hence:
$$E_k = \prod_{i=0}^{i=k} (x_i \leftrightarrow y_i) \tag{9.70}$$

Consider next the variable S'_k such that:

S'_k	$X(k) - Y(k)$
0	< 0
1	≥ 0

(9.71)

then:
$$S'_k = S_k + E_k = x_k \bar{y}_k + \overline{(x_k \oplus y_k)} \, S_{k-1} + \overline{(x_k \oplus y_k)} \, E_{k-1} \tag{9.72}$$
$$= x_k \bar{y}_k + \overline{(x_k \oplus y_k)} \, S'_{k-1} \tag{9.73}$$

Finally putting:
$$I_k = \overline{S'_k} \tag{9.74}$$

it follows that:
$$I_k = \overline{S'_k} = (\bar{x}_k + y_k)\left[(x_k \oplus y_k) + \overline{S'_{k-1}}\right]$$
$$= \bar{x}_k y_k + \bar{x}_k \overline{S'_{k-1}} + \bar{x}_k y_k + y_k \overline{S'_{k-1}}$$
$$= \bar{x}_k I_{k-1} + \bar{x}_k y_k + y_k I_{k-1}$$
$$= \bar{x}_k I_{k-1} \oplus \bar{x}_k y_k \oplus y_k I_{k-1} \tag{9.75}$$

(carry of a subtraction)
$$I_k = \bar{x}_k y_k \oplus (\bar{x}_k \oplus y_k) I_{k-1} = \bar{x}_k y_k \oplus \overline{(x_k \oplus y_k)} \, I_{k-1} \tag{9.76}$$

an expression whose interpretation is obvious, and which can also be determined directly, as for S_k or S'_k.

9.4.8 V-Scan Encoder Errors

9.4.8.1 Error Pattern

Let X and Y be 2 binary words:
$$\begin{cases} X = (x_n x_{n-1} \cdots x_k \cdots x_0) \\ Y = (y_n y_{n-1} \cdots y_k \cdots y_0) \end{cases} \tag{9.77}$$

for which a comparison is required. They could be considered as being binary numbers and their difference could be taken: dependent on whether it be null or different from O, X and Y are, or are not, identical words. X and Y could also be considered as 2 (binary) vectors whose vector difference is defined by:

$$E = (e_n e_{n-1} \cdots e_k \cdots e_1 e_0) \tag{9.78}$$

with

$$e_k = x_k \oplus y_k \quad (k = 0, 1, \ldots, n). \tag{9.79}$$

The vector E represents the vector error (digit by digit) conferred on X when it is replaced by Y and vice versa. When there is no ambiguity the "error pattern" will be known simply as the error.

We have the relations:

$$\begin{cases} E = X \oplus Y \\ Y = X \oplus E \\ X = Y \oplus E \end{cases} \tag{9.80}$$

9.4.8.2 Principle of the V-Scan Encoder

Each track of the encoder with the exception of the outer track carries 2 brushes B and B' ("leading" and "trailing", respectively) (Fig. 9.20). The spacing between the 2 brushes on a single track is equal to half the length of the corresponding sector. The brushes B and B' are

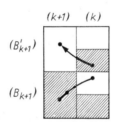

Fig. 9.19 Choice of a brush on a track — principle.

symmetric with respect to the radius joining the centre of the disc to the brush (unique) of the track for the least significant digit. The binary number corresponding to this radius, and which is to be read, will be designated by x. It is represented by the vector

$$X = (x_n x_{n-1} \cdots x_k \cdots x_0). \tag{9.81}$$

The method of V-decoding is as follows: if the value 0 has been read out at rank k, the (leading) brush B_{k+1} is used to read-out the next higher rank.

9.4 Examples

If the value 1 has been read out at rank k, the "trailing" brush B'_{k+1} is used (Figs. 9.19 and 9.20).

This method corresponds to the recurrence relation:

$$x^*_{k+1} = \bar{x}_k^* B_{k+1} + x_k^* B'_{k+1} \qquad (9.82)$$

x_k^* being the digit read for x at rank k).

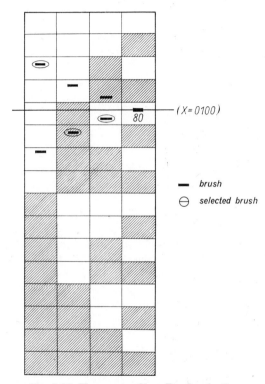

Fig. 9.20. V-scan encoding. Brush selection.

4.8.3 Relation Between the Value Available at a Brush and the Digits of the Number x to be Read Out

Leading brush B_k

here are 2^{n+1} quanta zones. Of these, the sector corresponding to rank ($k \geq 1$) comprises

$$\frac{2^{n+1}}{2^{n-k+1}} = 2^k$$

9. Applications and Examples

B_k is thus the digit of rank k of the number $x + 2^k/4 = x + 2^{k-2}$:

$$\left.\begin{aligned}X &= x_n \cdots x_k x_{k-1} x_{k-2} \cdots x_2 x_1 x_0, x_{-1} \cdots \\ 2^{k-2} &= 0 \cdots 0\ 0\ 1\ \cdots 0\ 0\ 0,\ 0\ \cdots \\ \overline{X + 2^{k-2}} &= s_{n+1} s_n \cdots s_k s_{k-1} s_{k-2} \cdots s_2 s_1 s_0,\ s_{-1}\end{aligned}\right\} \quad (9.8)$$

giving:

$$s_k = x_k \oplus x_{k-1} x_{k-2} \qquad (k \geq 1) \quad (9.8)$$

$$B_k = x_k \oplus x_{k-1} x_{k-2}$$

Trailing brush B_k'

Similarly, subtraction gives,

$$B'_k = x_k \oplus \bar{x}_{k-1} \bar{x}_{k-2} \quad (k \geq 1) \quad (9.8)$$

It is verified that this gives x correctly. If $x_k^* = x_k$,

$$x^*_{k+1} = \bar{x}_k (x_{k+1} \oplus x_k x_{k-1}) + x_k (x_{k+1} \oplus \bar{x}_k \bar{x}_{k-1}) \quad (9.8)$$
$$= \bar{x}_k x_{k+1} + x_k x_{k+1} = x_{k+1} \quad (9.8)$$

9.4.8.4 Error Caused by the Failure of a Single Brush

Problem: The number read-out from this type of encoder results from a sequential analysis of the brush system. By reason of the recurrent character of this analysis, an error, even a single error, in the read-out from a brush can produce errors in several other digits of superior rank.

Assume that the brush at rank p gives an erroneous reading $\bar{X}$ instead of X_p. Let X be the correct word, Y the word actually read. Then:

$$\begin{cases} X_{i+1} = B_{i+1} \bar{X}_i + B'_{i+1} X_i \\ Y_{i+1} = B_{i+1} \bar{Y}_i + B'_{i+1} Y_i \end{cases} \quad \text{(V-scan encoding)} \quad (9.8)$$

with:

$$i < p : Y_i = X_i \quad (9.8)$$
$$i = p : Y_i = \bar{X}_i$$

Y_i is required for $i > p$. We have

$$X_{i+1} \oplus Y_{i+1} = X_{i+1} \bar{Y}_{i+1} + \bar{X}_{i+1} Y_{i+1} \quad (9.9)$$

Moreover:

$$\bar{X}_{i+1} = \overline{B_{i+1} \bar{X}_i + B'_{i+1} X_i} = (\bar{B}_{i+1} + X_i)(\bar{B}'_{i+1} + \bar{X}_i)$$
$$= \bar{B}_{i+1} \bar{B}'_{i+1} + X_i \bar{B}'_{i+1} + \bar{X}_i \bar{B}_{i+1} \quad (9.9)$$

Similarly:
$$\overline{Y}_{i+1} = \overline{B}_{i+1}\overline{B}'_{i+1} + Y_i\overline{B}'_{i+1} + \overline{Y}_i\overline{B}_{i+1} \qquad (9.92)$$
$$X_{i+1} \oplus Y_{i+1} = \overline{X}_i Y_i B_{i+1} \overline{B}'_{i+1} + X_i \overline{Y}_i \overline{B}_{i+1} B'_{i+1}$$
$$\qquad + X_i \overline{Y}_i B_{i+1} \overline{B}'_{i+1} + \overline{X}_i Y_i \overline{B}_{i+1} B'_{i+1} \qquad (9.93)$$
$$X_{i+1} \oplus Y_{i+1} = (X_i \oplus Y_i)(B_{i+1} \oplus B'_{i+1}) \qquad (9.94)$$

and, consequently:
$$X_{i+j} \oplus Y_{i+j} = (X_i \oplus Y_i) \prod_{k=1}^{k=j} (B_{i+k} \oplus B'_{i+k}) \qquad (9.95)$$

The relations (9.94) and (9.95) translate the following phenomenon:

The error in the read-out at rank i produces an error in the following rank only if the 2 brushes are in contact with different sectors. The read-out error thus propagates so long as the brushes are in contact with different sectors. The errors cease as soon as the brushes make contact with a single sector.

The resultant total error in X, can be related to the value of X. We have:
$$B_{i+1} \oplus B'_{i+1} = (X_{i+1} \oplus X_i X_{i-1}) \oplus (X_{i+1} \oplus \overline{X}_i \overline{X}_{i-1})$$
$$= X_i X_{i-1} + \overline{X}_i \overline{X}_{i-1} = \overline{X_i \oplus X_{i-1}} \qquad (9.96)$$
$$B_{i+1} \oplus B'_{i+1} = \overline{X_i \oplus X_{i-1}} \qquad (9.97)$$

whence:
$$X_{i+j} \oplus Y_{i+j} = (X_i \oplus Y_i) \prod_{k=1}^{k=j} \overline{(X_{i+k-1} \oplus X_{i+k-2})} \qquad (9.98)$$

9.4.8.5 The Length of an Error Burst

1. If at rank $i+j$, the error
$$e_{i+j} = X_{i+j} \oplus Y_{i+j}$$
is null, it is also null for all higher ranks.

2. In order that the error may propagate beyond the brush i for which the read-out is defective, it is necessary that:
$$B_{i+1} \oplus B'_{i+1} = 1 \qquad (9.99)$$
i.e.
$$X_i \oplus X_{i-1} = 0 \qquad (9.100)$$

whence:
— if $X_i \oplus X_{i-1} = 1$, i.e. that the digit X_i is the first of a block of digits of the same nature as the number x to be read-out, for every $k > 0$, $e_{i+k} = 0$: the only erroneous digit in the read-out is the digit of rank i;

— if $X_i \oplus X_{i-1} = 0$, i.e. if the rank i is to be found in the interior of a block of digits of the same nature, $B_{i+1} \oplus B'_{i+1} = 1$.

Then the error propagates beyond the first erroneous digit, to a rank $i+j-1$ (inclusive) such that:

$$\begin{cases} X_{i+j-1} \oplus X_{i+j-2} = 1 & (9.101) \\ X_{i+j-2} \oplus X_{i+j-3} = 0 & (9.102) \end{cases}$$

If the initial error of rank i is not produced at the commencement of a block of digits of the same nature, then the length of the error is equal to the number of digits of the same nature commencing at rank i (inclusive) augmented by 1. Otherwise the error is limited to rank i.

9.4.8.6 Numerical Value of the Error: $e = y - x$ (x, y, e being the values of the numbers represented by X, Y, E)

First case: $X_i \oplus X_{i-1} = 0$. (9.103)

Determination of $y - x$

If j is the smallest integer such that $e_{i+j} = 0$ then:

$$X_{i+k-1} \oplus X_{i+k-2} = \begin{cases} 0 & \text{for } 2 \leq k < j \\ 1 & \text{for } k = j \end{cases} \quad (9.104)$$

which gives a situation of the following type:

$$\begin{cases} X = X_n X_{n-1} \cdots X_{i+j}\, 0\, 1\, 1 \cdots 1\, 1\ X_{i-1} \cdots X_0 \\ Y = X_n Y_{n-1} \cdots Y_{i+j}\, 1\, 0\, 0 \cdots 0\, 0\ X_{i-1} \cdots X_0 \end{cases} \quad (9.105)$$

$$E = 0\ \ 0\ \ \cdots 0\ \ \underbrace{1\, 1\, 1 \cdots 1\, 1}_{\text{erroneous block}}\, 0\ \ \cdots 0$$

$$y - x = \left(2^{i+j-1} - \sum_{k=i}^{k=i+j-2} 2^k\right) = 2^i \quad (9.106)$$

The other case (block of 0's) would give:

$$y - x = -2^i \quad (9.107)$$

Thus:

$$\begin{cases} X_i = 1 & y - x = 2^i \\ X_i = 0 & y - x = -2^i \end{cases} \quad (9.108)$$

which may be written:

$$y - x = (-1)^{|X_i|+1}\, 2^i \quad (9.109)$$

9.4 Examples

Second case:
$$X_i \oplus X_{i-1} = 1 \tag{9.110}$$

If $X_i = 1$, then $Y_i = 0$, whence
$$x - y = 2^i$$
$$y - x = -2^i \tag{9.111}$$

In the contrary case:
$$y - x = 2^i \tag{9.112}$$

$y - x$ is expressed in every case by means of the unique formula:
$$e = y - x = (-1)^{|x_i| + |x_i \oplus x_{i-1}|} \cdot 2^i$$
$$= (-1)^{|x_{i-1}|} 2^i = (-1)^{(1-x_{i-1})} \cdot 2^i \tag{9.113}$$

whence:

The absolute value of the error is always equal to the value of sector of the track with which the defective brush is in contact. The sign of the error is given by (9.113).

By way of an exercise the reader will be able to examine the case for multiple failures, first by the use of reasoning, then by means of Boolean algebra, and to establish the formula which gives the error pattern.

9.4.8.7 Examples

An arrow indicates the rank i at which the error commences (variable t)

$$
\begin{array}{c|l}
X & \quad\;\; — X_{i+j}\;0\;1\;1\;1\;1\;1\;0\; — \\
B \oplus B' & \quad\;\; — \quad\;\; 0\;\;\;1\;1\;1\;1\;0 \quad — \\
t & \quad\;\; 0\;\;\;0\;\;\;0\;0\;0\;0\;1\;0\;0\; — \\
y & \quad\;\; — X_{i+j}\;1\;0\;0\;0\;0\;1\;0\; — \\
e & \quad\;\; 0\;\;\;0\;\;\;1\;1\;1\;1\;1\;0\;0\; —
\end{array} \tag{9.114}
$$

$$
\begin{array}{c|l}
X & \quad — X_{i+j}\;0\;1\;1\;1\;1\;1\;0 \\
t & \quad 0\;\;\;0\;\;\;0\;0\;0\;0\;0\;1\;0 \\
e & \quad 0\;\;\;0\;\;\;0\;0\;0\;0\;0\;1\;0
\end{array} \tag{9.115}
$$

$$
\begin{array}{c|l}
X & \quad — X_{i+j}\;0\;1\;1\;1\;1\;1\;0 \\
t & \quad 0\;\;\;0\;\;\;0\;1\;0\;0\;0\;0\;0 \\
e & \quad 0\;\;\;0\;\;\;1\;1\;0\;0\;0\;0\;0
\end{array} \tag{9.116}
$$

9.4.9 Errors of a Reflected Binary Code Encoder

The read-out from the encoder gives a word:

$$R = R_n R_{n-1} \cdots R_k \cdots R_0 \qquad (9.117)$$

which is the representation of the number N in the Gray code, for which the pure binary representation is:

$$B = B_n B_{n-1} \cdots B_k \cdots B_0 \qquad (9.118)$$

Assuming the read-out R to be affected by an error and, instead of R supplies a vector R'. R' can be expressed as:

$$R' = R \oplus r, \quad r = (r_n r_{n-1} \cdots r_k \cdots r_0) \qquad (9.119)$$

Furthermore, assuming that R and R' be converted to pure binary:

$$\begin{cases} R \to B \\ R' \to B' \end{cases} \qquad (9.120)$$

Put

$$B' = B \oplus b, \quad b = (b_n b_{n-1} \cdots b_k \cdots b_0) \qquad (9.121)$$

Relation between b and r:
We have:

$$B_k = \bigoplus \sum_{i=n}^{i=k} R_i \qquad (9.122)$$

$$B_k' = \bigoplus \sum_{i=n}^{i=k} R_i'$$

$$= \bigoplus \sum_{i=n}^{i=k} (R_i \oplus r_i) = \bigoplus \sum_{i=n}^{i=k} R_i \oplus \sum_{i=n}^{i=k} r_i \qquad (9.123)$$

Whence:

$$B'_k = B_k \oplus \sum_{i=n}^{i=k} r_i \qquad (9.124)$$

Thus it follows that:

$$b_k = \bigoplus \sum_{i=n}^{i=k} r_i \qquad (9.125)$$

i.e.:

The error pattern on the binary equivalents is the binary equivalent of the error pattern on the representations in reflected code.

Examples:

$$\begin{array}{ll} R = 0\ 1\ 1\ 0 \to & B = 0\ 1\ 0\ 0 \\ R' = 1\ 0\ 1\ 0 \to & R' = 1\ 1\ 0\ 0 \\ \hline r = 1\ 1\ 0\ 0 & b = 1\ 0\ 0\ 0 \end{array}$$

9.4.10 Detector of Illicit Combinations in the Excess-3 Code

Consider the binary coded decimal (BCD) excess-3 code (9.126)

$$
\begin{array}{r|l}
 & 0\ 0\ 0\ 0 \\
 & 0\ 0\ 0\ 1 \\
 & 0\ 0\ 1\ 0 \\
\hline
0 & 0\ 0\ 1\ 1 \\
1 & 0\ 1\ 0\ 0 \\
2 & 0\ 1\ 0\ 1 \\
3 & 0\ 1\ 1\ 0 \\
4 & 0\ 1\ 1\ 1 \\
5 & 1\ 0\ 0\ 0 \\
6 & 1\ 0\ 0\ 1 \\
7 & 1\ 0\ 1\ 0 \\
8 & 1\ 0\ 1\ 1 \\
9 & 1\ 1\ 0\ 0 \\
\hline
 & 1\ 1\ 0\ 1 \\
 & 1\ 1\ 1\ 0 \\
 & 1\ 1\ 1\ 1 \\
\end{array}
\qquad (9.126)
$$

The 10 central binary combinations represent the words for the digits 0 to 9. Of the 16 combinations of 4 digits 6 are unused. It is required to determine a 4-input/1-output combinatiorial circuit, for

Table 9.21

A B C D	$f(A, B, C, D)$
0 0 0 0	1
0 0 0 1	1
0 0 1 0	1
0 0 1 1	0
0 1 0 0	0
0 1 0 1	0
0 1 1 0	0
0 1 1 1	0
1 0 0 0	0
1 0 0 1	0
1 0 1 0	0
1 0 1 1	0
1 1 0 0	0
1 1 0 1	1
1 1 1 0	1
1 1 1 1	1

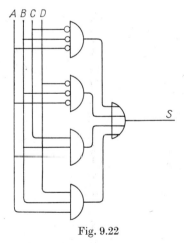

Fig. 9.22

which the output will have the value 1 when one of the 6 unused combinations is presented at the input; the 4 digits which comprise the word to be examined are presented in parallel.

This provides the truth table (Table 9.21).

The canonical form:

$$f(A,B,C,D) = \bar{A}\bar{B}\bar{C}\bar{D} + \bar{A}\bar{B}\bar{C}D + \bar{A}\bar{B}C\bar{D} + AB\bar{C}D$$
$$+ ABC\bar{D} + ABCD \tag{9.127}$$

which, after simplification can be expressed in the form:

$$f(A, B, C, D) = \bar{A}\bar{B}\bar{C} + \bar{A}\bar{B}\bar{D} + ABC + ABD \tag{9.128}$$

Which, with the AND, OR gates and inverters produces the layout of Fig. 9.22.

Chapter 10

The Simplification of Combinational Networks

10.1 General

One of the advantages of Boolean algebra is the ease with which diagrams for networks corresponding to a given function can be established. Once in possession of an algebraic expression for the function in question the functional diagram is easily produced.

Thus the three variable function

$$f = \sum (1, 2, 3, 4, 5)$$

takes the canonical form

$$f(x, y, z) = \bar{x}\bar{y}z + \bar{x}yz + x\bar{y}\bar{z} + x\bar{y}z + \bar{x}\bar{y}z,$$

from which the layout of Fig. 10.1 is immediately obtained. The network represented consists of the gates AND, OR and possesses two logic stages (disregarding the inversions).

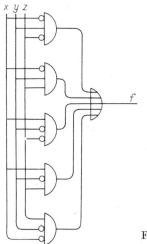

Fig. 10.1

From this point of view it will be seen that Boolean algebra provides a particularly convenient notation with which to represent a network. But, as is already known, a Boolean function may be represented by a variety of algebraic expressions (of which there is an infinity), leading to as many different logic networks. Thus the above function can still be expressed by the expression $x\bar{y} + \bar{x}y + \bar{x}z$. It will be conceded that the network represented by the second expression has a greater chance of being more 'economical' than the first. This represents an extremely general condition and, in practice, amongst all the expressions which represent a given function, it is convenient to look for that expression which minimizes a parameter associated with the network. Even more generally, in the case of the multi-output network, it is equally the search for a minimum network which will constitute the most important step in the synthesis. The parameter to be minimized will, in general, be known as the *'cost' function*, or the *'cost'* of the system.

10.2 Simplification Criterion. Cost Function

How is the criterion to be used to minimise or, at least, to simplify the network (i.e. the cost function) to be defined?

Evidently its definition will have a technological origin; by way of example consideration must be given to financial costs, weight, power dissipation, variables which, moreover, are not independent.

In fact the definition of a function, like that of cost, is extremely complex and, to the extent that the network in question be integrated into an even larger system, it generally affects the cost definition. Meanwhile the cost concept of a network remains a useful working tool. The above three criteria lead to a consideration of the network as being composed of individual elements $(E_1, \ldots, E_n)$, (transistors, diodes, resistors, capacitors or, at higher levels of complexity, gates, amplifiers, flip-flops or even modules or cards). The cost f of the network often takes the form

$$f = \sum_1^n f_i n_i,$$

where the coefficients are the individual costs of the elements retained for the evaluation of the cost, and n_i the number of elements of type E_i. It should be noted that, to a certain extent, f_i can include the cost of wiring and inter-element connections (multi-layer cards, connections between cards).

Examples:

1. Network of Fig. 10.1.

The constituent parts retained are the gates and the inverter, the cost is that for 5 gates AND, 1 gate OR and for 8 inverters.

$$f_1 = 5f_{\text{AND}} + f_{\text{OR}} + 8f_{\text{INV}} \tag{10.1}$$

10.2 Simplification Criterion. Cost Function

2. Network of Fig. 10.2.

Under the same conditions the cost is:

$$f_2 = 3f_{AND} + f_{OR} + 3f_{INV} \qquad (10.2)$$

Particular cases:

a) The cost is the total number of gates or inverters.

$$(f_{AND} = f_{OR} = f_{INV} = 1).$$

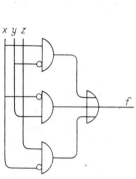

Fig. 10.2

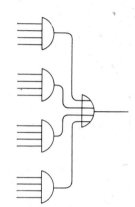

Fig. 10.3. Example of functional block.

Giving, for the above examples:

$$f_1 = 14, \qquad f_2 = 7 \qquad (10.3)$$

b) The cost of the total number of gates ($f_{INV} = 0$). In this case:

$$f_1 = 6, \qquad f_2 = 4$$

3. The constituant element is a functional block, constructed once and for all, which the circuit designer must consider as being indivisible (Fig. 10.3). The implementation corresponding to the canonical form would be particularly expensive to achieve (3 blocks) compared with that given for the simplified form (1 block) (Fig. 10.4). In general, it will be noted that to minimise the number of *blocks* leads to the use of algebraic expressions which differ from those associated with the minimum number of *gates*.

4. Another cost function: the number of symbols (direct or complemented variables).

Example:

$$f = x\bar{y} + \bar{x}y + \bar{x}z$$

Cost of f: 6.

10. The Simplification of Combinational Networks

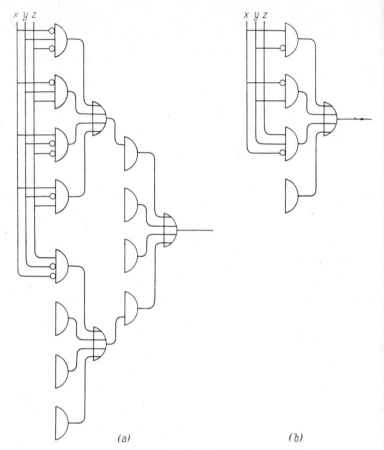

Fig. 10.4.

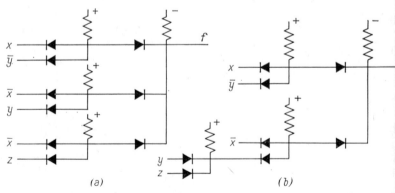

Fig. 10.5. 2 Practical Applications (2 and 3 levels) of Diode Logic.

5. The logic is a discrete diode logic and the cost function is equal to the number of diodes. For a two-level logic circuit of gates AND and OR, the sum of the products is the number of symbols (barred letters, non-barred letters) appearing in the monomials of more than one symbol, augmented by the number of monomials, if this is greater than 1. The circuits associated with the expressions $x\bar{y} + \bar{x}y + \bar{x}z$ and $x\bar{y} + \bar{x}(y+z)$ have the same number of gates, but do not require the same number of diodes (Fig. 10.5a, b). The first (2 levels) comprises 9, the second (3 levels) 8.

From the above examples are retained:
- the importance of the definition of cost,
- the role played by the number of logic levels. This number is a minimum of 2 for no matter what function (disregarding the inversions). Most often, considerations of speed and system clock frequency also impose an upper limit on the number of levels of the networks. (In particular, it is necessary that the propagation time across the network be less than the clock-time period of the system.) Between these two limits, the cost of the network depends on the number of levels. Under these conditions, with an arbitrary, or even a simple linear cost function (of the type described above) the problem of minimisation remains extremely complex, and is only partially resolved, as will later be seen. The reason for this is easily seen: the chosen cost criterion is of technological origin and, contrary to the majority of purely mathematical definitions, has little chance of generating notable mathematical results. Finally, the difficulty of this problem is worsened by the extreme variety of logic circuits and functional blocks supplied by the components industry. At best, a resonable simplification of the circuits will be proposed. This said, several methods are possible.

10.3 General Methods for Simplification

10.3.1 Algebraic Method

This consists in transforming the Boolean expressions describing the system by means of the diverse axioms and theorems of Boolean algebra in an effort to simplify the circuit (with respect to a given criterion). This method can be used for any number of outputs, for an arbitrary number of logic levels, and for logic elements of absolutely any type (AND, NOR, majority, etc.) It does, however, have the following disadvantages:
- in general, beyond 3 or 4 variables, the number of terms to be mani-

pulated is likely to grow very rapidly and the expressions then become very complicated;
— the method relies largely on intuition and 'know-how';
— in general, it is very difficult to demonstrate that a minimum expression has been obtained and that it is un-necessary to continue

10.3.2 Karnaugh Diagram Method

Basically this consists in replacing algebraic calculations on the functions by manipulation on sets. It makes use of the aptitude of the human brain to identify certain structures or patterns. It is applicable to networks possessing 1 or more outputs, two or more levels, and easily visualised functions. However, it is convenient only for a number of variables not exceeding 5, or, at the very most for $n = 6$. Beyond this the geometric aspect and intuitive character fade progressively in face of the systematic procedure which must be employed and, inevitably this leads to the use of algorithmic methods (3rd type of method).

10.3.3 Quine-McCluskey Algorithm

This method is chiefly based on the works of *Quine, McCluskey* and *Roth*. It can be used for networks having 1 or more outputs and any number of variables. It can be programmed on a computer. It provides results in the form of two-level expressions (summations of products or products of sums).

10.3.4 Functional Decomposition Method (Ashenhurst, Curtis, Povarov

This method seeks to represent a Boolean function as a function of function, by repeating this operation as many times as possible. In this way are obtained circuits of more than two levels.

Summarising, the following strategy will be used in practice:

1. For 2-variable problems, or problems with more than two variables but comprising few terms in canonical form: method 1 will be used (algebraic manipulation).
2. For problems with 3, 4 or 5 variables: the Karnaugh diagram method will be used.
3. For problems with n variables $(n \geq 6)$. For $n = 6$ either the Karnaugh diagram, or the Quine-McCluskey algorithm, or the functional decomposition method will be used. For $n \geq 6$ only the last two methods will be used.

In the case where the Quine-McCluskey method is used an expression which is a sum of products is obtained. From this (but without any guarantee of minimality), algebraic manipulations can be made in an attempt further to simplify the function (by factorisation).

Chapter 8 described the Karnaugh diagram method. The Quine-McCluskey method will now be described, This will be considered in two parts — circuits with one output and those with several outputs.

10.4 The Quine-McCluskey Algorithm

10.4.1 Single-Output Networks

10.4.1.1 Statement of the Problem

The method to be given below allows for the minimisation of the sums of products or the products of sums: it will explain the case for the sums of products and, finally, will indicate how this is used for the products of sums.

It will be assumed, therefore, that the function is provided by its truth table or by its disjunctive canonical form. Next it is proposed to find an algebraic form for the function which will be both a sum of products and give the minimum form for a certain cost function. It will be assumed that the cost is an increasing function of the number of symbols appearing in the expression. By way of an exercise, the reader could examine the (less restrictive) case with a simple non-decreasing function.

10.4.1.2 Geometrical Interpretation of the Minimisation Problem

As has already been done by various authors (*Urbano, Muller, Roth*) the problem of minimisation will be interpreted in geometrical form. Taking the function which, for the time being, is assumed from its characteristic set to be completely specified and which, it must be remembered, is a set of vertices of an n-dimensional cube. As has been indicated, this geometrical interpretation will necessarily preserve an abstract character, by reason of the impossibility conveniently to represent a n-dimensional cube. However, the theory which follows will be illustrated by particular representations which constitute Karnaugh diagrams.

10.4.1.3 'n'-Dimensional Cubes and Sub-Cubes

An n-dimensional cube possesses sub-cubes of lesser dimensions. For example, the 3-dimensional cube "cube" possesses faces which are "cubes" of 2 dimensions and 4 vertices, edges: which are cubes of dimension 1 with 2 vertices and, finally, vertices which may be considered to be cubes of zero dimension.

It must be remembered (chap. 8) that:

1. For a cube of dimension 0 (vertex) all the coordinates are fixed.
2. For a cube of dimension 1 (edge), the different points of the cube are obtained by causing *one* coordinate to vary, to which the two possible values (0 and 1) are ascribed.
3. For a cube of dimension 2 (square) the different points are obtained by causing *two* coordinates to vary.

In a general manner a sub-cube having k dimensions (and 2^k vertices) is obtained by allocating to k among the coordinates the values 0 and 1. A k-dimensional sub-cube can then be represented by an n-position combination of which $n - k$ positions are of value 0 or 1, and k being represented by a dash (—). In replacing the k dashes by all the 2^k possible combinations of 0 and 1, are obtained the 2^k vertices of the sub-cube in question.

For a given set E of the vertices of a cube, consideration can be given to all the cubes which can be formed with the aid of these points. A cube of dimension k and contained in E will be known as a k-cube. The following operations can then be defined:

Symmetrisation

With two 0-cubes (vertices) as (0010) and (0110) can be formed a 1-cube (0—10).

With two 1-cubes such as (0—10) and (1—10) can be formed one 2-cube: (——10) etc. More generally, given two k-cubes which possess the following properties:

— dashes (—) situated in the same positions,
— all other positions ("numerical" positions of value 0 or 1) identical with one exception,

a $(k + 1)$-cube is formed by putting a dash in the (unique) position which is 0 for one of the cubes and 1 for the other.

From a given set E, one can form the sets:

$$E^0, E^1, E^2, \ldots, E^k, \ldots, E^n,$$

where E^k is the set of k-cubes. Moreover, certain of these E^k may, beyond a certain rank, be empty. This results in a set of cubes:

$$\mathscr{E} = E^0 \cup E^1 \cup E^2 \cup \cdots \cup E^k \cup \cdots \cup E^n \qquad (10.4)$$

10.4 The Quine-McCluskey Algorithm

Equipped with the operation defined above and, with the converse operation which consists of separating a cube into 2 symmetrical faces and of allocating to a dash the values 0 and 1, the set $\mathscr{E}$ constitutes what is known as a 'complex' associated with the function f_E.

10.4.1.4 Maximal k-Cubes

A k-cube M possessing the following two properties is known as a maximal k-cube, or cocycle:

. It belongs to the characteristic set of f:

$$M \subset E_f.$$

. For any cube M' of greater dimension and containing M there exist points of M' which do not belong to E_f.

In a more visual way it can be said that a maximal k-cube is a -cube which is not a face of a $(k+1)$ cube.

Note: In logic the corresponding algebraic terms (products) are given the names *prime terms* or *prime implicants*.

10.4.1.5 Statement of the Problem of Minimisation

For a given set E of points defining the function, it is required to find a cover (R) for these points by means of k-cubes, such that the cost function $C(R)$ is at a minimum.

It is assumed that this is a function increasing with the number of symbols. (Particular case:

$$f = \sum_{k=0}^{k=n} a_k(n-k),$$

where a_k is the number of k-cubes of R; the cost is then the number of symbols, considering the direct and complemented variables as distinct).

10.4.1.6 Properties of the Minimal Cover

We have the following fundamental property:
every minimal cover R is necessarily formed of maximal k-cubes.

Unless this were true there would then exist at least one k-cube M belonging to R which would not be a maximal k-cube. From this, therefore, could be formed a cube M' of higher order which would contain M. It would then be possible to replace M by M' in the cover and the cost would be reduced since:

$$C(R') = C(R) + [C(M') - C(M)] < C(R) \qquad (10.5)$$

and, consequently, R would not be a minimum cost cover.

10.4.1.7 Essential Maximal k-Cubes

A maximal k-cube is said to be essential if it contains a vertex which is not included in any other maximal k-cube.

In these conditions it will be seen that every minimal cover necessarily comprises *all* the essential maximal k-cubes. Indeed, such a cover can only be formed from maximal k-cubes. If an essential k-cube had not been included, the vertex, only contained therein, would have been lost to the cover.

Note: It may happen that the set of essential k-cubes would suffice to cover the characteristic set E of the function. This, however, is not the general situation.

It can now be seen that the minimisation of a function of the form of sum of products (i.e. preserving two levels of logic) will comprise 3 phases

1. The determination of the maximal k-cubes.
2. The isolation of the essential k-cubes included in the above.
3. The determination of whether or not the essential k-cubes cover the given function. If they do, the set comprises the minimal form sought.

If they do not, other maximal k-cubes must be added to them in order to obtain a minimum cover.

For phase 1 use is made of the Quine-McCluskey algorithm, for phase there is also a tabular method available. For the third step, in the case where the essential k-cubes do not suffice for the cover, there is, in general, no non-exhaustive procedure which will allow the choice of the other maximal k-cubes necessary to complete the cover.

10.4.1.8 Quine-McCluskey Algorithm (geometric interpretation)

1. Write down the binary expressions for all the terms of disjunctive canonical form, i.e. the binary expressions for all the 0-cubes comprising the graph of the given function.
2. Arrange these terms in increasing order of weight (i.e. the number of 'ones') and form the sets $E_1^0, E_2^0, \ldots E_k^0, \ldots$ within which all terms have the same weight k and arrange the sets themselves in ascending order of weight of the terms of which they are composed. (Certain sets E_k^0 may be empty).
3. Form all possible 1-cubes endeavouring to associate the E_k^0 terms with those of E_{k+1}^0. Check whether they differ in 1 single position, i.e. check that the modulo 2 sum of the corresponding vectors has 1 as its weight.[1]

Each time that two terms can be combined, put a sign (V) against them and put the new term thus obtained in a column E^1. The term

---

[1] Note that for the distance of 2 vectors to be 1, it is necessary that the difference of their weights be 1.

10.4 The Quine-McCluskey Algorithm

not marked in this way at the end of this phase are the maximal cubes. They constitute a set C^0.

4. Arrange all the 1-cubes in columns such that, within a column, all terms have a constant weight. Carry out the same operation on the 0-cubes by combining the terms each time the dashes are in the same locations and the numerical positions (other than (−)) differ in 1 position and in one position only. Thus is obtained a set E^2 and those terms not marked are the maximal cubes which form a set C^1.

Repeat the same operations for E^2 (comparisons are effected on terms having all dashes (−) in identical positions and differing otherwise in 1 single position) etc. Thus are formed the sets $E^1, E^2, ..., E^k$, E^k being such that no cube of the order of $k+1$ can be formed from the cubes E^k.

This then gives rise to the set of maximal cubes, viz.:

$$C = C^0 \cup C^1 \cup \cdots \cup C^k \qquad (10.6)$$

$$C = \bigcup_0^k C^i$$

Example 1: minimisation of the expression:

$$f(x, y, z, t) = \sum (1, 2, 3, 4, 5, 7, 9, 10, 11, 13, 14, 15)$$

Here it will be assumed, as in the examples which follow, that the cost is equal to the number of symbols.

This function of 4 variables, whose form comprises 12 min-terms, corresponds to the following truth table (Table 10.6):

Table 10.6

	x y z t	$f(x, y, z, t)$
0	0 0 0 0	0
1	0 0 0 1	1
2	0 0 1 0	1
3	0 0 1 1	1
4	0 1 0 0	1
5	0 1 0 1	1
6	0 1 1 0	0
7	0 1 1 1	1
8	1 0 0 0	0
9	1 0 0 1	1
10	1 0 1 0	1
11	1 0 1 1	1
12	1 1 0 0	0
13	1 1 0 1	1
14	1 1 1 0	1
15	1 1 1 1	1

1) *Arrangement by weight*

The various terms whose function has the value 1 are taken from the truth table and arranged in increasing order of weight.

Thus are obtained the 4 groups of (Table 10.7)

Table 10.7

$$\text{Weight} = 1 \begin{cases} 0\ 0\ 0\ 1 \\ 0\ 0\ 1\ 0 \\ 0\ 1\ 0\ 0 \end{cases}$$

$$\text{Weight} = 2 \begin{cases} 0\ 0\ 1\ 1 \\ 0\ 1\ 0\ 1 \\ 1\ 0\ 0\ 1 \\ 1\ 0\ 1\ 0 \end{cases}$$

$$\text{Weight} = 3 \begin{cases} 0\ 1\ 1\ 1 \\ 1\ 0\ 1\ 1 \\ 1\ 1\ 0\ 1 \\ 1\ 1\ 1\ 0 \end{cases}$$

$$\text{Weight} = 4 \{\ 1\ 1\ 1\ 1$$

2) *Comparison by pairs*

By taking the term at the top of the above table, i.e. 0001, a check is made to determine whether this can be combined with one of the terms in the subset of weight group 2. It will be seen that this can be combined with 0011 from the scheme

$$\left. \begin{matrix} 0\ 0\ 0\ 1 \\ 0\ 0\ 1\ 1 \end{matrix} \right\} \rightarrow 0\ 0 - 1 \qquad (10.8)$$

This term $0\ 0 - 1$ is then placed at the top of a second table (which, in fact, is that for cubes of the order 1 (Table 10.8).

The term 0 0 0 1 again combines with 0 1 0 1:

$$\left. \begin{matrix} 0\ 0\ 0\ 1 \\ 0\ 1\ 0\ 1 \end{matrix} \right\} \rightarrow 0 - 0\ 1 \qquad (10.9)$$

and with the term 1 0 0 1

$$\left. \begin{matrix} 0\ 0\ 0\ 1 \\ 1\ 0\ 0\ 1 \end{matrix} \right\} \rightarrow -\ 0\ 0\ 1 \qquad (10.10)$$

these two new terms are then placed, below that previously determined in the new table.

10.4 The Quine-McCluskey Algorithm

Next, the following term in the group of terms of weight 1 is taken, e. 0 0 1 0 and a new search is made to determine whether it can be associated with one of the group of terms of weight group 2. It will be found that it can effectively be combined with the term 0 0 1 1 and with the term 1 0 1 0 as shown:

$$\left. \begin{array}{l} 0\ 0\ 1\ 0 \\ 0\ 0\ 1\ 1 \end{array} \right\} \to 0\ 0\ 1\ - \qquad (10.11)$$

and

$$\left. \begin{array}{l} 0\ 0\ 1\ 0 \\ 1\ 0\ 1\ 0 \end{array} \right\} \to -\ 0\ 1\ 0 \qquad (10.12)$$

0 0 1 — and — 0 1 0 will be the fourth and fifth terms in table 10.8. Finally, the term 0 1 1 0, the last of the group of weight 1, combines with 0 1 0 1 to give 0 1 0 —.

Thus ends the search for combinations between the group of weight 1 and the group of weight 2.

Each time that a term has been used it has been ticked (v) (cf. recapitulation — Table 10.11). This indicates that it is not a maximal cube but does not prevent its re-use in other comparisons.

This first phase having been completed a line is drawn below the terms already obtained in Table 10.8 and the procedure is recommenced in an endeavour to combine 0 0 1 1, the first term in the group of weight 2, with the terms of the group of weight 3, which gives 0 — 1 1 and — 0 1 1, etc.

Before continuing, the following remark must be made:

Table 10.8, produced from Table 10.7, has been divided into subsets by the horizontal lines as indicated, at the end of each comparison of terms of a group of Table 10.7, with the terms of the next higher weight group (e.g. at the end of the comparisons between the group of weight 1 and that of group 2 etc). It can easily be checked that the terms included in Table 10.8 are to be found again classified by weight as in Table 10.7, it being understood that these are the weights from consideration of positions other than those which correspond with the dashes (—). In other words, a procedure analogous to that used for Table 10.7 is utilised, the different horizontal bars of Table 10.8 playing the same role as those for Table 10.7 which serve to classify the terms by weight.

Based on Table 10.8 below, an analogous procedure will be resumed. The original terms are to be entered in the left-hand column.

For this table the procedure is a little different in the sense that the terms combine when:

1. The dashes are situated in the same positions in the word.

10. The Simplification of Combinational Networks

2. The other positions differ in 1 single place.

Thus the term 0 0 — 1 will be seen to combine with the term 0 1 — 1 to give 0 — — 1 which is then to be entered at the top of a new column

Table 10.8

1, 3	0 0 − 1
1, 5	0 − 0 1
1, 9	− 0 0 1
2, 3	0 0 1 −
2, 10	− 0 1 0
4, 5	0 1 0 −
3, 7	0 − 1 1
3, 11	− 0 1 1
5, 7	0 1 − 1
5, 13	− 1 0 1
9, 11	1 0 − 1
9, 13	1 − 0 1
10, 11	1 0 1 −
10, 14	1 − 1 0
7, 15	− 1 1 1
11, 15	1 − 1 1
13, 15	1 1 − 1
14, 15	1 1 1 −

The terms used are ticked as before (cf. 10.11) and so on. Thus is obtained Table 10.9

Table 10.9

1, 3, 5, 7	0 − − 1
1, 3, 9, 11	− 0 − 1
1, 5, 3, 7	0 − − 1
1, 5, 9, 13	− − 0 1
1, 9, 3, 11	− 0 − 1
1, 9, 5, 13	− − 0 1
2, 3, 10, 11	− 0 1 −
2, 10, 3, 11	− 0 1 −
3, 7, 11, 15	− − 1 1
3, 11, 7, 15	− − 1 1
5, 7, 13, 15	− 1 − 1
5, 13, 7, 15	− 1 − 1
9, 11, 13, 15	1 − − 1
9, 13, 11, 15	1 − − 1
10, 11, 14, 15	1 − 1 −
10, 14, 11, 15	1 − 1 −

By intention, the table has been reproduced just as it has been derived from the progressive application of the algorithm.

Several terms have appeared twice in the table (this being due to the fact that the same square can be obtained in two different manners, a

10.4 The Quine-McCluskey Algorithm

the combination of the two opposite edges); in practice, it is convenient to retain these terms only in 1 single place in the table in order to avoid uselessly complicating the procedure.

Finally, from Table 10.9, assumed thus to have been reduced, Table 10.10 can be formulated:

Table 10.10

1, 3, 5, 7, 9, 11, 13, 15	– – – 1
1, 3, 9, 11, 5, 7, 13, 15	– – – 1
1, 5, 9, 13, 3, 7, 11, 15	– – – 1

Thus the same term has been obtained three times (by different combinations of faces) of which only one will be retained. The application of the algorithm terminates at this point. Table (10.11) reproduces the set of Tables 10.7 to 10.10 obtained successively. Each term appears once only (even though it may have been obtained in several ways). Also represented are the ticks (V) with which the terms have been marked off, as also are the original terms. The various tables are set out side-by-side reading from the left, which is the procedure adopted in practice.

3. Maximal Cubes

The following terms are not ticked and thus they represent maximal cubes.

$$\begin{cases} 0\ 1\ 0\ - \\ -\ 0\ 1\ - \\ 1\ -\ 1\ - \\ -\ -\ -\ 1 \end{cases} \tag{10.13}$$

4. Selection of maximal cubes — Selection table

For this purpose a *selection table* is drawn up in the following manner (Table 10.12). Each column corresponds to a vertex of E. Each line corresponds to one of the maximal cubes found. A mark is drawn at the intersection of a line and a column each time that the vertex associated with the column is contained by the cube (eventually reducing to one point) associated with the line, i.e. each time that the vertex symbols are included in those of the maximal cube, according to the rule:

$$\begin{cases} 0,1 \subset - \\ 0 \quad \subset 0 \\ 1 \quad \subset 1 \end{cases} \tag{10.14}$$

10. The Simplification of Combinational Networks

Table 10.11

	$x\ y\ z\ t$				
1	0 0 0 1	✓	1, 3	0 0 - 1	✓	1, 3, 5, 7	0 - - 1	✓	1, 3, 5, 7, 9, 11, 13, 15	- - - 1
2	0 0 1 0	✓	1, 5	0 - 0 1	✓	1, 3, 9, 11	- 0 - 1	✓		
4	0 1 0 0	✓	1, 9	- 0 0 1	✓	1, 5, 9, 13	- - 0 1	✓		
3	0 0 1 1	✓	2, 3	0 0 1 -	✓	2, 3, 10, 11	- 0 1 -			
5	0 1 0 1	✓	2, 10	- 0 1 0	✓	3, 7, 11, 15	- - 1 1	✓		
9	1 0 0 1	✓	4, 5	0 1 0 -	✓	5, 7, 13, 15	- 1 - 1	✓		
10	1 0 1 0	✓	3, 7	0 - 1 1	✓	9, 11, 13, 15	1 - - 1	✓		
7	0 1 1 1	✓	3, 11	- 0 1 1	✓	10, 11, 14, 15	1 - 1 -			
11	1 0 1 1	✓	5, 7	0 1 - 1	✓					
13	1 1 0 1	✓	5, 13	- 1 0 1	✓					
14	1 1 1 0	✓	9, 11	1 0 - 1	✓					
15	1 1 1 1	✓	9, 13	1 - 0 1	✓					
			10, 11	1 0 1 -	✓					
			10, 14	1 - 1 0						
			7, 15	- 1 1 1	✓					
			11, 15	1 - 1 1	✓					
			13, 15	1 1 - 1	✓					
			14, 15	1 1 1 -						

10.4 The Quine-McCluskey Algorithm

Table 10.12

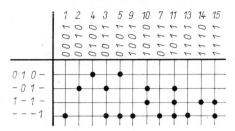

— The quest for essential maximal cubes.

For this it is sufficient to consult the table. In fact, the marks situated in a column corresponding to a vertex indicate to which maximal cubes the vertex belongs and, in particular, to how many of the cubes it belongs. It suffices, therefore, to look for those columns containing only one mark; the corresponding cube is essential since it contains a vertex which is not contained in any other maximal cube maximal.

It will thus be seen that the columns 1, 2, 4, 7, 9, 13, 14 carry only a single mark. It is then easy to deduce that the terms:

$$\begin{cases} ---1 \\ -0\,1\,- \\ 0\,1\,0\,- \\ 1\,-1\,- \end{cases} \quad (10.15)$$

are essential maximal cubes. Thus they constitute the minimum cover.

The following correspondence with the algebraic terms gives the minimal sum:

$$\begin{array}{c|c} ---1 & t \\ -0\,1\,- & \bar{y}z \\ 0\,1\,0\,- & \bar{x}y\bar{z} \\ 1\,-1\,- & xz \end{array} \quad (10.16)$$

whence, returning to the algebraic expressions:

$$f(x, y, z, t) = t + \bar{y}z + \bar{x}y\bar{z} + xz \quad (10.17)$$

which is the minimum sum of products sought.

Geometric interpretation.

The example which has been studied is that for a 4-variable function for which the use of a geometrical representation is both possible and easy. On Figure 10.13 have been represented the different maximal cubes

262   10. The Simplification of Combinational Networks

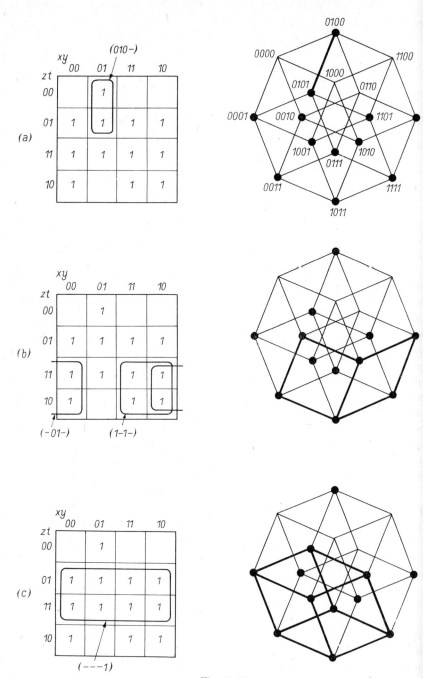

Fig. 10.13

10.4 The Quine-McCluskey Algorithm

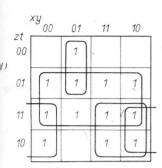
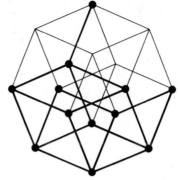

Fig. 10.13. (Continuation)

of different orders, which are all essential, as well as the minimal collection which was sought.

At the same time has been shown a 4-dimensional cube by means of the usual representation in order the better to show the maximal cubes of the different orders. Thus can be seen the correspondence of, on the one hand, the different squares and sets of squares on the Karnaugh map, and the points, edges, faces and cubes of the 4-dimensional cube, and on the other hand, the algebraic representation of these elements. It will be noted that, in this particular case, the Karnaugh map is the best way of highlighting directly the above geometric elements. It will also be noted that for one, two, three or four dimensions (at the most 5 or 6), the Karnaugh diagram (among other things) permits the maximal cubes to be found directly; the Quine-McCluskey algorithm and the selection tables permit them to be found in the general case (for any number of variables). It follows, then, that for 3 or 4 variables it is convenient to follow the diagram method.

The example above has been used only to emphasise:
1) the geometric elements defined by the general theory,
2) the correspondence with the Karnaugh diagram.

Example 2 — Minimisation of the function.

$$f(x, y, z, t, u) = \sum (0, 1, 3, 8, 9, 13, 14, 15, 16, 17, 19, 25, 27, 31)$$

The graph of the function f comprises 14 points (0-cubes) and by application of the Quine-McCluskey algorithm are found (Table 10.14) 10 maximum cubes (cocycles) among which are 4 essential maximal cubes and must, therefore, be included in the cover of the graph of the function f.

10. The Simplification of Combinational Networks

Table 10.14

	$x\ y\ z\ t\ u$				
0	0 0 0 0 0 √	0, 1	0 0 0 0 – √	0, 1, 8, 9	0 – 0 0 –
		0, 8	0 – 0 0 0 √	0, 1, 16, 17	– 0 0 0 –
1	0 0 0 0 1 √	0, 16	– 0 0 0 0 √	1, 3, 17, 19	– 0 0 – 1
8	0 1 0 0 0 √			1, 9, 17, 25	– – 0 0 1
16	1 0 0 0 0 √	1, 3	0 0 0 – 1 √	17, 19, 25, 27	1 – 0 – 1
		1, 9	0 – 0 0 1 √		
3	0 0 0 1 1 √	1, 17	– 0 0 0 1 √		
9	0 1 0 0 1 √	8, 9	0 1 0 0 – √		
17	1 0 0 0 1 √	16, 17	1 0 0 0 – √		
13	0 1 1 0 1 √	3, 19	– 0 0 1 1 √		
14	0 1 1 1 0 √	9, 13	0 1 – 0 1		
19	1 0 0 1 1 √	9, 25	– 1 0 0 1 √		
25	1 1 0 0 1 √	17, 19	1 0 0 – 1 √		
		17, 25	1 – 0 0 1 √		
15	0 1 1 1 1 √				
27	1 1 0 1 1 √	13, 15	0 1 1 – 1		
		14, 15	0 1 1 1 –		
31	1 1 1 1 1 √	19, 27	1 – 0 1 1 √		
		25, 27	1 1 0 – 1 √		
		15, 31	– 1 1 1 1		
		27, 31	1 1 – 1 1		

Thes election of the maximal cubes necessary to complete the cover is made with the aid of Table 10.15 which, in particular, highlights those which are essential.

The maximal cubes are the following terms:

$$\begin{cases} 0\ 1 - 0\ 1 \\ 0\ 1\ 1 - 1 \\ 0\ 1\ 1\ 1 - \\ 1\ 1 - 1\ 1 \\ -1\ 1\ 1\ 1 \\ 0 - 0\ 0 - \\ -0\ 0 - 1 \\ -0\ 0\ 0 - \\ --0\ 0\ 1 \\ 1 - 0 - 1 \end{cases} \qquad (10.18)$$

— The quest for the non-essential maximal cubes needed to complete the cover.

It is necessary to seek a set of lines such that these lines possess at least one mark for every column and also such that the number of

Table 10.15

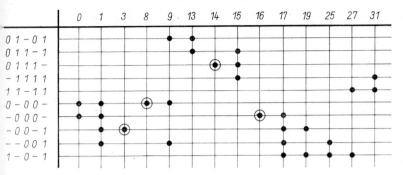

numerical positions of maximal cubes corresponding to these lines be minimal. This is the part of the problem for which no general algorithm exists. In certain cases this problem can be resolved easily by 'consulting' the table. In other cases the best solution is far from obvious and necessitates certain reasoning (profiting from the particular nature of the table) for the determination of the minimal cover.

The terms $0\,1\,1\,1\,-$, $0-0\,0-$, $-0\,0\,0-$, $-0\,0-1$, are essential (indicated by ⊙) and cover the vertices 0, 1, 3, 8, 9, 14, 15, 16, 17, 19. It remains to cover the vertices 13, 25, 27 and 31. For each vertex there are 2 covers with 2 or 4 points. In addition, no cube can cover more than 2 of the 4 missing vertices unaided. The term chosen to cover 13 will certainly not influence the total cost of the cover, since for the two possible terms $(0\,1-0\,1)$ and $(0\,1\,1-1)$ the remaining vertex belongs to the essential cubes, to the exclusion of the other vertices 25, 27 and 31.

From the foregoing, it will be seen that at least 2 cubes are needed to cover the 4 vertices. But with 2, one vertex always escapes. Thus at least 3 are necessary. But, it is easy to find a cover with 3 cubes. It is, therefore, with the cover by 3 cubes that the search for the minimum cover will commence.

There are only two 2-cubes among the candidates. If both are taken, the cover cannot be completed with a single term. Among the three term covers, the most that can be hoped for is, then, to have one 2-cube and two 1-cubes (cost: 11).

There are, effectively, 6 combinations of this type:

$$\begin{cases} 1-0-1 & 1\,1-1\,1 & 0\,1-0\,1 \\ 1-0-1 & -1\,1\,1\,1 & 0\,1-0\,1 \\ 1-0-1 & 1\,1-1\,1 & 0\,1\,1-1 \\ 1-0-1 & -1\,1\,1\,1 & 0\,1\,1-1 \\ --0\,0\,1 & 1\,1-1\,1 & 0\,1-0\,1 \\ --0\,0\,1 & 1\,1-1\,1 & 0\,1\,1-1 \end{cases}$$

266   10. The Simplification of Combinational Networks

Added to the essential cubes, any one of these 6 sets of cubes thus supplies a minimal collection. (There are no solutions of 4 or more cubes which supply other minimal collections, given that the minimum cost of 4 cubes is $4 \times 3 = 12 > 11$).
Hence the 6 minimal expressions:

$$f(x, y, z, t, u) = \bar{x}yzt + \bar{x}\bar{z}\bar{t} + \bar{y}\bar{z}\bar{t} + \bar{y}\bar{z}u + \begin{cases} x\bar{z}u + xytu + \bar{x}y\bar{t}u \\ x\bar{z}u + xztu + \bar{x}y\bar{t}u \\ x\bar{z}u + xytu + \bar{x}yzu \\ x\bar{z}u + yztu + \bar{x}yzu \\ \bar{z}\bar{t}u + xytu + \bar{x}y\bar{t}u \\ \bar{z}\bar{t}u + xytu + \bar{x}yzu \end{cases}$$

(10.19)

Example 3: Minimisation of the 6-variable function:

$$(x, y, z, t, u, v) = \sum (1, 3, 5, 13, 15, 17, 23, 24, 25, 26, 27, 28, 30, 32, 33,$$
$$34, 35, 36, 37, 38, 40, 42, 44, 46, 48) \qquad (10.20)$$

The graph of the function includes 25 points. The different cubes resulting from the application of the Quine-McCluskey algorithm are represented in Table 10.16. Finally there remain 13 maximal cubes of which 7 are essential (e). Hence the minimal covers: these 7 cubes leave 3 non-covered vertices (5, 17 and 37) which belong to 6 non-essential cubes. None of these cubes suffice to cover the 3 vertices, but the 2-cube combinations (and these only) $-0\,0-0\,1$ and $0-0\,0\,0\,1$ or $0\,1-0\,0\,1$ permit this cover, (cost of these 2-cubes: 9) of 2 cubes. Additionally, every 3-cube combination has a minimal cost of $5 + 2 \times 4 = 13$. There are, thus, two minimal collections here and the corresponding algebraic expression is written:

$$f(x, x, z, t, u, v) = \bar{x}y\bar{z}tuv + x\bar{z}\bar{t}u\bar{v} + \bar{x}\bar{y}ztv + \bar{y}\bar{z}\bar{t}v + \bar{x}yz\bar{t} + \bar{x}yz\bar{v}$$
$$+ x\bar{y}\bar{v} + \bar{y}\bar{z}\bar{u}v + \bar{x}\bar{z}\bar{t}\bar{u}v \quad (\text{or } \bar{x}y\bar{t}\bar{u}v) \qquad (10.21)$$

(the first 7 terms are essential).

Example 4: non-simplifiable function.

$$f(x_1, x_2, \ldots, x_n) = x_1 \oplus x_2 \oplus \cdots \oplus x_i \oplus \cdots \oplus x_n$$

This function cannot be simplified as a logical sum of logical products. Indeed, it takes the value 1 for all of the combinations containing an odd number of ones. Two cases can be considered in the endeavour to form a cube of order 1 with two points of the characteristic set of the function f.

10.4 The Quine-McCluskey Algorithm

Table 10.16

x y z t u v			
0 0 0 0 0 1 √	0 0 0 0 - 1 √	- 0 0 0 - 1 e	1 0 - - - 0 e
1 0 0 0 0 0 √	0 0 0 - 0 1 √	- 0 0 - 0 1	
	0 - 0 0 0 1	1 0 0 0 - -	
0 0 0 0 1 1 √	- 0 0 0 0 1 √	1 0 0 - - 0 √	
0 0 0 1 0 1 √	1 0 0 0 0 - √	1 0 0 - 0 -	
0 1 0 0 0 1 √	1 0 0 0 - 0 √	1 0 - 0 - 0 √	
0 1 1 0 0 0 √	1 0 0 - 0 0 √	1 0 - - 0 0 √	
1 0 0 0 0 1 √	1 0 - 0 0 0 √		
1 0 0 0 1 0 √	1 - 0 0 0 0 e	0 1 1 0 - - e	
1 0 0 1 0 0 √		0 1 1 - - 0 e	
1 0 1 0 0 0 √	- 0 0 0 1 1 √	1 0 - - 1 0 √	
1 1 0 0 0 0 √	- 0 0 1 0 1 √	1 0 - 1 - 0 √	
	0 0 - 1 0 1	1 0 1 - - 0 √	
0 0 1 1 0 1 √	0 1 - 0 0 1		
0 1 1 0 0 1 √	0 1 1 0 0 - √		
0 1 1 0 1 0 √	0 1 1 0 - 0 √		
0 1 1 1 0 0 √	0 1 1 - 0 0 √		
1 0 0 0 1 1 √	1 0 0 0 - 1 √		
1 0 0 1 0 1 √	1 0 0 - 0 1 √		
1 0 0 1 1 0 √	1 0 0 0 1 - √		
1 0 1 0 1 0 √	1 0 0 - 1 0 √		
1 0 1 1 0 0 √	1 0 - 0 1 0 √		
	1 0 0 1 0 - √		
0 0 1 1 1 1 √	1 0 0 1 - 0 √		
0 1 0 1 1 1 e	1 0 - 1 0 0 √		
0 1 1 0 1 1 √	1 0 1 0 - 0 √		
0 1 1 1 1 0 √	1 0 1 - 0 0 √		
1 0 1 1 1 0 √			
	0 0 1 1 - 1 e		
	0 1 1 0 - 1 √		
	0 1 1 0 1 - √		
	0 1 1 - 1 0 √		
	0 1 1 1 - 0 √		
	1 0 - 1 1 0 √		
	1 0 1 - 1 0 √		
	1 0 1 1 - 0 √		

First case: the 2 binary combinations have the same weight. In this case their distance is an even number $2k (k \geq 1)$ and consequently they cannot be associated since it is necessary and sufficient that the distance be equal to 1.

Second case: the 2 binary combinations have different weights. In this case the weights differ by a multiple of $2 (2k', k' \geq 1)$ and conse-

Table 10.17. *Selection Table.*

quently they, too, cannot be associated since it is *imperative* that the difference in weights be 1.

Consequently, if it is required to state the function $f(x_1, x_2, \ldots x_i \ldots x_n)$ in the form of a sum of products, the minimal form is the first canonical form.

No adjacent squares are to be found in a Karnaugh diagram representative of the function (Fig. 8.36).

10.4.1.9 *Incompletely Specified Functions*

The following procedure to be adopted constitutes a generalisation of that discussed above. The reader may seek the justification thereof by way of an exercise, or turn to reference [10].

Method

- Find the maximal cubes for the "maximum function", i.e. that obtained by replacing all un-specified values by 1.
- Refer these cubes to a selection table in an endeavour to cover the "minimum function", i.e. that obtained by replacing all un-specified values by 0.

10.4.1.10 *Product of Sums*

Here the principle of duality is applied twice: the dual function f is minimised as above, and applied again to the new dual form which will then supply the required form.

10.4.2 Multi-Output Circuits

In the case of a network having n inputs and m outputs $(m > 1)$ it will not suffice separately to simplify the algebraic expressions associated with the different outputs: it is necessary, in general, to take account of terms which may be common to two or more among them. This is a general observation, valid irrespective of the number of levels. In the particular case of 2-level circuits, the Quine-McCluskey method can be generalised (*Bartee*). This present work will be limited to a single example of the method which is fairly intuitive; the reader will find detailed explanations in the works of *Bartee, Lebow, Reed* [10] among others.

Consider, for example, the 3 functions of Table 10.18.

The maximal cubes are determined not merely for the 3 functions f_1, f_2, f_3, but also for their pairwise products and the products of all 3.

270   10. The Simplification of Combinational Networks

Table 10.18

x y z t	f_1	f_2	f_3	$f_1 \cdot f_2$	$f_1 \cdot f_3$	$f_2 \cdot f_3$	$f_1 \cdot f_2 \cdot f_3$
0 0 0 0	0	0	0	0	0	0	0
0 0 0 1	0	0	0	0	0	0	0
0 0 1 0	0	0	0	0	0	0	0
0 0 1 1	0	0	0	0	0	0	0
0 1 0 0	1	0	1	0	1	0	0
0 1 0 1	1	1	1	1	1	1	1
0 1 1 0	0	0	0	0	0	0	0
0 1 1 1	1	0	0	0	0	0	0
1 0 0 0	0	0	0	0	0	0	0
1 0 0 1	0	0	0	0	0	0	0
1 0 1 0	0	0	0	0	0	0	0
1 0 1 1	1	0	0	0	0	0	0
1 1 0 0	1	0	1	0	1	0	0
1 1 0 1	1	1	0	1	0	0	0
1 1 1 0	0	0	0	0	0	0	0
1 1 1 1	0	0	0	0	0	0	0

(In the general case of n functions, the pairwise products, three-by-three etc., must be considered). The products of the functions for this present example are shown in Table 10.18. Thus is established a selection table completely analogous to that used in the case of an unique function with, however, the difference that for all the maximal cubes of the function products, the summits covered are indicated for every function of the product. From the example in question is obtained Table 10.19.

Table 10.19

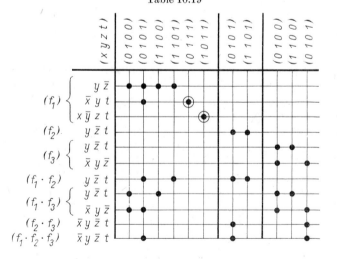

10.5 Functional Decompositions

From which is derived the cover:

$$f_1 = x\bar{y}zt + \bar{x}yt + (y\bar{z}t + y\bar{z}\bar{t})$$
$$f_2 = y\bar{z}t \qquad (10.22)$$
$$f_3 = \bar{x}y\bar{z} + y\bar{z}\bar{t}$$

For f_2 and f_3, the above expressions are identical with those which would have been obtained from treating f_2 and f_3 in isolation. For f_1 this is not the case: the term $y\bar{z}$ (which the minimisation of f_1 alone would yield) is replaced by the terms $y\bar{z}t$ and $y\bar{z}\bar{t}$ which appear in f_2 and f_3 respectively. The reader is left the task of evaluating the gain thus realised in relation to the case where, for f_1, the expression (minimum for f_1 only) had been used:

$$f_1 = x\bar{y}zt + \bar{x}yt + y\bar{z}$$

taking as the cost:
- the total number of letters for the monomials;
- the number of distinct monomials.

The case for un-specified values

The procedure is as that above, the determination of the maximum cubes for the maximal functions, and the covering of the minimal functions.

10.5 Functional Decompositions (Ashenhurst, Curtis, Povarov)

The Quine-McCluskey algorithm leads to 2-level networks of AND and OR gates. If the restriction on the number of levels be lifted, the circuit thus obtained may well no longer be minimum: by factorising (leading to more than 2 levels) the circuits then obtained are often less costly. However, there is nothing that prevents the execution of such operations of factorisation starting from a 2-level expression. If reference back to the parallel binary adder of § 9.4.2 be made it will be seen that, in its iterative form, it supplies a "carry" r_k after passing through a number of levels which grows with k (e.g. $2k + 1$) if the elementary adders are of two levels) and the problem presented is that of the time of propagation of information. Conversely, in the case of a "carry look ahead", it will be seen that for all 'bit' positions, or at least for whole groups of bits, there is a carry network which is only a 2-level[1] version of the preceding one and that, this time, the reduction of the number of levels

or 4, dependent on the design.

to a minimum results in a much more complicated circuit than that for the iterative circuit. Analogous consideration could be given to the subject of counters. This then gives a glimpse of the theory of decompositions: if the cost of the circuit decreases rapidly with the number of variables n (e.g. exponentially, frequent case) the design of circuits in the form of function of function (with few variables if possible) is likely to be particularly economical. It is with this theory of *functional decomposition*, due to the works of *Ashenhurst, Curtis* and *Povarov* in particular, that the most important results are to be found in the text which follows.

10.5.1 Simple Disjoint Decompositions

10.5.1.1 Definition

It is said that a Boolean function with n variables $f(x_1, ..., x_i, ... x_n)$ admits a *simple disjoint decomposition* (or that it is "*functionally separable*") if:

1. There exists a *partition* of the n variable set $\{x_1, ..., x_n\}$ into two subsets $\{x_{i_1}, ... x_{i_s}\}$ and $\{x_{i_{s+1}}, ..., x_{i_n}\}$ such that $f(x_1, ..., x_i, ..., x_n)$ can be expressed as a compound function by a relation of the form:

$$f(x_1, ..., x_i, ..., x_n) = g[h(x_{i_1} ... x_{i_s}), x_{i_{s+1}} ... x_{i_n}] \qquad (10.23)$$

2. $2 \leq s \leq n - 1$

The pair of functions g and h is known as the decomposition of f. This definition gives rise to the following observations:

1. g is a Boolean function with $s - n + 1$ variables.
2. h is a Boolean function with s variables.
3. The condition $s \neq 0$ states that in order for there to be a true decomposition, it must, effectively, be a function of function; if not, it is possible only to state that:

$$f(x_1, ... x_i, ... x_n) = g(x_{i_1}, ... x_{i_n}),$$

a relation resulting from a simple permutation of the variables and which, consequently, presents no interest from the viewpoint of circuit simplification.

4. The condition $s \neq 1$ eliminates the two decompositions of type:

$$g[h(x_{i_k}), x_{i_1}, ..., x_{i_{k-1}}, x_{i_{k+1}}, ..., x_{i_n}]$$

with

$$h(x_{i_k}) = x_{i_k} \quad \text{or} \quad h(x_{i_k}) = \bar{x}_{i_k}$$

10.5 Functional Decompositions

which can be considered to be trivial: it can easily be verified that these two decompositions exist for every variable:

$$x_{i_k} \ (k = 1, 2, \ldots, n)$$

5. The condition $s \neq n$ eliminates the case where h contains the n variables. In this case:

$$f(x_1, \ldots, x_n) = g[h(x_1, \ldots, x_n)]$$

where g would be the function of a single variable. This would give, if $f(x_1, \ldots, x_n)$ is not constant, either $g(h) = h$ or $g(h) = \bar{h}$ and, consequently, would leave only two possiblities:

$$h(x_1, \ldots, x_n) = f(x_1, \ldots, x_n) \quad \text{and} \quad g = h \qquad (10.24)$$

or

$$h(x_1, \ldots, x_n) = \bar{f}(x_1, \ldots, x_n) \quad \text{and} \quad g = \bar{h} \qquad (10.25)$$

The circuit for the first solution is in no way different from that for $f(x_1, \ldots, x_n)$. The second solution necessitates 2 complementations which, from the logic viewpoint at any rate, are useless.

6. To simplify the notation it will be necessary to use the now traditional notation for the variables of h.

$$Y = \{y_1, \ldots, y_s\} = \{x_{i_1}, \ldots, x_{i_s}\}$$

and for the others:

$$Z = \{z_1, \ldots, z_{n-s}\} = \{x_{i_{s+1}}, \ldots, x_{i_n}\}.$$

This allows us to write:

$$f(x_1, \ldots, x_n) = g[h(y_1, \ldots, y_s), z_1, \ldots, z_{n-s}] = g[h(Y), Z] \quad (10.26)$$

Also putting

$$X = \{x_1, \ldots, x_n\} \qquad (10.27)$$

then

$$\begin{cases} Y \cap Z = \emptyset \\ Y \cup Z = X \end{cases} \qquad (10.28)$$

Finally the conditions $s \neq 0$ and $s \neq n$ are equivalent to: $Y \neq \emptyset$; $Z \neq \emptyset$, the condition $s \neq 1$ to $Y \neq \{x_i\}$ for every i. The notation to be used is:

$g[h(Y), Z]$: function (compound) of n variables,

$g[h, Z]$: function of $n - s + 1$ variables, where h is a variable.

For greater convenience a function and its characteristic vector will, generally, be designated by the same letter.

7. The interpretation in terms of circuits of the type (10.23) is evident: among the variables $x_1, \ldots, x_n$, s among them, $y_1, \ldots y_s$ are the inputs to a circuit performing the function h, for which the output, together with the remaining variables $z_1, \ldots, z_{n-s}$, serve as an input to a second circuit performing the function $g[h, z_1, \ldots, z_{n-s}]$ (cf. Fig. 10.20 and, as a supplementary example, Figs. 9.16 and 9.17).

8. In the text which follows the name non-trivial[1] function will be given to a function which is not constant (with respect to 1 or more variables).

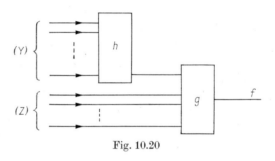

Fig. 10.20

10.5.1.2 Number of Partitions (YZ) of Variables

A partition of variables is defined here by one of the 2 (complementary) sets Y and Z, i.e. by a knowledge of one of the 2^n subsets of set X. By eliminating from these subsets the empty set $\varnothing$, the set X and the n sets containing 1 single variable, it will be seen that there are $2^n - n - 2$ possible partitions (YZ) of the set X having input variables $x_1, \ldots, x_n$.

Note: Decomposition is a property of functions. For a given decomposition the functions g and h can have various algebraic expressions.

10.5.1.3 Necessary and Sufficient Condition for the Existence of a Simple Disjoint Decomposition

From consideration of a partition $\{y_1, y_2, \ldots, y_s\}$, $\{z_1, z_2, \ldots z_{n-s}\}$ of n variables of f, a *rectangular truth table* can be constructed as follows: in abscissa will be shown the 2^s possible binary combinations

---

[1] (or non-degenerate).

10.5 Functional Decompositions

$y_1, y_2, \ldots, y_s$, in ordinate will be shown the 2^{n-s} possible binary combinations $z_1, z_2, \ldots z_{n-s}$. At the intersection of line L and column C is to be found a position corresponding to a certain binary combination of *all* the variables $x_1, x_2, \ldots, x_n$ at which point will be inscribed the value corresponding to the function. In theory this procedure is applicable to any n and s as was seen in Chapter 7. Moreover for (small) n and s ($n \leq 5$ or 6) the truth table so compiled can be presented in the form of a Karnaugh diagram, maintaining the same partition. In the general case the truth table corresponding to the partition of (YZ) is given the name of *partition table* or *partition matrix*. This leads to the following (Ashenhurst) theorem:

— *Line criterion*

In order that the function $f(x_1, \ldots, x_n)$ shall admit a simple disjoint decomposition, it is necessary and sufficient that the lines of the partition matrix be recruited from among the following 4 types, and from these only.

— the trivial vector (0)
— the trivial vector (1)
— a non-trivial vector h
— the vector $\bar{h}$

Proof

1. If $f(x_1, \ldots, x_n)$ can be decomposed, then:

$$f(x_1, \ldots, x_n) = g[h(y_1, \ldots, y_s) z_1, \ldots, z_{n-s}] \tag{10.29}$$

which is equivalent to (Shannon theory):

$$f(x_1, \ldots, x_n) = h(y_1, \ldots y_s) g[1, z_1, \ldots, z_{n-s}] + \bar{h}(y_1, \ldots, y_s) g[0, z_1, \ldots, z_{n-s}] \tag{10.30}$$

which may be shortened to:

$$f(x_1, \ldots, x_n) = h g_1 + \bar{h} g_0 \tag{10.31}$$

Although h may be a function of s variables only, and g_1 and g_0 of the $n - s$ others, there is nothing to prevent consideration of these functions as functions of n variables. The truth table for g_0, for example, is then made up of 2^s identical columns. The same applies to g_1 (cf. Fig. 10.21 b).

In one or other of these tables there are but 2 types of horizontal vectors, the vectors (0) and (1), thus all the components are respectively 0 and 1. As for h, its truth table comprises 2^{n-s} identical lines; thus this horizontal vector will again be indicated by h. It can then be stated as:

$$f = A + B \tag{10.32}$$

with

$$A = h g_1, \qquad B = \bar{h} g_0. \tag{10.33}$$

The vector lines of A can only be h or 0 (depending on whether $g_1 = 1$ or 0), those for B, $\bar{h}$ or 0, and consequently $A + B$ can only be

$$h + 0 = h, \qquad \bar{h} + 0 = \bar{h}, \qquad 0 + 0 = 0, \qquad h + \bar{h} = 1.$$

(Table 10.22 shows an example).

Table 10.21. (n_C distinct columns n_L distinct lines)

$(f(Y))$ $\begin{cases} n_C = 2 \\ n_L = 1 \end{cases}$

$Y \rightarrow$
$\downarrow Z$

| 1 0 1 1 0 1 0 1 |
| 1 0 1 1 0 1 0 1 |
| 1 0 1 1 0 1 0 1 |
| 1 0 1 1 0 1 0 1 |

$\downarrow$

| 1 0 1 1 0 1 0 1 |

(a)

| 0 |
| 1 |
| 1 |
| 0 |

$\rightarrow$

| 0 0 0 0 0 0 0 |
| 1 1 1 1 1 1 1 |
| 1 1 1 1 1 1 1 |
| 0 0 0 0 0 0 0 |

$(f'(Z))$ $\begin{cases} n_C = 1 \\ n_L = 2 \end{cases}$

(b)

Table 10.22

| 1 0 1 1 0 1 0 1 |
| 1 0 1 1 0 1 0 1 |
| 1 0 1 1 0 1 0 1 |
| 1 0 1 1 0 1 0 1 |

$\cdot$

| 0 0 0 0 0 0 0 0 |
| 1 1 1 1 1 1 1 1 |
| 1 1 1 1 1 1 1 1 |
| 0 0 0 0 0 0 0 0 |

$+$

| 0 1 0 0 1 0 1 0 |
| 0 1 0 0 1 0 1 0 |
| 0 1 0 0 1 0 1 0 |
| 0 1 0 0 1 0 1 0 |

$\cdot$

| 0 0 0 0 0 0 0 0 |
| 0 0 0 0 0 0 0 0 |
| 1 1 1 1 1 1 1 1 |
| 1 1 1 1 1 1 1 1 |

$=$

h g_1 $\bar{h}$ g_0

| 0 0 0 0 0 0 0 0 |
| 1 0 1 1 0 1 0 1 |
| 1 0 1 1 0 1 0 1 |
| 0 0 0 0 0 0 0 0 |

$+$

| 0 0 0 0 0 0 0 0 |
| 0 0 0 0 0 0 0 0 |
| 0 1 0 0 1 0 1 0 |
| 0 1 0 0 1 0 1 0 |

$=$

0	0	0	0	0	0	0	0	(0)
1	0	1	1	0	1	0	1	(L)
1	1	1	1	1	1	1	1	(1)
0	1	0	0	1	0	1	0	($\bar{L}$)

$(c_1)(c_2)(c_1)(c_1)(c_2)(c_1)(c_2)(c_1)$

$h_0 g_1$ $\bar{h} g_0$

2. Conversely, should there be a partition (YZ) of the variables such that the line vectors of the truth table $f(x_1, \ldots, x_n)$ callup only from among the types (0), (1), h, $\bar{h}$, the function f is then decomposable.

10.5 Functional Decompositions

It should be noted that there exists at least one non-trivial line differing from (0) and (1)), since f depends on the variables Y in a non-trivial way. If a non-trivial line vector be taken at random it can de described in two ways: L or $L' = \bar{L}$ and in two ways only. If the complement of the vector considered also appears in the table it automatically takes the name $\bar{L}$ or $\bar{L}'$ respectively. With the vectors L, $\bar{L}$, (0) and (1) are associated functions of (Z): if these are grouped in the sets E_L, $E_{\bar{L}}$, $E_{(0)}$ and $E_{(1)}$ (certain of which may be empty), the characteristic functions of these sets of vectors:

$$f_L(Z), f_{\bar{L}}(Z), f_{(0)}(Z) \text{ and } f_{(1)}(Z)$$

will, by definition, also be equal to 1 since, for the combination Z, the corresponding vector line appertains to the set with the same index.

$$(f_L + f_{\bar{L}} + f_{(0)} + f_{(1)} = 1) \tag{10.34}$$

giving:
$$(x_1, \ldots, x_n) = g(L, Z) = \bar{L} g_0 + L g_1 \tag{10.35}$$

with

$$\begin{cases} g_1 = f_L + f_{(1)} \\ g_0 = f_{\bar{L}} + f_{(1)} \end{cases} \tag{10.36}$$

The pair (L, F) is thus properly determined. If in taking the convention $L = \bar{L}'$ and endeavouring to write $f(x_1, \ldots, x_n) = g'(L', Z)$ it will be seen that:

$$g'(L', Z) = \bar{L}' g'_0 + L' g'_1 = L g'_0 = \bar{L} g'_1 \tag{10.37}$$

whence
$$\begin{cases} g'_0 = g_1 \\ g'_1 = g_0 \end{cases} \tag{10.38}$$

Within the terms of the hypothesis, it will be seen that there are only two decompositions (L, g) and $(\bar{L}, g')$, with the relations (10.38) between g' and g.

Particular cases:

Consider the case where the table possesses only two types of vector line, of which at least one is non-trivial.
To emphasise the concept consider the following case:

1. $f_{(1)} = f_{(0)} = 0, \quad f_L, f_{\bar{L}} \neq 0$ \hfill (10.39)

We have:
$$f_L + f_{\bar{L}} = 1 \tag{10.40}$$

But:
$$f_L \cdot f_{\bar{L}} = 0 \tag{10.41}$$
therefore
$$f_L = \bar{f}_{\bar{L}} \tag{10.42}$$
and
$$\begin{cases} g_0 = f_{\bar{L}} \\ g_1 = f_L = \bar{g}_0 \end{cases} \tag{10.43}$$

Thus giving:
$$g = \bar{L}g_0 + Lg_1 = \bar{L}f_{\bar{L}} + L\bar{f}_{\bar{L}} = \bar{L}\bar{f}_L + Lf_L \tag{10.44}$$
i.e.
$$g = L \oplus f_{\bar{L}} = \bar{L} \oplus f_L = \overline{L \oplus f_L} \tag{10.45}$$

If, however, the contrary convention had been used, by postulating $L' = \bar{L}$ the corresponding function g' would have given:
$$g' = L'g_0 + \bar{L}'g_1 = L'f_{\bar{L}} + \bar{L}'f_L = L'f_{L'} + \bar{L}'f_{\bar{L}'} \tag{10.46}$$
$$g' = L' \oplus \bar{f}_{L'} = \bar{L}' \oplus f_{L'} = \overline{L' \oplus f_{L'}} \tag{10.47}$$

2.
$$f_{(1)} = f_{\bar{L}} = 0, \quad f_{(0)}, f_L \neq 0 \tag{10.48}$$
thus:
$$\begin{cases} g_0 = 0 \\ g_1 = f_L \end{cases} \tag{10.49}$$
$$g = Lf_L \tag{10.50}$$

If we put $L' = \bar{L}$, then:
$$g' = \bar{L}'g'_0 + L'g'_1 = \bar{L}'f_L = \bar{L}'f_{\bar{L}'} \tag{10.51}$$

3.
$$f_{(0)} = f_{\bar{L}} = 0, \quad f_{(1)}, f_L \neq 0 \tag{10.52}$$
giving:
$$f_L + f_{(1)} = 1 \tag{10.53}$$
whence:
$$\begin{cases} g_0 = f_{(1)} \\ g_1 = f_L + f_{(1)} = 1 \end{cases} \tag{10.54}$$
and
$$g = \bar{L}f_{(1)} + L = f_{(1)} + L \tag{10.55}$$

With the convention $L' = \bar{L}$, g will be replaced by:
$$g' = \bar{L}' + L'f_{(1)} = \bar{L}' + f_{(1)} \tag{10.56}$$

10.5.1.4 Other Form of Criterion of Decomposability — Column Criteria

The types of lines and of columns are not independent. Let n_L and n_C be the numbers of vector lines and distinct columns respectively. Consider the following particular case where there are at most two distinct columns C and C' ($n_C \leq 2$). The different cases of column-line dependence can be summarised convincingly by Table 10.23b with 2 column vectors of 4 components and 4 line vectors of 2 components. From this "reduced" partition matrix can be constructed Table 10.23a which recapitulates the correspondence between the types of line and the types of column for $n_C \leq 2$. The variables t_Y and t_Z take the value 1 when f does not depend on Y or Z respectively. The absence or presence

Table 10.23

$C_1 C_1' C_2 C_2' C_3 C_3' C_4 C_4'$	C	L	n_C	n_L	$t_Y t_Z$
0 0 0 0	—	—	—	—	— —
0 0 0 1	(1)	(1)	1	1	1 1
0 0 1 0	(1), (0)	(L)	2	1	0 1
0 0 1 1	(1), (C)	(1), (L)	2	2	0 0
0 1 0 0	(0), (1)	(L)	2	1	0 1
0 1 0 1	(1), (C)	(1), (L)	2	2	0 0
0 1 1 0	(C) $(\bar{C})$	(L), $(\bar{L})$	2	2	0 0
0 1 1 1	(C_1), (C_2)	(1), (L), $(\bar{L})$	2	3	0 0
1 0 0 0	(0)	(0)	1	1	1 1
1 0 0 1	(C)	(0), (1)	1	2	1 0
1 0 1 0	(0), (C)	(0), (L)	2	2	0 0
1 0 1 1	(C_1), C_2	(0), (1), (L)	2	3	0 0
1 1 0 0	(0), (C)	(0), (L)	2	2	0 0
1 1 0 1	(C_1), (C_2)	(0), (1), (L)	2	3	0 0
1 1 1 0	(C_1), (C_2)	(0), (L), $(\bar{L})$	2	3	0 0
1 1 1 1	(C_1), (C_2)	(0), (1), (L) $(\bar{L})$	2	4	0 0

(a)

$(C)\ (C')$

$C_1 C_1'$	0	0
$C_2 C_2'$	0	1
$C_3 C_3'$	1	0
$C_4 C_4'$	1	1

(b)

of type $C_i C'_i$ is indicated by 0 or 1. The columns C and L indicate the types of vector columns and lines which exist at the same time.

From this table it is deduced that:
— for $n_C = 1$, $n_L \leq 2$,
— for $n_C = 2$, $n_L \leq 4$,
— the criterion of decomposability can be re-formulated in terms of columns:

Column criteria

In order that a function $f(x_1, \ldots, x_n)$ be decomposable in accordance with the partition (YZ) it is necessary and sufficient that the corresponding separation table should have a number of distinct columns ≤ 2 ($n_C \leq 2$).

Example: Consider the function f given in Table 10.24a (where the order of the binary combinations is that of a Karnaugh diagram). The truth table 10.24b for the partition $[(x_1 x_3 x_4), (x_2 x_5)]$ bears witness to 3 types of line: (0), (1) and a non-trivial line denoted by L. There are two types of column (cf. also Table 10.22 for another example).

Table 10.24

x_1 0 1 1 0 0 1 1 0 x_1 0 1 0 1 0 1 0 1
x_2 0 0 1 1 0 0 1 1 x_2 0 0 1 1 0 0 1 1
x_3 0 0 0 0 1 1 1 1 x_3 0 0 0 0 1 1 1 1

$x_4 x_5$
0 0   0 0 1 0 0 0 0 1
0 1   0 1 1 1 1 0 1 1
1 1   1 0 1 1 0 1 1 1
1 0   0 0 0 1 0 0 1 0

(a)

$x_2 x_5$
0 0   0 0 0 0 0 0 0 0   (0)
0 1   0 1 1 0 1 0 0 1   } L
1 1   0 1 1 0 1 0 0 1
1 0   1 1 1 1 1 1 1 1   (1)

(b)

we have:

$$\begin{cases} f_L = 0\,1\,1\,0 \\ f_{\bar L} = 0\,0\,0\,0 \\ f_{(0)} = 1\,0\,0\,0 \\ f_{(1)} = 0\,0\,0\,1 \end{cases} \quad (10.57)$$

Whence:
$$g_0 = f_{\bar L} + f_{(1)} = 0\,0\,0\,1 = x_2 x_5 \quad (10.58)$$

$$g_1 = f_L + f_{(1)} = 0\,1\,1\,1 = x_2 + x_5 \quad (10.59)$$

and:
$$\begin{cases} L(x_1, x_3, x_4) = x_1 \oplus x_2 \oplus x_4 \\ f(x_1, x_2, x_3, x_4, x_5) = \bar L x_2 x_5 + L(x_2 + x_5) \end{cases} \quad (10.60)$$

10.5 Functional Decompositions

Before replacing L by a specific algebraic expression, it should be noted that:

$$x_2 x_5 + L(x_2 + x_5) = L(x_2 + x_5) + x_2 x_5 = L \# x_2 \# x_5 \qquad (10.61)$$

whence:

$$f(x_1, x_2, x_3, x_4, x_5) = (x_1 \oplus x_3 \oplus x_4) \# x_2 \# x_5. \qquad (10.62)$$

Proceeding to the following theorems (*Ashenhurst*):

10.5.1.5 Theorem

If, for a given function $f(x_1, x_2, \ldots, x_n)$ there exists a partition A, B, C such that:

$$f(A, B, C) = g[h(A, B), C] = k[l(A), B, C] \qquad (10.63)$$

then

$$f(A, B, C) = g[u[l(A), B], C] \qquad (10.64)$$

where u is a function of l and B is determined in a unique manner by the relation:

$$u[l(A), B] = h(A, B). \qquad (10.65)$$

Indeed, N_C being the number of elements of C,

$$f(A, B, C) = \sum_{i=0}^{i=N_c} g_i[h(A, B)]\, m_i(C) = \sum_{i=0}^{i=N_c} k_i[l(A), B]\, m_i(C) \qquad (10.66)$$

where for every i:

$$g_i[h(A, B)] = k_i[l(A), B]. \qquad (10.67)$$

Every line, for which $g_i(h) \neq (0) \neq (1)$ (these can exist unless $g(h, C)$ depends only on C), can be written:

$$g_i(h) = h^{e_i} \quad (e_i = 0 \text{ or } 1) \qquad (10.68)$$

Whence for every line of this type:

$$h^{e_i}(A, B) = k_i[l(A), B] \qquad (10.69)$$

or

$$h(A, B) = k_i^{e_i}[l(A), B]. \qquad (10.70)$$

Thus $h(A, B)$ possesses a decomposition. The function l being chosen, u is determined in a unique manner

$$h = u[l(A), B]. \qquad (10.71)$$

Example:

Tables 10.25 bear witness to two decompositions which satisfy the assumptions of the statement.

Table 10.25

$x_1 x_2 x_3$

$x_6 x_4 x_5$

```
0 0 0 0 0 0 0 0
0 1 1 0 1 0 0 1
0 1 1 0 1 0 0 1
1 1 1 1 1 1 1 1
1 1 1 1 1 1 1 1
1 0 0 1 0 1 1 0
1 0 0 1 0 1 1 0
0 0 0 0 0 0 0 0
```
$\}\, l$

(b)

$x_1 x_2 x_3 x_4 x_5$

x_6

```
0 0 0 0 0 0 0 0 0 1 1 0 1 0 0 1 0 1 1 0 1 0 0 1 1 1 1 1 1 1 1 1
1 1 1 1 1 1 1 1 1 0 0 1 0 1 1 0 1 0 0 1 0 1 1 0 0 0 0 0 0 0 0 0
```
$\}\, h$

(a)

Table (a) gives:
$$f = h(x_1, x_2, x_3, x_4, x_5) \oplus x_6. \qquad (10.72)$$

Table (b) gives:
$$\begin{cases} l = x_1 \oplus x_2 \oplus x_3 \\ k_0 = 0\,0\,0\,1\,1\,1\,1\,0, \quad k_1 = 0\,1\,1\,1\,1\,0\,0\,0 \end{cases} \qquad (10.73)$$

i.e.
$$\begin{cases} k_0 = \bar{x}_6 x_4 x_5 + x_6 \overline{x_4 x_5} = x_6 \oplus x_4 x_5 \\ k_1 = \bar{x}_6 (x_4 + x_5) + x_6 \overline{(x_4 + x_5)} \\ \quad\; = x_6 \oplus (x_4 + x_5) \end{cases} \qquad (10.74)$$

whence:
$$\begin{aligned} f &= l[x_6 \oplus (x_4 + x_5)] + \bar{l}(x_6 \oplus x_4 x_5) \\ &= l[x_6 \oplus (x_4 + x_5)] \oplus \bar{l}(x_6 \oplus x_4 x_5) \\ &= x_6(l \oplus \bar{l}) \oplus \bar{l}(x_4 + x_5) \oplus l x_4 x_5 \qquad (10.75) \\ &= x_6 \oplus l(x_4 + x_5) + x_4 x_5 \\ &= x_6 \oplus (l \# x_4 \# x_5). \end{aligned}$$

It can be verified directly that:

$$x_6 \oplus h = x_6 \oplus (l \# x_4 \# x_5) \tag{10.76}$$

$$h(x_1, x_2, x_3, x_4, x_5) = (x_1 \oplus x_2 \oplus x_3) \# x_4 \# x_5 \tag{10.77}$$

which gives:

$$f(x_1, x_2, x_3, x_4, x_5, x_6) = [(x_1 \oplus x_2 \oplus x_3) \# x_4 \# x_5] \oplus x_6 \tag{10.78}$$

The function between brackets (h) is, after abstraction of the order of variables, that of § 10.5.1.4 denoted as f).

10.5.1.6 Theorem

If, for a given function $f(x_1, x_2, \ldots, x_n)$ there exists a partition (A, B) such that:

$$f(A, B) = g[h(A), B] = k[l(B), A] \tag{10.79}$$

then:

$$f(A, B) = u[h(A), l(B)] \tag{10.80}$$

where u is a function of two variables $u(h, l)$, determined in a unique manner by the functions $g(h, B)$, $h(A)$, $k(l, A)$ and $l(B)$.

Indeed, if it is assumed that the conditions (10.79) are satisfied, the partition table of f according to AB possesses two types of column and two types of line.

Additionally, the configurations of lines and columns are not independent. It is easily verified that this gives only the configurations:

$$[L, \bar{L} \text{ and } C, \bar{C}], [L, (1) \text{ and } C, (1)], [L, (0) \text{ and } C, (0)] \tag{10.81}$$

which gives the following cases:

1. *Configuration* $[L, (0) - C, (0)]$

It is necessary to have (§ 10.5.1.3)

$$g = L(A) f_L(B) = C(B) f_C(A) \tag{10.82}$$

which leads to:

$$\begin{cases} L(A) = f_C(A) \\ C(B) = f_L(B) \end{cases} \tag{10.83}$$

Assume that, in fact, one of the above equations, the first (for the purpose of the discussion) had not been satisfied. There would exist a combination A_0 such that $L(A_0) = 0$ and $f_C(A_0) = 1$ and, by taking a value B_0

such that $f_L(B_0) = 1$ would have given:

$$g = 1 \cdot 1 = 1 = C(B_0) \cdot 0 = 0,$$

which is a contradiction. Therefore:

$$g = L(A) \cdot C(B). \tag{10.84}$$

More generally, replacing L by L^e ($e = 0$ or 1), C by $C^{e'}$ ($e' = 0$ or 1) results in:

$$g_{(ee')} = L^e(A) \cdot C^{e'}(B).$$

2. *Configuration* $[L, (1) - C, (1)]$

In an analogous manner:

$$g = L(A) + f_{(1)}(B) = C(B) + f'_{(1)}(A). \tag{0.85}$$

By reasoning analogous to that for the preceding example is obtained:

$$g = L(A) + C(B). \tag{10.86}$$

More generally, replacing L by L^e, C by $C^{e'}$ will give:

$$g_{(ee')} = L^e(A) + C^{e'}(B). \tag{10.87}$$

3. *Configuration* $\left[L, \bar{L} - C, \bar{C}\right]$

we have:
$$g = L \oplus f_L = C \oplus f_C \tag{10.88}$$

Whence:
$$L \oplus f_C = C \oplus f_L \tag{10.89}$$

but, $L \oplus f_C$ and $C \oplus f_L$, are functions of different variables. To be equal they can only be constants. Whence:

$$L \oplus f_C = C \oplus f_L = K \quad (K = 0 \text{ or } 1) \tag{10.90}$$

whence:
$$\begin{cases} f_C = K \oplus L \\ f_L = K \oplus C \end{cases} \tag{10.91}$$

and
$$g = L \oplus C \oplus K. \tag{10.92}$$

There remains, in this case, a constant K to be determined: to dispense with the uncertainty a point in the truth table for which $L = 1$, $C = $ can be considered. Thus we have $K = f = g$. More generally, if L is

10.5 Functional Decompositions

replaced by L^e, C by $C^{e'}$ then:

$$g_{(ee')} = L^e \oplus C^{e'} \oplus K. \tag{10.93}$$

The value of the constant K being:

$$g_{(ee')}(A_0 B_0) \oplus L^e(A_0) \oplus C^{e'}(B_0),$$

where $A_0 B_0$ is a particular combination of variables.

Examples:

For each of the Tables 10.26 a, b, c, a non-trivial vector line and a non-trivial column have been chosen arbitrarily, respectively denoted by L and C. The designation for the other lines and columns has, therefore, been determined in a unique manner.

Table 10.26

$C\overline{C}\ x_1 x_2 x_3$ (1) $C\ x_1 x_2 x_3$ C (0) $x_1 x_2 x_3$

$x_4 x_5$
```
1 0 0 1 0 1 1 0  }L̄    x_4 x_5  1 0 0 1 0 1 1 0  }L    x_4 x_5  1 0 0 1 0 1 1 0  }L
0 1 1 0 1 0 0 0  }L             1 0 0 1 0 1 1 0                   1 0 0 1 0 1 1 0
0 1 1 0 1 0 0 0                 1 1 1 1 1 1 1 1  }(1)             0 0 0 0 0 0 0 0  }(0)
1 0 0 1 0 1 1 0                 1 0 0 1 0 1 1 0                   1 0 0 1 0 1 1 0
```
     (c)   (b)   (a)

— Table a $\begin{cases} L = 1\ 0\ 0\ 1\ 0\ 1\ 1\ 0 \\ C = 1\ 1\ 0\ 1 \end{cases}$

Taking

$$\begin{cases} h = L \\ l = C \end{cases} \tag{10.94}$$

which, from the foregoing, gives:

$$u = L \cdot C. \tag{10.95}$$

Hence:

$$(x_1, x_2, x_3, x_4, x_5) = L \cdot C = h \cdot l \tag{10.96}$$

$$= (\overline{x}_1 \overline{x}_2 \overline{x}_3 + \overline{x}_1 x_2 x_3 + x_1 \overline{x}_2 x_3 + x_1 x_2 \overline{x}_3)(\overline{x_4 \overline{x}_5}). \tag{10.97}$$

— Table b $\begin{cases} L = 1\ 0\ 0\ 1\ 0\ 1\ 1\ 0 \\ C = 0\ 0\ 1\ 0 \end{cases}$

which, with convention (10.94) gives:

$$f = L + C \tag{10.98}$$
$$= (\bar{x}_1\bar{x}_2\bar{x}_3 + \bar{x}_1 x_2 x_3 + x_1 \bar{x}_2 x_3 + x_1 x_2 \bar{x}_3) + x_4 \bar{x}_5. \tag{10.99}$$

$$- \text{ Table } c \quad \begin{cases} \bar{L} = 1\ 0\ 0\ 1\ 0\ 1\ 1\ 0 \\ C = 1\ 0\ 0\ 1 \end{cases}$$

we have:
$$u = L \oplus C \oplus K.$$
for
$$x_1 x_2 x_3 x_4 x_5 = 0\ 0\ 1\ 0\ 0, \quad f = 0, \quad L = 1, \quad C = 1, \quad K = 0.$$

therefore:
$$f = L \oplus C.$$

10.5.1.7 Theorem

If a function $f(x_1, x_2, \ldots, x_n)$ admits two decompositions of the type

$$f(x_1, x_2, \ldots, x_n) = g[h(A), B, C] = k[l(B), A, C] \tag{10.100}$$

then:
$$f(x_1, x_2, \ldots, x_n) = u[h(A), l(B), C]. \tag{10.101}$$

In the development with respect to the variables C it will be seen that for any combination i of these variables:

$$g_i[h(A), B] = k_i[l(B), A] \tag{10.102}$$

which, from the preceding theorem, leads to:

$$f_i = u_i[h, l].$$

Hence:
$$f = \sum_i f_i m_i(C) = \sum_i u_i[h(A), l(B)]\, m_i(C) \tag{10.103}$$
$$= \sum_i u[h, l, i]\, m_i(C) = u[h, l, C]$$

Example:

The function

$$f(x_1, x_2, x_3, x_4, x_5) = \sum (1, 2, 4, 5, 6, 7, 13, 14, 21, 22, 29, 30)$$

is represented in the partition Tables 10.27 a and 10.27 b. It can be seen that it can be decomposed in both cases.

10.5 Functional Decompositions

Table 10.27

<pre>
        x₁x₂   x₃x₄
        0 1 0 1   0 1 0 1
x₅x₄x₃ 0 0 1 1   x₁x₂x₅   0 0 1 1

0 0 0   0 0 0 0 (0)   0 0 0   0 1 1 0   (l)
0 0 1   1 0 0 0 (h)   0 0 1   0 0 0 0   (0)
0 1 0   1 0 0 0   0 1 0   0 0 0 0
0 1 1   0 0 0 0   0 1 1   0 0 0 0
1 0 0   1 0 0 0   1 0 0   1 1 1 1   (1)
1 0 1   1 1 1 1 (1)   1 0 1   0 1 1 0
1 1 0   1 1 1 1   1 1 0   0 1 1 0
1 1 1   1 0 0 0   1 1 1   0 1 1 0

         (a)   (b)
</pre>

we have:

$$\begin{cases} h(x_1, x_2) = \bar{x}_1\bar{x}_2 = x_1 \downarrow x_2 \\ l(x_3, x_4) = x_3 \oplus x_4 \end{cases} \quad (10.104)$$

With respect to function g.

$$\begin{cases} g_0 = 0\,0\,0\,0\,0\,1\,1\,0 \\ g_1 = 0\,1\,1\,0\,1\,0\,0\,1 + 0\,0\,0\,0\,0\,1\,1\,0 = 0\,1\,1\,0\,1\,1\,1\,1. \end{cases} \quad (10.105)$$

Hence:

$$g = \bar{h}\,x_5(x_3 \oplus x_4) + h[x_5 + (x_3 \oplus x_4)]$$
$$= x_5(x_3 \oplus x_4) + h[x_5 + (x_3 \oplus x_4)]$$
$$= h \,\#\, (x_3 \oplus x_4) \,\#\, x_5. \quad (10.106)$$

There fore:

$$f = (x_1 \downarrow x_2) \,\#\, (x_3 \oplus x_4) \,\#\, x_5. \quad (10.107)$$

10.5.1.8 Theorem

If a function admits 2 decompositions of the type:

$$f(x_1, \ldots, x_n) = g[h(A, B)C] = k[l(A, C), B] \quad (10.108)$$

then it can be expressed in the form:

$$f(x_1, \ldots x_n) = u[r(A), s(B), t(C)]. \quad (10.109)$$

The function $u(r, s, t)$ is one of the three following types: $r + s + t$, $r \cdot s \cdot t$, or $r \oplus s \oplus t$ and is determined in an unique manner from the decompositions (10.108) as also are r, s and t.

288   10. The Simplification of Combinational Networks

Envisage, now, a 3-dimensional truth table for the partition (ABC) formed from the horizontal matrices AB, stacked vertically as a function of the combinations of C.

By the same reasoning as for 10.5.1.6 the reader will convince himself that:

1. there are at most 4 types of horizontal planes taken from h, $\bar{h}$, (0) and (1);

2. there are at most 4 types of vertical planes taken from l, $\bar{l}$, (0) and (1).

The simultaneous existence of conditions 1 and 2 above, as the reader will see from an examination of the different cases possible, leads to the conclusion that:

3. to a truth table satisfying the conditions of (10.108) there can only exist the conditions of vertical and horizontal planes of table 10.28 which recapitulates the different cases which can be produced, including the degenerate cases. It will be seen that there are only 3 types of configuration: $[h, \bar{h} - l, \bar{l}]$, $[h, (0) - l, (0)]$ and $[h, (1) - l, (1)]$, h and $\bar{h}$ being the horizontal planes associated with the functions of (10.108). (t_A, t_Z, t_C have the meaning of § 10.5.1.4, the presence or absence of the

Table 10.28

(V)				(H)	t_A	t_B	t_C
P	$\bar{P}$	(0)	(1)				
0	0	0	0	—	—	—	—
0	0	0	1	(1)	1	1	1
0	0	1	0	(0)	1	1	1
0	0	1	1	(Q_1)	1	0	1
0	1	0	0	$(Q_2), (\bar{Q}_2)$, (0), (1)	0	1	0
0	1	0	1	(Q_3), (1)	0	0	0
0	1	1	0	(Q_4), (0)	0	0	0
0	1	1	1	(Q_5)	0	0	1
1	0	0	0	$(Q_6), (\bar{Q}_6)$, (0), (1)	0	1	0
1	0	0	1	(Q_7), (1)	0	0	0
1	0	1	0	(Q_8), (0)	0	0	0
1	0	1	1	(Q_9)	0	0	1
1	1	0	0	$(Q_{10}), (\bar{Q}_{10})$	0	0	0
1	1	0	1	(Q_{11})	0	0	1
1	1	1	0	(Q_{12})	0	0	1
1	1	1	1	(Q_{13})	0	0	1

10.5 Functional Decompositions

'vertical" types $P, \overline{P}$ (0), or (1) being indicated, as in Table 10.23, by the binary variables; (V) and (H) indicate the possible selections of vertical and horizontal planes, certain of which may be degenerate).

Taking the foregoing into account, it will be observed that the planes h and l are associated with decomposable functions for the partition AB:

$$\begin{cases} h = m[p(A), B] \\ l = n[p(A), B] \end{cases}$$

Table 10.29 summarises, as a function of the 3 possible different configurations, the types of function which can be obtained — (cf. § 10.5.1.3).

Table 10.29

$h, \overline{h}$	$l, \overline{l}$	$p, \overline{p}$	$g_0 = \overline{g}_1$	$k_0 = \overline{k}_1$	$g = h \oplus g_0$	$k = l \oplus k_0$	$f = p \oplus g_0 \oplus k_0$
$h, 0$	$l, 0$	$p, 0$	$g_0 = 0$	$k_0 = 0$	$g = h \cdot g_1$	$k = l \cdot k_1$	$f = p \cdot g_1 \cdot k_1$
$h, 1$	$l, 1$	$p, 1$	$g_1 = 0$	$k_1 = 0$	$g = h + g_0$	$k = l + k_0$	$f = p + g_0 + k_0$

Example:

Table 10.30

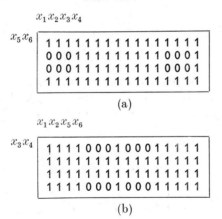

(a)

(b)

From Table 10.30a, is derived:

$$f = [x_1, x_2 + (x_3 \oplus x_4)] + \overline{x_5 \oplus x_6}. \qquad (10.110)$$

Table 10.30b corresponds to the decomposition:

$$f = [x_1 x_2 + \overline{(x_5 \oplus x_6)}] + (x_3 \oplus x_4) \qquad (10.111)$$

i.e.:

$$f = x_1 x_2) + (x_3 \oplus x_4) + \overline{(x_5 \oplus x_6)}. \qquad (10.112)$$

10.5.1.9 Theorem

If a function admits 2 decompositions of the type:

$$f(x_1, x_2, \ldots, x_n) = g[h(A, B), C, D] = k[l(A, C), B, D] \quad (10.113)$$

then:

$$f(x_1, x_2, \ldots, x_n) = v\{u[r(A), s(B), t(C)], D\}. \quad (10.114)$$

For each combination i of the variables D there exists a function $f_i(A, B, C)$ which, as before, can be associated with a 3-dimensional truth table.

It is easily verified that the preceding theorem is applicable and that for every i, $f_i = u_i[r, s, t]$.

But, for every truth table, the functions h and l being the same, the resulting vertical or horizontal types of plane are the same for all the tables and, consequently, all the functions u_i belong to one and the same category within the 3 possible categories. If u is one of these functions then all others take the form $a_i u + b_i \bar{u}$. From which:

$$f = \sum_i (a_i u + b_i \bar{u}) m_i(D) = \left[\left(\sum_i a_i\right) u + \left(\sum_i b_i\right) \bar{u}\right] m_i(D) \quad (10.115)$$

$$= v(u, D) \quad (10.116)$$

Example:

Consider the table for an 8-variable function $x_1, x_2, x_3, \ldots, x_7, x_8$, identical with table 10.30 for $x_7 x_8 = 00$ or 11 and identical with the complementary table for $x_7 x_8 = 01$ or 10. We have:

$$\begin{cases} f'_{00}(x_1, x_2, \ldots, x_6) = f \\ f'_{01}(x_1, x_2, \ldots, x_6) = \bar{f} \\ f'_{10}(x_1, x_2, \ldots, x_6) = \bar{f} \\ f'_{11}(x_1, x_2, \ldots, x_6) = f \end{cases} \quad (10.117)$$

whence:

$$f(x_1, x_2, \ldots, x_8) = f \bar{x}_7 \bar{x}_8 + \bar{f} x_7 x_8 + \bar{f} x_7 \bar{x}_8 + f x_7 x_8$$

$$= f \overline{(x_7 \oplus x_8)} + \bar{f}(x_7 \oplus x_8)$$

$$= f \oplus (x_7 \oplus x_8)$$

$$= \left[x_1 x_2 + (x_3 \oplus x_4) + \overline{x_5 \oplus x_6}\right] \oplus x_7 \oplus x_8$$

10.5 Functional Decompositions

10.5.2 Simple Disjoint Decompositions and Maximal Cubes

The following notations will be used:

a) $\begin{cases} Y = x_{i_1}, \ldots, x_{i_s} \\ Z = x_{i_{s+1}}, \ldots, x_{i_n} \end{cases}$ (10.118)

(sets of variables comprising n and $s - n + 1$ elements)

b) $\begin{cases} (Y) = (x_{i_1}, \ldots, x_{i_s}) \\ (Z) = (x_{i_{s+1}}, \ldots, x_{i_n}) \end{cases}$ (10.119)

$(YZ) = (x_{i_1}, \ldots, x_{i_s}, x_{i_{s+1}}, \ldots, x_{i_n})$ $\quad (x_{i_j} = 0, 1 \text{ or } -)$

$(hZ) = (h, x_{i_{s+1}}, \ldots, x_{i_n})$

(cubes with dimensions: s, $n - s$, $n - s + 1$)

c) *Vertices of* $(Y), (Z)$

These are obtained by allocating to the free variables (dashes: —) the fixed values 0 or 1.

$\begin{cases} A = (a_{i_1}, \ldots, a_{i_s}) \\ B = (b_{i_{s+1}} \ldots b_{i_n}) \end{cases}$ $(a_{i_k}, b_{i_j} = 0, 1)$ $(k = 1, \ldots, s;\ j = s+1, \ldots, n)$ (10.120)

d) *Sub-cubes of* (YZ) *and* (hZ)

It will be necessary to consider the sub-cubes obtained by allocating to the free variables (Y), or of (Z) or of both, the values 0 and 1; examples:

$$(AZ), (YB), (AB) \text{ or } AB$$
$$\begin{cases} (AZ), (YB) \subset (YZ) \\ (AB), AB \in (YZ) \end{cases}$$
$$(aZ), (hB), (aB) \text{ or } aB \qquad (10.121)$$
$$\begin{cases} aZ, hB \subset (hZ) \\ (aB), aB \in (hZ). \end{cases}$$

Particular cases:

$\begin{cases} (-\cdots-Z) & (\text{dimension } (Y) = s) \\ (Y-\cdots-) & (\text{dimension } (Z) = n - s) \\ (-Z) & (\text{dimension } (h) = 1) \end{cases}$ (10.122)

10. The Simplification of Combinational Networks

e) E_f, E_g, E_h: *characteristic sets of* f; g; h

For a given function possessing a single disjoint decomposition

$$f(x_1, x_2, \ldots, x_n) = g[h(Y), Z]$$

are obtained the following properties (Pichat)[1]

Theorem: *If a cube* (YZ) *is a maximal cube of* f, *and if the cube* (Y) *is of dimension less than* s, (Y) *is a maximal cube, either for* h, *or for* $\bar{h}$.
Indeed, the cube (YZ) being maximal, by definition:

1. $(YZ) \subset E_f$
2. For every cube $K \supset (YZ)$, $K \not\subset E_f$ [2]

Assuming that there are 2 vertices,

$$A = (a_{i_1}, \ldots, a_{i_s}) \quad \text{and} \quad A' = (a'_{i_1} \ldots a'_{i_s}),$$

of (Y) such that:
$$h(A) = b, \quad h(A') = \bar{b}. \tag{10.123}$$

Given that:
$$(AZ) \subset (YZ) \subset E_f \text{ and } (A'Z) \subset (YZ) \subset E_f, \tag{10.124}$$

for every vertex B of Z
$$f(AB) = f(A'B) = 1 \tag{10.125}$$

whence:
$$h(A)g(1, B) + \bar{h}(A)g(0, B) = h(A')g(1, B) + \bar{h}(A')g(0, B)$$
$$bg_1(B) + \bar{b}g_0(B) = \bar{b}g_1(B) + bg_0(B), \tag{10.126}$$

which results in:
$$g_0(B) = g_1(B) = 1$$

and, consequently, for every point $YB (B \in Z)$, we have

$$f(YZ) = h(Y)g(1, B) + \bar{h}(Y)g(0, B) \tag{10.127}$$
$$= g(1, B) = g(0, B) = 1, \tag{10.128}$$

which will, therefore, give

$$(YZ) \subset (-\cdots - Z) \subset E_f \tag{10.129}$$

---

[1] A geometric presentation has been adopted in this instance, by utilising the n-dimensional cube.

[2] 1. $\not\subset$ signifies: is not included in.
2. $\supset$ is, for the cubes considered here, a strict inclusion (true subsets).

10.5 Functional Decompostiions

which contradicts the hypothesis which states that (YZ), where (Y) is of dimension $\leq s - 1$, is the maximal cube of f. Therefore:

$$Y \subset E_h \quad \text{or} \quad Y \subset E_{\bar{h}}. \tag{10.130}$$

In addition, for every cube (Y') containing (Y) and of superior dimension,

$$(Y'Z) \supset (YZ) \tag{10.131}$$

(YZ) being maximal, there exists $A'B(A' \in Y', B \in Z)$, such that

$$A'B \in (Y'Z) \tag{10.132}$$

and

$$f(A'B) = h(A')g(1, B) + \bar{h}(A')g(0, B) = 0. \tag{10.133}$$

But, for every vertex AB where $A \in (Y)$, we have

$$f(AB) = h(A)g(1, B) + \bar{h}(A)g(0, B) = 1. \tag{10.134}$$

Consequently, if for $A \in (Y)$, $h(A) = b$, (b being a constant value 0 or 1) there exists, therefore, $A' \in Y'$ such that $h(A') = \bar{b}$. In short, (Y) is a maximal cube of h or of $\bar{h}$.

Consider next the function $f(x_1, \ldots, x_n)$, assumed decomposable for the separation YZ. The maximal cubes of f can be classified into two categories: those of form $(Y_j Z_j)$ where (Y_j) has a dimension $\leq s - 1$, and those of form $(Y_i Z_i)$, where (Y_i) has a dimension s, i.e. the cubes which it has been agreed to denote as $(- \cdots - Z_i)$. In their turn the first may be divided into two groups: those denoted as $(Y_j{}^1 Z_j{}^1)$ for which $(Y_j{}^1)$ a maximal cube of h or $\bar{h}$ by the preceding theorem, is more precisely a maximal cube of $h^1 = h$, and those denoted by $(Y_k{}^0 Z_k{}^0)$, for which $(Y_k{}^0)$ is a maximal cube of $h^0 = \bar{h}$: The set of these cubes can, therefore, be expressed:

$$B = \bigcup_i (- \cdots - Z_i) \cup \bigcup_j (Y_j{}^1 Z_j{}^1) \cup \bigcup_k (Y_k{}^0 Z_k{}^0) \tag{10.135}$$

we then have the theorem (Pichat).

Theorem: *The sets of cubes $\bigcup_j (Y_j{}^1)$ and $\bigcup_k (Y_k{}^0)$, so long as they are not empty, constitute the maximal cubes of h and $\bar{h}$ respectively.*

From the preceding theorem, and the definitions of $(Y_j{}^1)$ and $(Y_k{}^0)$, every $(Y_j{}^1)$ is a maximal cube of $h' = h$, and every $Y_k{}^0$ a maximal cube of $h^0 = \bar{h}$.

Conversely, consider a maximal cube of h or $\bar{h}$, say h for instance, which will be denoted by (Y'). For every $A \in Y'$, and fro every B, we have:

$$f(AB) = g[h(A), B] = g(1, B). \tag{10.136}$$

In order that a term $(Y'Z')$ may be a maximal cube of f, it is necessary and sufficient that:

(1) $(Y'Z') \subset E_f$.
(2) For every $(Z'') \supset (Z')$, $(Y'Z'') \not\subset E_f$.
(3) For every $(Y'') \supset (Y')$, $(Y''Z') \not\subset E_f$.

The conditions (1) and (2) ensure that (Z') is a maximal cube of $g(1, Z)$ (cf. Fig. 10.33 by way of an example). Given such a maximal cube (Z'), condition (3) ensures that:

$$A \in (Y''), \quad B \in (Z') \tag{10.137}$$

such that
$$f(AB) = 0.$$

Since $g(1, B) = 1$ for every B belonging to Z', there exists $B \in (Z')$ such that $g(0, B) = 0$. There exists at least one cube (Z') so that this condition can be fulfilled. Indeed, assuming that this was not true: for every maximal cube (Z') and for every $B \in (Z')$ there would exist $g(0, B) = 1$. Since the set of maximal cubes of $g(1, Z)$ covers the characteristic set $E_{g(1,Z)}$ of $g(1, Z)$, there would consequently be for every $B \in E_{g(1,Z)}$, $g(0, B) = 1$, i.e. $B \in E_{g(0,Z)}$, or, equivalently, $E_{g(1,Z)} \subset E_{g(0,Z)}$, which could be written:

$$g(0, Z) = g(1, Z) + A(Z), \quad \bigl(A(Z) \not\subset g(1, Z)\bigr) \tag{10.138}$$
$$\bigl(\text{e.g.:} \ A(Z) \cdot g(1, Z) = 0, \ A(Z) \neq 0\bigr)$$

which would give:

$$g(h, Z) = h g(1, Z) + \bar{h} g(0, Z) = (h + \bar{h}) g(1, Z) + \bar{h} A(Z)$$
$$= g(1, Z) + \bar{h} A(Z). \tag{10.139}$$

The function $f = g(h, Z)$ would then have no maximal cube of form $(Y_1 Z_1)$, where (Y_1) is of dimension $\leq s-1$. Indeed, for every $AB \in Y_1 Z_1$, since $Y_1 \subset E_h$, there would be $\bar{h}(A) = 0$ and $f = g(1, B) = 1$, which would then give:

$$(Y_1 Z_1) \subset (-\cdots - Z_1) \subset E_f \tag{10.140}$$

and $Y_1 Z_1$ would not be a maximal cube of f.

Summarizing, if for a given maximal cube (Y') of h, there had not been a term of form $(Y'Z')$ in the set of maximal cubes of f, then there would be *no* maximal cube of the type $(Y_j^1 Z_j^1)$ therein, which is a contradiction of the hypothesis which states that there must be at least one. Consequently, for every maximal cube Y_i^1 of h, there exists a term $Y_i^1 Z_i^1$ in B.

This leads to the following theorem (Pichat).

10.5 Functional Decompositions

Theorem: The set of cubes

$$B' = \bigcup_i (-Z_i) \cup \bigcup_j (1, Z_j^1) \cup \bigcup_k (0, Z_k^0) \qquad (10.141)$$

is the set of maximal cubes of $g(h, Z)$.

Firstly, it must be shown that if a cube of form $(- \cdots - Z_i)$ is a maximal cube for f, $(-Z_i)$ is a maximal cube for $g(h, Z)$. Indeed, if h is not identically 1, there exist AB such that $h(A) = 0$, AB' such that $h(A') = 1$. The property $f = 1$ for every $YB \in (- \cdots - Z_i)$ implies, therefore, that for every $B \in Z_i$, $g(1, B) = 1$ and that $g(0, B) = 1$, thus that for every $aB \in (-Z_i)$, $g(aB) = 1$. Additionally, if $Z' \supset Z_i$, there exists AB such that $f(AB) = 0$, i.e. by posing $a = h(A)$, there exists aB such that $g(aB) = 0$.

Next it must be shown that for every maximal cube $(Y_j^1 Z_k^1)$ or $(Y_k^0 Z_k^0)$, such that $Y_j^1 (Y_k^0)$ is of dimension $\leq s - 1$, the cube $1 Z_j^1 (0 Z_k^0)$ is the maximal cube of $g(h, Z)$.

Consider a cube of the type $y_j^1 Z_j^1$. Since

$$(Y_j^1) \subset E_h \quad \text{and} \quad (Y_j^1 Z_j^1) \subset E_f,$$

we have, for every $B \in (Z_j^1)$, $g(1, B) = 1$. Thus $1 Z_j^1 \subset E_{g(h,Z)}$. Moreover, no maximal cube of form $(- Z_j^1)$ is contained in $E_{g(h,Z)}$, since this would give for every $B \in Z_j^1$, $g(1, B) = 1$ and $g(0, B) = 1$, and so

$$Y_j^1 Z_j^1 \subset Z_j^1 \subset E_f, \qquad (10.142)$$

which contradicts the hypothesis which states that $X_j^1 Z_j^1$ is a maximal cube of f. From what has already been seen, neither is there a maximal cube of form $1 Z'(Z' \supset Z_j^1)$ contained in $E_{g(h,Z)}$. There is thus no maximal cube containing $1 Z_j^1$ and contained in $E_{g(h,Z)}$; in short, $1 Z_j^1$ is a maximal cube of $g(h, Z)$.

Conversely:

— If $(- Z_i)$ is a maximal cube of $g(h, Z)$, the cube $(- \cdots - Z_i)$ is so, too, for f.

Indeed, for every $B \in (Z_i)$, $g(1, B) = 1$ and $g(0, B) = 1$, whence for every $B \in (Z_i)$

$$f = h g(1, B) + \bar{h} g(0, B) = h + \bar{h} = 1.$$

In addition, for every (Z') such that $(- Z') \supset (- Z_i)$, there exists $aB \in (- Z_i)$ such that $g(a, B) = 0$. If $a = 1, h$ not being constant, there exists (A) such that $h(A) = 1$, and $B \in Z'$ such that $g(1, B) = 0$. If $a = 0$; there exist A such that $h(A) = 0$ and $B \in Z'$ such that $g(0, B) = 0$. In any case, there exists $AB \in (- \cdots - Z_i)$ such that $f(AB) = 0$.

— For every maximal cube $1 Z_j^1(0, Z_k^0)$ of $g(h, Z)$, there exists a term $(Y_j^1 Z_j^1)$, $[(Y_k^0 Z_k^0)]$, which is maximal cube of f. Indeed, let us consider, for example, the case of a cube of type $1 Z_j^1$ and try for Y_j^1 a maximal cube of h. For every AB, such that $A \in Y_j^1$, $B \in Z_j^1$, $f = g(1, B) = 1$. For every $Y_j^1 Z'$, where $Z' \supset Z_j^1$, there exists $Z' \not\subset E_f$, since there exists $B \in Z'$, such that $g(1, B) = 0$, and A such that $A \in Y_j^1$ $(h(A) = 1)$; there exists, therefore, $AB \in Y_j^1 Z'$ such that $f = 1 \cdot g(1, B) = 0$. On the other hand, for every $Y' Z_j^1$, where $Y' \supset Y_j^1$, there exists $A \in Y'$, $B \in Z_j^1$, such that $f = 0$. Otherwise there would exist (Y') such that for every $A \in Y'$ and every $B \in Z_j^1$, we would have

$$g(1, B) = 1, \qquad g(0, B) = 1$$
$$1 Z_j^1 \subset (- Z_j^1) \subset E_{g(h, Z)} \tag{10.143}$$

and $1 Z_j^1$ would, therefore, not be a maximal cube, contradicting the hypothesis.

For a given partition (YZ) of variables, the set B of maximal cubes f can always be re-written in the form:

$$B'' = \overbrace{\bigcup_i (- \cdots - Z_i)}^{A} \cup \Big[\overbrace{\bigcup_j (Y_j^1 Z_j^1) \cup \bigcup_k (Y_k^0 Z_k^0)}^{B'''} \Big] \cup \overbrace{\bigcup_l (Y_l - \cdots -)}^{C} \tag{10.144}$$

by bringing into play the cubes $(Y_l - \cdots -)$, where the Z term is of dimension $n - s$, the terms $(Y_j^1 Z_j^1)$, $(Y_k^0 Z_k^0)$ being such that the (Y_j^1), (Y_k^0) are of dimension $\leq s - 1$, the (Z_j^1), (Z_k^0) of dimension $\leq n - s - 1$, some of the three sets A, B''', C may well be empty. One then has the following theorem (Pichat):

1. If the 3 sets A, B''', C are non-empty, the function f cannot be decomposed according to the partition YZ in question.

2. If the set $C = \bigcup_l (Y_l - \cdots -)$ is empty, a necessary and sufficient condition that f be decomposable with the partition in question is that it shall be possible to write:

$$\bigcup_j (Y_j^1 Z_j^1) \cup \bigcup_k (Y_k^0 Z_k^0) = \bigcup_{j'} \bigcup_{j''} Y_{j'}^1 Z_{j''}^1 \cup \bigcup_{k'} \bigcup_{k''} Y_{k'}^0 Z_{k''}^0 \tag{10.145}$$

and that we have:

$$\bigcup_{j'} Y_{j'}^1 = \varnothing \tag{10.146}$$

or

$$\bigcup_{k'} Y_{k'}^0 = \varnothing$$

or
$$\bigcup_j E_{Y_j^1} = \overline{\bigcup_k E_{Y_k^0}}. \tag{10.147}$$

then:
$$\begin{cases} E_h = \bigcup_j E_{Y_j^1} \\ E_{\bar h} = \bigcup_k E_{Y_k^0} \end{cases} \tag{10.148}$$

Finally, if there are no terms in B''', f can be decomposed.

First part — Assume that f is decomposable, and that there is a term of form $(Y_k - \cdots -)$ with $Y_k \in E_h$. (It could be reasoned in an analogous way if there were no term Y_h in E_h, but only in $E_{\bar h}$). Thus, for every $A \in Y_k$ and for every B,

$$f(AB) = g(1, B) = 1$$

which would give
$$f(YZ) = g(1, Z) h(Y) + g(0, Z) \bar h(Y) = h(Y) + \bar h(Y) g(0, Z)$$
$$= h(Y) + g(0, Z). \tag{10.149}$$

It is easily verified that the maximal cubes of f are not of form $(Y_j^1 Z_j^1)$ or $(Y_k^0 Z_k^0)$ (terms where the Y_j, Y_k are of dimension $< s$, and where the Z_j^1, Z_j^0 are of dimension $< n - s$), which contradicts the hypothesis (B''' non-empty). Consequently, if there are 3 types of terms $(Y_k - \cdots -)$, $(- \cdots - Z_i)$, and $(Y_j^e E_j^e)$, then f cannot be decomposed.

Second part — From the preceding theorems and their proofs it follows that if a term $(Y_j^1 Z_{jl})$ is a maximal cube, and if there exists at least one term commencing with $Y_{j'}^1$, $(Y_{j'}^1 Z')$ which is also a maximal cube, then $(Y_{j'}^1 Z_{jl})$ is also a maximal cube. Consequently the terms $Z_{j''}^1$ which can be combined with $Y_{j'}^1$ are the same for every $Y_{j'}^1$. This result can be transposed in terms of Y_k^0 and Z_k^0. Hence the relationship (10.145). As for the relations (10.146-7), they result immediately from the preceding theorems.

Conversely, the relations (10.145-8) imply that f can be decomposed. Indeed, by designating by ΣZ_j^1, ΣZ_k^0, ΣZ_i ΣY_k^0 and ΣY_j^1, the functions having for their characteristic sets the unions of cubes (Z_j^1), (Z_k^0), (Z_i) (Y_k^0) and (Y_j^1) respectively, it can be said that:

$$f = \left(\sum_{j'} Y_{j'}^1\right)\left(\sum_{j''} Z_{j''}^1\right) + \left(\sum_{k'} Y_{k'}^0\right)\left(\sum_{k''} Z_{k''}^0\right) + \sum_i Z_i$$
$$= h \left(\sum_{j''} Z_{j''}^1\right) + \bar h \left(\sum_{k''} Z_{k''}^0\right) + \sum_i Z_i$$
$$= h \left(\sum_{j''} Z_{j''}^1 + \sum_i Z_i\right) + \bar h \left(\sum_{k''} Z_{k''}^0 + \sum_i Z_i\right)$$
$$= h g_1(Z) + \bar h g_0(Z) \tag{10.150}$$

The last part of the theorem (the case where the set B''' is empty) is immediate.

In the example of Fig. 10.33, we have, $x_2, x_1, x_5, x_4, x_3, x_6$ being arranged as A, B, C, D, E, F in Fig. 8.13 and the diagram f being repeated in the interests of clarity:

— 4 cubes of form $(Y_h^0 Z_h^0)$,
— 8 cubes of form $(Y_j^1 Z_j^1)$,
— 1 cube of form $(- \cdots - Z_i)$ (not shown).

The set of these cubes is to be found in Table 10.31 below.

Table 10.31

Maximal cubes of $f(x_1, x_2, x_3, x_4, x_5, x_6)$	Maximal cubes of $h(x_1, x_2, x_3)$ or $\bar{h}(x_1, x_2, x_3)$
1. $\bar{x}_1 \ - \ \bar{x}_3 \ \bar{x}_4 \ - \ \bar{x}_6$	$\left. \begin{array}{l} \bar{x}_1 - \bar{x}_3 \\ x_1 \bar{x}_2 x_3 \end{array} \right\} (\bar{h})$
2. $\bar{x}_1 \ - \ \bar{x}_3 \ \bar{x}_4 \ \bar{x}_5 \ -$	
3. $x_1 \ \bar{x}_2 \ x_3 \ \bar{x}_4 \ - \ \bar{x}_6$	
4. $x_1 \ \bar{x}_2 \ x_3 \ \bar{x}_4 \ \bar{x}_5 \ -$	
5. $x_1 \ - \ \bar{x}_3 \ x_4 \ - \ -$	$\left. \begin{array}{l} x_1 - \bar{x}_3 \\ -\ x_2 x_3 \\ x_1 x_2 - \\ \bar{x}_1 - x_3 \end{array} \right\} (h)$
6. $- \ x_2 \ x_3 \ x_4 \ - \ -$	
7. $x_1 \ x_2 \ - \ x_4 \ - \ -$	
8. $\bar{x}_1 \ - \ x_3 \ x_4 \ - \ -$	
9. $- \ x_2 \ x_3 \ - \ \bar{x}_5 \ \bar{x}_6$	
10. $\bar{x}_1 \ - \ x_3 \ - \ \bar{x}_5 \ \bar{x}_6$	
11. $x_1 \ x_2 \ - \ - \ \bar{x}_5 \ \bar{x}_6$	
12. $x_1 \ - \ \bar{x}_3 \ - \ \bar{x}_5 \ \bar{x}_6$	
13. $- \ - \ - \ \bar{x}_4 \ \bar{x}_5 \ \bar{x}_6$	

Table 10.32

$x_4 x_5 x_6$ \ $x_1 x_2 x_3$	$\bar{x}_1 - \bar{x}_3$	$x_1 \bar{x}_2 x_3$	$x_1 - \bar{x}_3$	$- x_2 x_3$	$x_1 x_2 -$	$\bar{x}_1 - x_3$
$\bar{x}_4 \ - \ \bar{x}_6$	1	1				
$\bar{x}_4 \bar{x}_5 \ -$	1	1				
$x_4 \ - \ -$			1	1	1	1
$- \ \bar{x}_5 \bar{x}_6$			1	1	1	1

In practice, to determine whether a function can be decomposed, use can be made of a special table. On one axis are shown the Y parts of the prime term of the function, on the other axis are shown those of Z. The above theorem takes the form of a lines criterion (cf. 10.5.1.3): in order

10.5 Functional Decompositions

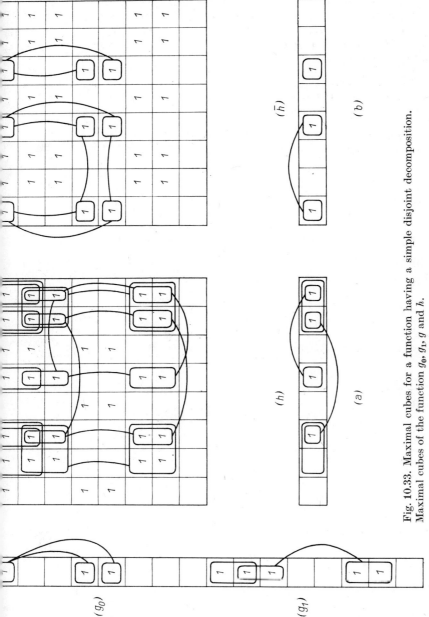

Fig. 10.33. Maximal cubes for a function having a simple disjoint decomposition. Maximal cubes of the function g_0, g_1, g and h.

that the function may be decomposed, it is necessary and sufficient that the table of prime terms possesses at most 3 lines recruted from the types L, $\bar{L}$ and 1.

Example: the function of the above example gives Table 10.32. From which the following functions can be deduced:

$$\begin{cases} h^\alpha(x_1, x_2, x_3) = \bar{\alpha}[\bar{x}_1\bar{x}_3 + x_1\bar{x}_2x_3] + \alpha[x_1\bar{x}_3 + x_2x_3 + x_1x_2 + x_2x_3] \\ g_{\bar{\alpha}} = \bar{x}_4\bar{x}_6 + x_4x_6 \\ g_\alpha = x_4 + \bar{x}_5\bar{x}_6, \qquad f = g_\alpha h^\alpha + g_{\bar{\alpha}} h^{\bar{\alpha}} \end{cases} \quad (10.115)$$

α being a Boolean parameter of value 0 or 1 at choice. These diverse functions are brought to light in Fig. 10.33 (with the exception of the term $\bar{x}_4\bar{x}_5\bar{x}_6$).

10.5.3 Simple Non-Disjoint Decompositions

A function $f(x_1, \ldots, x_n)$ $(n \geq 3)$ allows a simple, non-disjoint (non-trivial) decomposition if it can be expressed:

$$f(x_1, \ldots, x_n) = g[h(x_{i_1}, \ldots, x_{i_k} x_{i_{s+1}}, \ldots, x_{i_n}) \, x_{i_{k+1}}, \ldots, x_{i_s}, x_{i_{s+1}}, \ldots, x_{i_n}] \quad (10.152)$$

with $(1 < k < s)$.

The variables $x_{i_{s+1}}, \ldots, x_{i_n}$ are thus common to both functions g and h. In abridged notation this can be expressed:

$$f(x_1, \ldots, x_n) = f(X\,Y\,Z) = g[h(X, Z), Y, Z] \quad (X, Y, Z \neq \varnothing) \quad (10.153)$$

The reader will verify that the decompositions eliminated by the restrictions $(n \geq 3$ and $1 < k < s)$ are trivial.

A function can be represented by a 3-dimensional truth table corresponding to the 3 sets X, Y and Z for each combination $Z = j$ (height); there exists, therefore, a rectangular table which may be considered as the partition matrix of the function $f(X, Y, j) = f_j(X, Y)$. It takes the name of a sub-matrix relating to the combination $Z = j$. Then we have the following condition of decomposability.

Theorem: (*Ashenhurst*) — In order that a function may admit a simple non-disjoint decomposition with the cover YXZ it is necessary and sufficient that for every value j of $Z(0 \leq j \leq 2^{n-s} - 1)$, the function $f_j(X, Y)$ represented by the corresponding sub-matrix admit a simple disjoin decomposition of the type:

$$f_j(X, Y) = g_j[h_j(X), Y] \quad (10.154)$$

This condition is necessary.

10.5 Functional Decompositions

Envisage a 3-dimensional truth table for f. A function $h(X, Z)$ is represented by the horizontal vectors (0) or (1) associated with the combinations (XZ), a function of Z and Y is represented by the vectors 0) and (1) associated with the combinations (ZY). Consequently, in applying the relationship:

$$f(x_1, x_2, \ldots x_n) = h(X, Z) g_1(Y, Z) + \bar{h}(X, Z) g_0(Y, Z) \quad (10.155)$$

It will be seen that for $Z = a$ given j, will produce a plane comprising lines taken from among the following types:

$$h_j, \bar{h}_j, (0) \text{ and } (1)$$

The functions $f(X, Y, j) = f_j(X, Y)$ are, therefore, decomposable for every j with a decomposition of the form:

$$f_j(X, Y) = g_j[h_j(X), Y] \quad (j = 1, \ldots 2^{n-s}) \quad (10.156)$$

This condition is sufficient.

Indeed, if the relation (10.156) is true, for every j the functions h_j and g_j can be determined upto complementation. We may then put:

$$h_j(X) = h(X, j)$$

and

$$g_j(h, Y) = g(h, Y, j),$$

giving

$$h(X, Z) = \sum_{j=0}^{j=2^{n-s}} h(X, j) m_j(Z)$$

and

$$g(h, Y, Z) = \sum_{j=0}^{j=2^{n-s}} g(h, Y, j) m_j(Z)$$

whence:

$$f = \sum_j f_j(X, Y) m_j(Z) = \sum_j g_j[h_j(X), Y] m_j(Z)$$
$$= \sum_j g[h(X, Z), Y, j] m_j(Z) = g[h(X, Z), Y, Z]. \quad (10.157)$$

Example: consider the function $f(x_1, x_2, \ldots x_6)$ given by Table 10.34. For $x_3 = 0$ and $x_3 = 1$, the submatrices show the decompositions. By taking, for $x_3 = 0$,

$$h_0 = x_1 \bar{x}_2 + \bar{x}_1 x_2 = x_1 \oplus x_2$$

then:

$$\begin{cases} g_0^1 = \bar{x}_4 \bar{x}_5 \bar{x}_6 + x_4 x_5 \bar{x}_6 + \bar{x}_4 \bar{x}_5 x_6 + x_4 \bar{x}_5 x_6 \\ g_0^0 = x_4 \bar{x}_5 \bar{x}_6 + x_4 x_5 \bar{x}_6 + \bar{x}_4 \bar{x}_5 x_6 + \bar{x}_4 x_5 x_6 \end{cases} \quad (10.158)$$

Table 10.34

$x_6 x_5 x_4$	x_1 x_2 x_3	0 1 0 1 0 1 0 1 0 0 1 1 0 0 1 1 0 0 0 0 1 1 1 1	g_0^1	g_0^0	g_1^1	g_1^0
0 0 0		0 1 1 0 0 0 0 1	1	0	1	0
0 0 1		1 0 0 1 0 0 0 0	0	1	0	0
0 1 0		0 0 0 0 0 0 0 0	0	0	0	0
0 1 1		1 1 1 1 1 1 1 0	1	1	0	1
1 0 0		1 1 1 1 0 0 0 0	1	1	0	0
1 0 1		0 1 1 0 0 0 0 1	1	0	1	0
1 1 0		1 0 0 1 1 1 1 0	0	1	0	1
1 1 1		0 0 0 0 1 1 1 1	0	0	1	1

$$\begin{cases} h_0 = 0\ 1\ 1\ 0 \\ h_1 = 0\ 0\ 0\ 1 \end{cases}$$

and
$$g_0 = \bar{h}_0 g_0^0 + h_0 g_0^1 \tag{10.159}$$

and for $x_3 = 1$,
$$h_1 = x_1 x_2$$

then
$$\begin{cases} g_1^1 = \bar{x}_4 \bar{x}_5 \bar{x}_6 + x_4 \bar{x}_5 x_6 + x_4 x_5 x_6 \\ g_1^0 = x_4 x_5 \bar{x}_6 + \bar{x}_4 x_5 x_6 + x_4 x_5 x_6 \end{cases} \tag{10.160}$$

and
$$g_1 = \bar{h}_1 g_1^0 + h_1 g_1^1 \tag{10.161}$$

whence:
$$h(x_1, x_2, x_3) = \bar{x}_3 h_0(x_1, x_2) + x_3 h_1(x_1, x_2) \tag{10.162}$$
$$f = \bar{x}_3 g_0(h, x_4, x_5, x_6) + x_3 g_1(h, x_4, x_5, x_6). \tag{10.163}$$

Chapter 11

Concept of the Sequential Network

So far, studies have been made of networks which have been known as combinatorial networks,[1] i.e. those which associate to every binary input combination an unique binary output combination: they realise the mapping of the set of input combinations into that of the output combinations. With this type of function it happens, in particular, that every time a given combination C is presented at the input, at distinct instants, the same combination is seen to appear at the output. The input/output correspondence established by such a network is, therefore, "punctual", it is indifferent to what has happened between the different instants at which the combination C has been applied: and does not cause to intervene any hereditary phenomena. Now, with the help of a number of examples, we shall introduce a new type of network which does not fall into the above category and where, on the contrary, the history of the input is to play a rôle, that of the sequential network.

11.1 Elementary Example. The Ferrite Core

Consider for example the system consisting of a ferrite core (i.e. a torus-shaped ferrite element) used as a "memory unit" in a digital computer in accordance with the technique of coincident currents. It is equipped with 2 windings corresponding to the 2 coordinates X and Y of the core in a so-called core matrix (which will be associated with, for

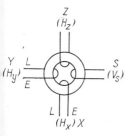

Fig. 11.1. Ferrite core with input and output windings

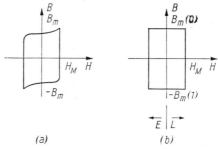

Fig. 11.2. Hysteresis cycle (a, real; b, simplified)

[1] or "combinational networks".

example, a 'bit' position in a binary word), a third winding, the 'inhibitor' winding Z (used for writing into the memory) and finally an output winding S, by means of which the memory is read (Fig. 11.1). In a much simplified manner, it will be assumed that the magnetic behaviour of the core will be manifested by an ideal rectangular hysteresis cycle (Fig. 11.2b). Under these conditions the functioning of the core during read-out can be described as follows:

If H_x and H_y are magnetic excitation fields, and H_z that furnished by the inhibitor winding, then by construction:

$$\begin{pmatrix} \varepsilon_X \ll H_M \\ \varepsilon_Y \ll H_M \end{pmatrix} \quad \begin{array}{c|c} \text{Reading phase} & \text{Writing phase} \\ \hline H_X = \begin{cases} 0 \\ \text{ou} \\ \frac{1}{2} H_M + \varepsilon_X \end{cases} & H_X = \begin{cases} 0 \\ \text{ou} \\ -\frac{1}{2} H_M - \varepsilon_X \end{cases} \\ H_Y = \begin{cases} 0 \\ \text{ou} \\ \frac{1}{2} H_M + \varepsilon_Y \end{cases} & H_Y = \begin{cases} 0 \\ \text{ou} \\ = \frac{1}{2} H_M - \varepsilon_Y \end{cases} \\ H_Z = 0 & H_Z = \frac{1}{2} H_M \end{array} \quad (11.1)$$

The selection of a core is made by the application of simultaneous current impulses to the lines X and Y, having the values given above. Introducing the binary variables X, Y, Z (Table 11.3):

Tableau 11.3 Binary variables associated with a core

H_X	X	H_Y	Y	H_Z	Z
0	0	0	0	0	0
$\neq 0$	1	$\neq 0$	1	$\neq 0$	1

and the variable S defined as:

$\begin{cases} S = 0 \text{ so long as the output voltage is zero (less than a certain threshold } V_m') \\ S = 1 \text{ so long as the output voltage is not zero (greater than the threshold } V_m') \end{cases}$

11.1 Elementary Example. The Ferrite Core

Finally, in the permanent case, the core possesses one of 2 states of magnetic induction, $B = B_m$, $B = -B_m$ (Fig. 11.4) which will be identified with the aid of the binary variable Q:

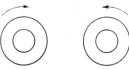

B	Q
B_m	0
$-B_m$	1

(11.2)

Fig. 11.4. The two states of magnetic induction of the core

11.1.1 Read-Out

Three phases can be distinguished with respect to the operation of the core as a read-out unit. (Fig. 11.5b)

(Φ) — permanent state before the application of the inputs X, Y, Z.

(Φ') — application of input and subsequent core 'swing' (transient case.

(Φ'') — permanent final state (after extinction of the transients).

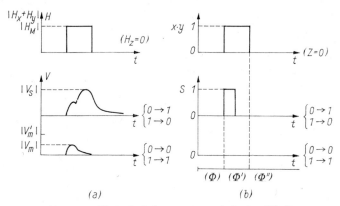

Fig. 11.5. 'Swing' of the core (a, real; b, simplified)

Schematically, the output S is read in the Φ' portion of the phase $\Phi' + \Phi''$), the state Q of the core has its definitive value in the portion Φ'' of the phase ($\Phi' + \Phi''$). These phases will be denoted by the attribution of the index n to $(\Phi + \Phi')$, the index $n+1$ to (Φ'') and expressed

$$\begin{cases} X_n = \text{maximum (binary) of } X \text{ in the phase } n \\ Y_n = \text{maximum (binary) of } Y \text{ in the phase } n \\ Z_n = \text{maximum (binary) of } Z \text{ in the phase } n \\ Q_n = \text{maximum (binary) of } Q \text{ in the phase } n \\ S_n = \text{maximum (binary) of } S \text{ in the phase } n \end{cases} \quad (11.3)$$

The operation, in read-out of the memory core is then as described in Table 11.6

Table 11.6

$Q_n X_n Y_n$	Q_{n+1}	S_n	
0 0 0	0	0	
0 0 1	0	0	
0 1 0	0	0	$(Z_n = Z_{n+1} = 0)$
0 1 1	0	0	
1 0 0	1	0	
1 0 1	1	0	
1 1 0	1	0	
1 1 1	0	1	

which can be represented by the expressions $(Z_n = Z_{n+1} = 0)$:

$$\begin{cases} Q_{n+1} = Q_n(\overline{X}_n \overline{Y}_n \overline{X}_n Y_n + X_n \overline{Y}_n) \\ \qquad\;\; = Q_n \overline{X_n Y_n} = Q_n(X_n / Y_n) \\ S_n \;\;\;\; = Q_n X_n Y_n \end{cases} \quad (11.4)$$

It can be verified that:

— the "new" magnetic state $(n+1)$ of the core depends on its "present" state (n) and on the present inputs.
— the output depends not only on the present inputs (n) but also on the present magnetic state.

Let us examine what happens if $Z_n = 1$.

If ε_x is an increase sufficient to ensure that $H_M + \varepsilon_x$ causes the core to swing, Table 11.6 can be completed as follows (Table 11.7). Thus we have:

$$\begin{cases} Q_{n+1} = Q_n(\overline{X}_n \overline{Y}_n + \overline{Z}_n \overline{X}_n + \overline{Z}_n \overline{Y}_n) = Q_n \overline{(X_n \# Y_n \# Z_n)} \\ S_n = Q_n(Z_n X_n + Z_n Y_n + X_n Y_n) = Q_n(X_n \# Y_n \# Z_n) \end{cases} \quad (11.5)$$

11.1 Elementary Example. The Ferrite Core

Table 11.7

Z_n	Q_n	X_n	Y_n	Q_{n+1}	S_n	
0	0	0	0	0	0	
0	0	0	1	0	0	
0	0	1	0	0	0	
0	0	1	1	0	0	normal read-out
0	1	0	0	1	0	
0	1	0	1	1	0	
0	1	1	0	1	0	
0	1	1	1	0	1	
1	0	0	0	0	0	
1	0	0	1	0	0	
1	0	1	0	0	0	
1	0	1	1	0	0	
1	1	0	0	1	0	
1	1	0	1	0	1	
1	1	1	0	0	1	
1	1	1	1	0	1	

11.1.2 Writing

A currently used procedure consists of storing within the core the contents of a flip-flop which, at this instant, contains either the information in the core read from the preceding cycle (this consists purely and simply of a re-write) or new information to be stored. In both cases it is to be assumed that the core is in condition 0 (it will have been brought back to this condition, if it was already not there, during the preceding cycle). There is then, no interest in the output S, but only in the storage of information.

Under these conditions it may be supposed that $X, Y, Z = 1$ since $I_X = H_Y = -1/2 H_M$, $H_Z = 1/2 H_M$ whence Table 11.8

It is interesting to complete this table by adding to it what happens when one is still in the storing condition, but where Q_n is not necessarily zero. Thus we have Table 11.9.

To sum up, the functioning of a coincident current memory core can be considered to be a 2 phase function $(n, n + 1)$ the Boolean variables in the phases n and $n + 1$ being linked by the above relations. The following observations can be made in this respect:

1. The physical time t can be divided into a certain number of phases, or periods of time, associated with the successive inter-states switchings in the manner shown above and can be numbered $0, 1, 2 \ldots n, n + 1$, etc., from on a permanent reference state. Each of the Boolean variables

11. Concept of the Sequential Network

Table 11.8

Q_n	Z_n	X_n	Y_n	Q_{n+1}
0	0	0	0	0
0	0	0	1	0
0	0	1	0	0
0	0	1	1	1
0	1	0	0	0
0	1	0	1	0
0	1	1	0	0
0	1	1	1	0

(Inscription of 1)

(Inscription of 0)

Table 11.9

Q_n	Z_n	X_n	Y_n	Q_{n+1}
0	0	0	0	0
0	0	0	1	0
0	0	1	0	0
0	0	1	1	1
0	1	0	0	0
0	1	0	1	0
0	1	1	0	0
0	1	1	1	0
1	0	0	0	1
1	0	0	1	1
1	0	1	0	1
1	0	1	1	1
1	1	0	0	1
1	1	0	1	1
1	1	1	0	1
1	1	1	1	1

describing the system therefore passes through a sequence of values X_n. They can, thus, be described as a function of "simplified" ("reduced") time, which is the discrete variable n. The equations which translate the sequence of the phases are known as recurrence equations defining the phase $n + 1$ as a function of the phase n. Such equations, analogous to differential equations, also carry the name of 'difference equations'.

2. The recurrent equations can be put in the form of "truth tables" where the inputs are the phase n variables, the outputs the time variables n and $n + 1$.

3. The variable Q_n defines the state (memory) of the core; it remains unchanged so long as the core $(X_n \cdot Y_n = 0)$ is not selected; this is the memory property of the core being considered.

4. An examination of Tables 11.6 and 11.9 reveals that the same input combination can give different output values; in table 11.6, for example, the equations $X_n, Y_n = 1$ give an output value S_n of 0 or 1 dependent on whether $Q_n = 0$ or 1. The output does not depend solely on the input variables but also on the variable Q_n, the variable for the "internal state" or "memory state".

The study of the core in the example given above appears, therefore as an elementary switching network subjected to the input signals possessing 2 memory states and endowed with an output permitting the state to be read-out.

A characteristic property of the new type of network, and of which the combinatorial networks are deprived, that of memory, is to be found, in the case of the core, linked to certain physical properties of the magnetic material of which it is composed. It can, as is known, be obtained with the aid of other systems, notably electronic systems thanks to the various arrangements of the elementary components which "simulate", in a way, the property of residual magnetism and memory furnished by the core. The following are cited as constituting a number of purely electronic "memory elements".

11.2 Eccles-Jordan Flip-Flop

The principles of this type of flip-flop are presented schematically in Fig. 11.10 — representing a transistorised flip-flop. It should be noted that, here, the layout has been much simplified. In fact the flip-flops actually used in electronic computers or logic networks, in general,

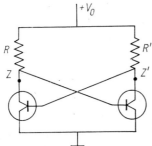

Fig. 11.10. transistorised flip-flop

correspond to a considerably more complex scheme where, in particular, the number of transistors greatly exceeds 2. This factual complexity of the flip-flop is due to the introduction of various refinements aimed at improving the electrical performance (switching time, switching threshold and noise sensitivity). However, the logic functioning of these diverse flip-flops can be boiled down to a small number of variants, among which are the following.

11.2.1 Type S-R Flip-Flop (Elementary Register)

From the functional viewpoint this is a box having 2 inputs denoted by S and R, and two complementary outputs Z and $\bar{Z}$. One of the outputs, Z for example, can permanently operate only in one of 2 states (dependent on whether the corresponding transistor is in the blocked or conducting state) which will be denoted by 0 and 1. The functioning is, therefore, as follows:

- If the value 1 be applied at the input S only, the output Z takes the value 1 and remains so, irrespective of its former value (recording of 1).
- If the value 0 be applied at the input S and 0 at R, the output Z retains an unchanged value (memory property).
- If the value 1 be applied at the input R only, Z takes the value 0, irrespective of its former value (recording of 0).
- The combination 11 applied at the input is inacceptable (from the electrical viewpoint, it introduces several undesirable features: e.g. random switching, amplification modes etc.).

This functioning can again be represented by a truth table linking the 2 consecutive phases (before and after application of an input).

The internal state of the memory unit is accessible by an examination of the output combination $Z\bar{Z}$ (for which there exist two possibilities: 01 and 10). In order to conform with the core model it is possible to introduce a variable for the internal state: for which there are 2 possible conventions:

$$Q = Z \quad \text{or} \quad Q = \bar{Z}. \tag{11.6}$$

If it is assumed that $Q = Z$ Table 11.11 is obtained:

Table 11.11 *Operation of the type S-R flip-flop*

Q_n S_n R_n	Q_{n+1}	Z_n
0 0 0	0	0
0 0 1	0	0
0 1 0	1	0
0 1 1	—	0
1 0 0	1	1
1 0 1	0	1
1 1 0	1	1
1 1 1	—	1

Z_n is thus the output at the moment of application of input $S_n R_n$, Q_n is the memory state in the phase which follows, after the swing (if any). This flip-flop can, therefore, be considered to be an elementary register (containing 1 item of information or 'bit') equipped with recording lines (SR) and read by sampling of Z or $\bar{Z}$. By the juxtaposition of n flip-flops of this type is obtained an n binary positions flip-flop register.

11.2.2 Type T Flip-Flop (Elementary Counter, Symmetric Flip-Flop)

This is a variant of the preceding unit obtained, from the electrical viewpoint, by the union of the two preceding inputs SR (based on diverse procedures). It possesses, therefore, 1 input E, 2 outputs $(Z, \bar{Z})$ and always contains one item or 'bit' of information.
Operation:

$$\begin{cases} \text{if } E = 1 & Z \text{ becomes } \bar{Z} \\ \text{if } E = 0 & Z \text{ is unchanged} \end{cases} \quad (11.7)$$

Once more taking a variable of internal state Q, gives Table 11.12:

Table 11.12. *Functioning of a T type flip-flop*

Q_n E_n	Q_{n+1}	Z_n
0 0	0	0
0 1	1	0
1 0	1	1
1 1	0	1

Equations:

$$\begin{cases} Q_{n+1} = \bar{Q}_n E_n + Q_n \bar{E}_n = Q_n \oplus E_n \\ Z_n = Q_n \end{cases} \quad (11.8)$$

The flip-flop returns to its original state at every second impulse; giving:

$$Q_{n+k} = Q_n \oplus \sum_{i=n}^{i=n+k-1} E_i.$$

It sums (modulo 2) the applied inputs. Should, for example, a return to the initial state be detected by a network producing a single pulse each time, this pulse will appear at every second input pulse and the T type flip-flop can be considered, in this respect, as a divisor by 2.

11.3 Dynamic Type Flip-Flop

11.3.1 Type S-R Flip-Flop

Consider the circuit of Fig. 11.13. The inputs S and R are the control lines, the input H is that of the time base: the signal applied at this input comprises a regular sequence of pulses of the form described in Table 11.15 which have the effect of validating the signals S and R only at the moment at which they occur.

At the output of the OR gate there is a delay Δ equal to the interval between clock pulses (clock period). Under these conditions the circuit of Fig. 11.13 is nothing more than a flip-flop possessing two states characterized by the absence or presence of a recirculation of pulses within the loop $abcda$. Consider for example an initial state characterised by the absence of an impulse in $abcda$. If an impulse be applied at S in synchronisation with H it will then re-circulate with a period

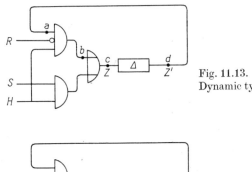

Fig. 11.13.
Dynamic type S-R Flip-flop

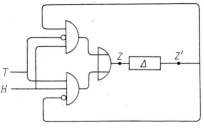

Fig. 11.14.
Dynamic type T Flip-flop

Δ so long as no impulse be applied at R (which would block the re-entry of the impulse) this, whether or not impulses are applied at S. On the other hand, if an impulse be applied at R in synchronisation with the time base the re-circulating impulse which appears at a cannot re-enter the loop and re-circulation is interrupted. This constitutes a return to the initial state. This function is, therefore, that of the type $S - R$ flip-flop described earlier, with the slight difference that the combination $SR = 11$ is permissible here. With n and $n + 1$ designating the present and future 'bit' times respectively can be obtained the expression:

$$Z_{n+1} = H_n(Z_n \overline{R}_n + S_n) \qquad (11.9)$$

Z_n being, at the same time, both the state variable and the output variable.

Again it can be said that such a flip-flop is of the dynamic type. It should be noted that the re-circulation of the impulse is possible only with an adequate regeneration of the signal. Within the loop $abcda$ must be included one or more elements (amplifiers) which can be either autonomous (they must be added into a detailed scheme), or be included in the gates or in the delay device, otherwise the re-circulating signal will rapidly decay.

11.3.2 Type T Flip-Flop (Fig. 11.14)

This time, with the same convention as for the type $S - R$ flip-flop,

$$Z_{n+1} = H_n(Z_n \overline{T}_n + \overline{Z}_n T_n) = H_n(Z_n \oplus T_n). \qquad (11.10)$$

This is the case of the type T flip-flop (cf. 11.8) analogous, from the functional viewpoint, to that of the Eccles-Jordan type T flip-flop. The output Z (or Z') resulting from the application of a series of impulses at T has the value 1 once only in every 2 impulses: the unit will permit the passage of only one impulse in every two (Fig. 11.15)

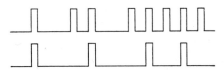

Fig. 11.15. Input and output sequences for a type T dynamic flip-flop

11.4 Some Elementary Counters

11.4.1 Dynamic Type Flip-Flop Pure Binary Counter

An example of the application of the type T flip-flop takes the form of a counter. For this it suffices to couple the output of the flip-flop at rank n (containing the binary figure of rank n) to the input T of the

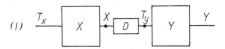

Fig. 11.16. Counter comprising 2 type T dynamic flip-flops

flip-flop at rank $n + 1$. Fig. 11.16 describes such a counter for the case of 2 flip-flops (of type T described in 11.3.2)

The input is T_0 the output $T_2 T_1$. The word $T_2 T_1$ is, in pure binary code, the residue modulo 4, $|N|_4$, of the number N of impulses applied at T_0 when $T_2 T_1$ has the initial state 00.

11.4.2 Gray Code Counter with Eccles-Jordan Type T Flip-Flops

The outputs $X\bar{X}$, $Y\bar{Y}$ (Fig. 11.17) are only sampled at clock times. Any modification in the state of the flip-flops X and Y produced at a time n, brings about, after a transient state, a new permanent voltage

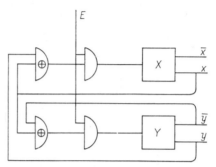

Fig. 11.17. Eccles-Jordan Flip-flop Counter

regime at the outputs which is not read-out (sampled) until clock time $n + 1$. Under these conditions the relations produced are:

$$\begin{cases} X_n \oplus X_{n+1} = E_n(\overline{X_n \oplus Y_n}) \\ Y_n \oplus Y_{n+1} = E_n(X_n \oplus Y_n) \end{cases} \quad (11.11)$$

corresponding to the truth table (11.18) which follows:

Table 11.18

X_n Y_n	$E_n = 0$		$E_n = 1$	
0 0	0	0	0	1
0 1	0	1	1	1
1 0	1	0	0	0
1 1	1	1	1	0

The content of the flip-flops YX starting from the condition 00 is, therefore, the representation in the Gray code of $|N|_4$ where N is the number of impulses having been applied at the input E.

11.4.3 Gray Code Continuous Level Counter

In the interests of simplicity it will be assumed that the delays Δ_x and Δ_y shown in Fig. 11.19 are equal ($\Delta_x = \Delta_y = \Delta$) (the contrary case will be dealt with later in this work). These delays can either be natural delays (more precisely they would be "time constants") due to the gates, or else delays expressly built-in, or a combination of both. The values Y and Y' on the one hand, X and X' on the other corresponding with 2 different instants (by an interval of time). Thus it can be said:

$$\begin{cases} X = EX' + \bar{E}\,Y' \\ Y = \bar{E}\,Y' + E\bar{X}' \end{cases} \tag{11.12}$$

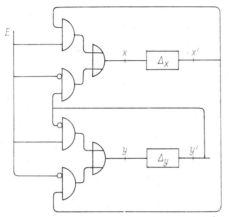

Fig. 11.19. Gray binary code continuous level counter

or, by assigning the index n to X', $\bar{X}'$, Y', $\bar{Y}'$ and E, and the index $n+1$ to X, $\bar{X}$, Y, $\bar{Y}$,

$$\begin{cases} X_{n+1} = E_n X'_n + \bar{E}_n Y'_n \\ Y_{n+1} = \bar{E}_n Y'_n + E_n \bar{X}'_n \end{cases} \tag{11.13}$$

which gives Table 11.20:

Table 11.20

$X_n\ Y_n$	X_{n+1}		Y_{n+1}	
	$E=0$		$E=1$	
0 0	0	0	0	1
0 1	1	1	0	1
1 1	1	1	1	0
1 0	0	0	1	0

If it is assumed (for simplification) that the delay Δ is negligible in comparison with the duration of the input square-wave 'castellations', the use of positive logic results in the input-output relation illustrated in

Fig. 11.21. It will be noted that there is no time signal and that the functioning of the system does not take into account the duration of the castellations (always assumed to be large in comparison with Δ):

Fig. 11.21. Input-output sequences for the counter shown in Fig. 11.19

the counter will indifferently accept the regular signal of Fig. 11.21a or the irregular signal of Fig. 11.21b. The following characteristics should also be noted:

— the fronts (ascending and descending) of Y are associated (only) with the ascending fronts of signal E;
— in an analogous manner the fronts of X are defined solely by the descending fronts of the signal E;
— each of the outputs X and Y therefore halves the number of fronts applied at the input. Equally, the number of square wave castellations presented is halved.

The ouput XY is renewed regularly with a periodicity of 4, during which time it passes successively through the values 00, 01, 11, 10, 00 etc.

— In the case of a periodic square-wave input, the lengths of the castellations X and Y is double that of the E castellations, the period T of the signal XY is equal to two periods of E.
— For a signal E applied to the counter initially in the state $XY = 00$,

XY subsequently represents (in Gray code) the value $|N_4|$ where N is the number of fronts E from the initial instant.

In passing, it will be noted that the table 11.20 differs from 11.18.

This circuit may be regarded as a "random castellation counter" (random both in respect of their initiation and of their duration).

11.4.4 Random Impulse Counter

The system as shown in Fig. 11.16 functions as follows: the flip-flop X is a type T Eccles-Jordan unit which completes the transition $1 \to 0$ every second impulse I. The block denoted by D supplies an impulse every time that X passes from 1 to 0. (The reader who is familiar with the

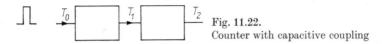

Fig. 11.22.
Counter with capacitive coupling

Eccles-Jordan flip-flop and frequency division will recognise that this circuit contains, essentially, a derivation network $R - C$ with rectification supplied by the input of flip-flop Y.) The flip-flop Y is of the same type as X. (Fig. 11.22)

The impulses I are very fine with steep fronts and arrive at the input at random instants, sufficiently spaced, however, to ensure that the transients have time to decay before the arrival of the following impulse. Under these conditions, the contents of the flip-flops XY, assuming that at the initial instant $XY = 0$ represents the remainder $|N_4|$ of the number N of impulses received, in pure binary code. If T_Y is considered as an output from the system, this output is made up of pulses obtained by the suppression of pulses I of even rank (initial state $XY = 00$ or 10) or of odd rank (initial state $XY = 01$ or 11). (Fig. 11.23)

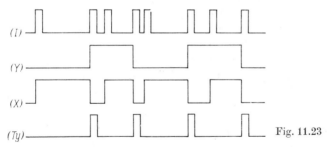

Fig. 11.23

To sum up, a number of elementary networks have now been examined. They possess certain common characteristics and, essentially:

an operation no longer 'punctual' but of a recurrent type, taking time into account by means of a memory.

However, these various networks present certain differences. In the chapter which follows an attempt will be made to determine to what extent these diverse types of circuit can be defined for one or more models, and to provide a general definition for sequential networks.

Chapter 12

Sequential Networks. Definitions and Representations

12.1 Quantization of Physical Parameters and Time in Sequential Logic Networks

A logic network cannot be other than an abstraction: the actual voltages, currents, positions and, more generally, any physical parameter used to convey information are not, themselves, binary but each extend over continuous ranges of values. There are two reasons for this:

— for a given network, the values of a parameter at a given point such as voltages, are not fixed but distributed statistically around a mean value. This scatter of values is due primarily to a statistical "spread" in the parameter values of the components themselves (diodes transistors). There are also the effects of fluctuations on the network during operations which affect the electrical characteristics (transistor loadings etc.) due to variations in power supply levels and in ambient conditions (notably temperature).

— the existence of *transient* periods of non-negligible duration which link, in continuous manner, the 'steady-state' periods during which the signals are in operation within the system.

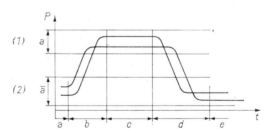

Fig. 12.1. Binary quantification

However, thanks to a careful control of component parameter values in production and to a knowledge of their distribution, it is possible to ensure, with the aid of various techniques (worst case design, statistical design), that these parameter values fall into two separate, *disjoint* domains (a', $\bar{a}'$ Fig. 12.1) during every steady-state period of operation of the network and to which a binary coding can be assigned.

It should be noted that, in order to achieve such a *'discrete'* coding of physical parameters, it is necessary to assume that time be divided into

12.1 Quantization of Physical Parameters

time intervals having special properties (steady-state periods etc.). (zones a, c, e, Fig. 12.1). These time intervals can be assigned integer numbers which, in a sense, amounts to defining a '*discrete*' time. Discrete (binary) parameters and discrete time are abstractions which appear together in the study of sequential networks.

12.1.1 Discrete Time

Time, as used in the theory of sequential networks, is not the continuous physical time which, for example, serves to describe the behaviour of a physical network (explicitly, or with the aid of differential equations, or with partial differential equations), but a *discrete* time only.

The correspondence between the two is as follows: ordinary time is broken down into a certain number of "*phases*" numbered from the origin by use of a variable T which takes the values $0, 1, 2, 3, \ldots n \ldots$ This variable is the discrete time. Later it will be seen that the phases can, in certain cases, be considered as of constant duration (synchronous networks).

In other cases, they are of variable physical duration, from a few nano or microseconds to several hours for example (asynchronous networks).

12.1.2 Operating Phases. Timing Signals

12.1.2.1 Synchronous Systems

This is the name given to networks where all operations are initiated by a *periodic* signal originating *outside* the network itself, known as the *synchronisation signal*, or *timing signal*, or *clock signal* which in the case of logic networks appears in the two forms depicted in Fig. 12.2a, Fig. 12.2b representing the idealised signal which will be used in the theory which follows:

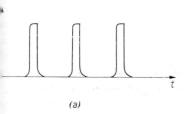
(a)

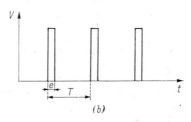
(b)

Fig. 12.2

The appearance of this latter is the same in both cases except for the width of the pulse.

From the *electronic* point of view, the triggering of the flip-flops by the pulses can be of various types. Certain flip-flops ("pulse-input flip-flops") must be triggered by a very narrow pulse (e.g. 0, 1 or 1 microsecond) having a very steep front, applied directly to the base or collector of the transistors (the condition relating to the width of the pulse is intended to ensure that there is no return to the initial state as the input pulse returns to zero).

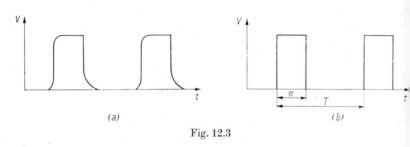

Fig. 12.3

Other types of flip-flop are controlled by one of the two fronts (always the same front) of the pulse or input square wave (e.g. the descending front). The difference is due to the input circuit of this type of flip-flop which, by various methods (in general derivation, rectification) "simulates" the operating conditions for the first type of flip-flop.

Finally, there is a third type of flip-flop for which the pulse (or square wave) must have a minimum width. However, these are merely internal differences and in each case there will be the same *logic* function and the symbolic representations to be used will be confined to only one type of flip-flop, it being understood that, from an electronic point of view, the pulses will remain subjected to various restrictions.

The ratio e/T is sometimes called the "form factor" of the *time signal*. T is the *time period* the *unit of time* for the system e is the width of the pulse.

The intervals of time of length e during which the clock signal is not zero are known as 'bit' times or "*clock pulse*" times.

12.1.2.2 Asynchronous Systems

There is no external time signal to control the evolution of the system. The time phases can only be defined with respect to the time behaviour of the inputs and of the internal functioning of the system in a manner to be examined later.

12.2 Binary Sequential Networks. General Model

By definition a binary sequential network is an abstract model of a logic network possessing the following properties:

12.2.1 Inputs

There are m binary inputs $(m \geq 1)$ e_i $(e_i = 0 \text{ or } 1)$ which form a binary vector $E = (e_1, e_2, \ldots, e_i, \ldots, e_m)$. To each input e_i there corresponds an *input line*. E will be known as the *input vector* (or *input state*), or more simply as the *input* so long as there is no ambiguity. The input E can, therefore, take a maximum of 2^m values in the binary code.

12.2.2 Outputs

Similarly, there are n $(n \geq 1)$ binary outputs S_j $(S_j = 0 \text{ or } 1)$ carried by as many lines, and forming the vector

$$S = (S_1, S_2, \ldots, S_j, \ldots, S_n)$$

known as the *output vector*, or *output state*, or simply as the *output* so long as there is no ambiguity. The output S has, therefore, a maximum of 2^n distinct values in the binary code.

There are thus $m + n$ external terminals to the network.

12.2.3 Internal States

The network also contains $r(r \geq 1)$ *memory points*, i.e. 2-state memory elements each having a capacity of 1 'bit' of information. The memory content will be $(q_1, q_2, \ldots, q_k, \ldots, q_r)$ $(q_k = 0 \text{ or } 1 \text{ for every } k)$. The vector $Q = (q_1, q_2, \ldots, q_k, \ldots q_r)$ is the *memory state* or "*internal state*" of the sequential network.

The number r is the memory capacity (expressed in memory points or in 'bits' of information).

12.2.4. Operation

The input, output and internal states vectors are defined during *phases* numbered $0, 1, 2, \ldots, n$, i.e. at discrete time values. Every binary component of these vectors is thus a function of n.

Let E_n, Z_n and Q_n be the input, output and internal state vectors at time n respectively. Then, starting from an initial state Q_0, the operation of the sequential network is defined as follows:

For every n $(n \geq 0)$ the network computes, by means of E_n and Q_n, a new vector to be assigned to the phase $n+1$, and denoted as Q_{n+1}.

As a function of E_n and Q_n it also computes a vector S_n which is the output during the phase n. This operation can be summarised by the equations:

$$\begin{cases} Q_{n+1} = F(E_n, Q_n) & (12.1) \\ S_n = G(E_n, Q_n) & (12.2) \end{cases}$$

(Huffmann-Mealy model)

The equations (12.1) and (12.2) define a method of operation which is that studied by *Huffmann* (asynchronous systems) and *Mealy* (synchronous systems). For this reason it is known as the Huffmann-Mealy model, or Mealy (when it concerns synchronous systems).

Particular case: Moore model

There exists another type of sequential network, that of Moore, which can be considered as a particular case of the preceding type: here the output is a function of the internal state only:

$$\text{(Moore model)} \begin{cases} Q_{n+1} = F(E_n, Q_n) & (12.3) \\ S_n = G(Q_n) & (12.4) \end{cases}$$

12.2.5 Present and Future States. Total State. Transitions

The states carrying the subscript n (E_n, Q_n, S_n) are designated *present input*, *present internal* and *present output* states respectively, the state Q_{n+1} the future ("next") state. The vector (EQ) formed by the juxtaposition of E and Q (vector having $m+r$ elements) given the name of *total state*. We write:

$$\begin{cases} Q_{n+1} = \Phi(EQ)_n \\ S_n = \Gamma[(EQ)_n] \end{cases} \quad (12.5)$$

Relations (12.1) and (12.2) or (12.3) and (12.4) specify:

— the way in which the network, based on the present internal state and the present input, determines the future state. This is the calculation of the *transitions* $Q_n \to Q_{n+1}$;
— the way in which the outputs are determined based on the present (Moore) state or on the present (Huffmann-Mealy) input — internal state combination.

12.2.6 Concept of a Finite Automaton

With the *sequential network* can be associated an abstract structure known as a *finite automaton* (or sequential machine). A *finite automaton* is a collection of 5 mathematical entities, denoted

$$\mathscr{A} = \langle \mathscr{E}, \mathscr{Q}, \mathscr{S}, F, G \rangle \tag{12.6}$$

where $\mathscr{E}$, $\mathscr{Q}$, $\mathscr{S}$ are finite sets, respectively known as the sets of input states (or input letters), of internal states and of output states. F and G are the mappings of $\mathscr{E} \times \mathscr{Q}$ in $\mathscr{Q}$ and $\mathscr{S}$ respectively and are known as the transition and output functions.

12.2.7 Synchronous and Asynchronous Sequential Networks

It should be noted that, mathematically speaking, the memory of the system is manifested by the recurrence relation (12.1). The physical memory only intervenes when the instants $0, 1, \ldots, n$ have a positive duration Δt_n: thus permitting the system to await the phase $n+1$ from the initiation of the phase n.

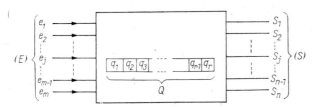

Fig. 12.4. Sequential network

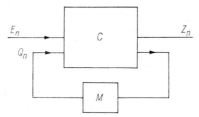

Fig. 12.5. Another symbolic representation of a sequential network

The circuit of Fig. 12.4 can thus be re-drawn as in Fig. 12.5. It will be seen that it is made up of a combinatorial network (C) and a memory M. At a given instant in time, the states of input and output of the memory M are respectively associated with the future state Q_{n+1} and the state Q_n (but not necessarily identical with them, as will be seen).

12.2.7.1 Synchronous Sequential Networks

Memory

In general, this consists of electronic flip-flops or, more rarely, of calibrated delays equal to one clock period. Thus from the functional point of view there are r closed paths, or feedback loops, each traversing a flip-flop (or a delay) and associated with r internal state variables q_1.

In the case of the delays each loop is represented physically by a line.

In the case of the flip-flops having, usually, 2 complementary outputs and, according to the type employed, 1, 2, 3 inputs or more, each loop can be associated with several lines.

In each case these lines convey the input control signals to each memory cell on the one hand, and the output values on the other.

Combinational networks (C). They can be of the following types where all the flip-flops are of the same type:

Flip-flop	Type of circuit C
T	$[m + 2r, r + n]$
$S - R$	$[m + 2r, 2r + n]$
$S - R - T$	$[m + 2r, 3r + n]$
etc.	

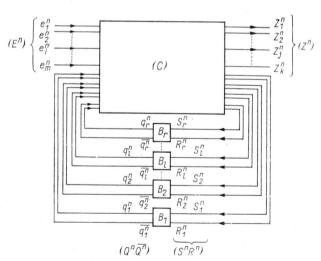

Fig. 12.6. Synchronous sequential network with S-R flip-flops

12.2 Binary Sequential Networks. General Model

If flip-flops of different types were used simultaneously there would result intermediate values ($[m + 2r, \alpha\, r]$). With the network C will, therefore, correspond $r + n$ Boolean functions for the $m + 2r$ variables.

By way of an example, for the case of the $S - R$ type flip-flops, there will be a network $[m + 2r,\ 2r + k]$, having the form represented in Fig. 12.6 which analytically translates into the system of equations which follow:

$$\begin{cases} S_1{}^n = s_1(e_1{}^n, e_2{}^n, ..., e_i{}^n, ..., e_m{}^n;\ q_1{}^n, q_2{}^n, ..., q_r{}^n;\ \bar{q}_1{}^n, \bar{q}_2{}^n, ..., \bar{q}_l{}^n, ..., \bar{q}_r{}^n) \\ R_1{}^n = r_1(e^n{}_1, e_2{}^n, ..., e_i{}^n, ..., e_m{}^n;\ q_1{}^n, q_2{}^n, ..., q_r{}^n;\ \bar{q}_1{}^n, \bar{q}_2{}^n, ..., \bar{q}_l{}^n, ..., \bar{q}_r{}^n) \\ S_2{}^n = s_2(e_1{}^n, e_2{}^n, ..., e_i{}^n, ..., e_m{}^n;\ q_1{}^n, q_2{}^n, ..., q_r{}^n;\ \bar{q}_1{}^n, \bar{q}_2{}^n, ..., \bar{q}_l{}^n, ..., \bar{q}_r{}^n) \\ R_2{}^n = r_2(e_1{}^n, e_2{}^n, ..., e_i{}^n, ..., e_m{}^n;\ q_1{}^n, q_2{}^n, ..., q_r{}^n;\ \bar{q}_1{}^n, \bar{q}_2{}^n, ..., \bar{q}_l{}^n, ..., \bar{q}_r{}^n) \\ \vdots \\ S_l{}^n = s_l(e_1{}^n, e_2{}^n, ..., e_i{}^n, ..., e_m{}^n;\ q_1{}^n, q_2{}^n, ..., q_r{}^n;\ \bar{q}_1{}^n, \bar{q}_2{}^n, ..., \bar{q}_l{}^n, ..., \bar{q}_r{}^n) \\ R_l{}^n = r_l(e_1{}^n, e_2{}^n, ..., e_i{}^n, ..., e_m{}^n;\ q_1{}^n, q_2{}^n, ..., q_r{}^n;\ \bar{q}_1{}^n, \bar{q}_2{}^n, ..., \bar{q}_l{}^n, ..., \bar{q}_r{}^n) \\ \vdots \\ S_r{}^n = s_r(e_1{}^n, e_2{}^n, ..., e_i{}^n, ..., e_m{}^n;\ q_1{}^n, q_2{}^n, ..., q_r{}^n;\ \bar{q}_1{}^n, \bar{q}_2{}^n, ..., \bar{q}_l{}^n, ..., \bar{q}_r{}^n) \\ R_r{}^n = r_r(e_1{}^n, e_2{}^n, ..., e_i{}^n, ..., e_m{}^n;\ q_1{}^n, q_2{}^n, ..., q_r{}^n;\ \bar{q}_1{}^n, \bar{q}_2{}^n, ..., \bar{q}_l{}^n, ..., \bar{q}_r{}^n) \end{cases} \quad (12.7)$$

$$\begin{cases} z_1{}^n = z_1(e_1{}^n, e_2{}^n, ..., e_i{}^n, ..., e_m{}^n;\ q_1{}^n, q_2{}^n, ..., q_r{}^n;\ \bar{q}_1{}^n, \bar{q}_2{}^n, ..., \bar{q}_l{}^n, ..., \bar{q}_r{}^n) \\ z_2{}^n = z_2(e_1{}^n, e_2{}^n, ..., e_i{}^n, ..., e_m{}^n;\ q_1{}^n, q_2{}^n, ..., q_r{}^n;\ \bar{q}_1{}^n, \bar{q}_2{}^n, ..., \bar{q}_l{}^n, ..., \bar{q}_r{}^n) \\ \vdots \\ z_j{}^n = z_j(e_1{}^n, e_2{}^n, ..., e_i{}^n, ..., e_m{}^n;\ q_1{}^n, q_2{}^n, ..., q_r{}^n;\ \bar{q}_1{}^n, \bar{q}_2{}^n, ..., \bar{q}_l{}^n, ..., \bar{q}_r{}^n) \\ \vdots \\ z_k{}^n = z_k(e_1{}^n, e_2{}^n, ..., e_i{}^n, ..., e_m{}^n;\ q_1{}^n, q_2{}^n, ..., q_r{}^n;\ \bar{q}_1{}^n, \bar{q}_2{}^n, ..., \bar{q}_l{}^n, ..., \bar{q}^n)_r \end{cases}$$

In the case of the electronic components they will, here (and for the remainder of this work), be assumed to be made up of electronic gates. In the case of diode type gates it will take the form of a diode matrix.

Synchronisation

The circuit functions in synchronisation with the periodic pulses from the clock base. It is only during these periods that the combinatorial network C is active and able to take account of the input, output and internal state variables (state of the feedback loop).

For the study of the *synchronous* operation of the network, what occurs between the time pulses plays no part and the values of the variables are ignored by the combinatorial network (C) at that time. The intermediate phenomena between two time pulses give rise to a more detailed study whose main features will be mentioned a little later during the discussion on asynchronous networks.

Various methods are available for the slaving of the network to the clock control. One, for example, consists of interposing AND gates at the inputs e_i and q_s of the network C.

As a result, the outputs Z_j and the flip-flop inputs will also be synchronised. This, however, is a costly method and often the synchronisation can be introduced directly at the gates comprising the network C thus avoiding the use of additional gates.

In some designs the time base can be introduced at a special input (a line specifically provided for this purpose) or as a normal input to the gate.

In the first instance the time input is usually distinct from the supply line, but in certain instances the latter is used directly to introduce the clock signal (the circuit only operating when the supply line is pulsed).

It can also be introduced at an input to the flip-flops. Whatever the procedure adopted, a study of the different synchronisation techniques will show that it is possible, at least for the study of the logic function of the network, to assume that the signals exist only during the clock pulses.

For these reasons the time base will, in general, not be illustrated, — once it is known that the system is synchronous a convention, analogous to that used for the supply to the logic circuits, will be adopted. The time base as well as the power supply will, therefore, be implicit in the diagrams. It must not, however, be forgotten that in certain cases (studies of time base, supply fault analysis, for example) it may be necessary to represent them explicitly as Boolean variables.

Notes:

— *use of delays*

In a synchronous sequential network, the use of flip-flops in order to obtain a delay of 1 clock period is not absolutely necessary. Delays if precisely calibrated to one period serve the purpose, as has been shown earlier. This method, for which the circuit C will be of type $[m + r, r + k]$ is rarely employed, however, since it requires that all the delays have the value of exactly one time period and, in particular, that they do not 'drift' under the influence of ambient conditions (e.g. heat) to which they may be exposed. In addition the clock period itself can vary.

12.2 Binary Sequential Networks. General Model

In both vases the delays can get out-of-step with reference to the clock period and functioning can be disturbed. On the other hand, in the case of the flip-flops, the delays automatically adjust themselves to the required values.

— *clock signal*

As has been seen, the output of the flip-flops is used for input control through the circuit C. As a consequence, the synchronising pulses must obey several restictions in order to ensure that there is only one transition ($Q_n \to Q_{n+1}$) per impulse.

In the case of pulse controlled flip-flops, for example, it is essential that the flip-flop output signals do not reach the input to (C) until after the extinction of the clock pulse, otherwise the network (C) will remain active and will utilise the new output S and may thus provoke a new transition. Thus, (Fig. 12.7), Q and Q' being signals changing with time n,

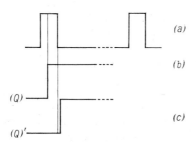

Fig. 12.7

the signal Q will again be taken into account during the same period o time. Alternatively, the signal Q' will arrive sufficiently late so as not to be available for use until the following time period, which is the desired objective.

Fig. 12.8 illustrates the type of phenomenon which can be produced when there is too rapid a return of the flip-flop outputs. This phenomenon will occur with the delay as indicated in the diagram both for pulse control and for control by an ascending front.

It will be seen that the control pulse E, instead of initiating the single transition $X_1 X_2 = 00 \to 10$, initiates a double transition: $00 \to 10 \to 11$ within a single clock cycle.

This second transition constitutes a defect, a synchronisation "lapse" which is undesirable in a synchronous network.

If, on the other hand, a control by means of the descending fronts is considered, the risk of parasitic operation of this type is greatly reduced. In fact, the flip-flops will arrive at their new state immediately

328   12. Sequential Networks. Definitions and Representations

after the impulse which terminates at this front. Even though these fronts are not the vertical fronts represented in the diagrams, control by means of the descending fronts (the end of the square wave) offers greater protection against the premature return of the information signals.

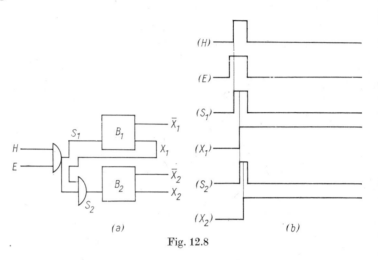

Fig. 12.8

Another condition to be satisfied is that of the minimum spacing of the time pulses: it is necessary that the period T be sufficient to ensure that the transition initiated by a pulse or front at time n be terminated before the pulse or the analogous front of rank $n+1$ be repeated.

In fact, the output from the flip-flops is not used until they have taken up a permanent state (p in Fig. 12.9a) which must, therefore, be attained after a time $t < T$. The signal of Fig. 12.9a can, for example, be the output signal from a Eccles-Jordan type T flip-flop, encountered directly by the time (clock) unit H (Fig. 12.9b).

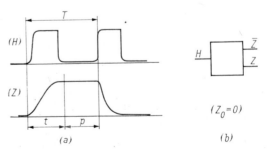

Fig. 12.9. Input/output of type T flip-flop. Transient/permanent states

Examples of synchronous sequential networks

Among the examples which follow reference will again be made to some already discussed in the preceding chapter but, this time, they will be re-written in the general notation defined above.

— $S - R$ *Dynamic flip-flop* (Fig. 11.13)

This a (rare) case where there is a calibrated delay T. The following convention will apply:

$Q_{n+1} = 1$: there is a recirculation pulse within the loop, i.e. (which constitutes the delay) between the time n and the time $n+1$.

$Q_{n+1} = 0$: there is no recirculation during this same period. Another equivalent convention:

$$\begin{cases} Q_{n+1} = 1 \text{ if } Z' = 1 \text{ at the instant } n+1 \\ Q_{n+1} = 0 \text{ if } Z' = 0 \text{ at the instant } n+1 \end{cases} \quad (12.8\,\text{a})$$

Alternatively:

$$\begin{cases} Z_n = 1 \text{ if } Z = 1 \text{ at the instant } n \\ Z_n = 0 \text{ if } Z = 0 \text{ at the instant } n \end{cases} \quad (12.8\,\text{b})$$

Then:

$$\begin{cases} Q_{n+1} = Q_n \bar{R}_n + S_n \\ Z_n = Q_n \end{cases} \quad (12.9)$$

(the time base being implicit only).

— Type T dynamic flip-flop

Using the same convention, this time:

$$\begin{cases} Q_{n+1} = Q_n \oplus T_n \\ Z_n = Q_n \end{cases} \quad (12.10)$$

— Type T dynamic flip-flop counter

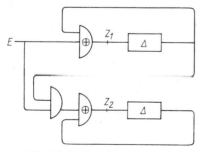

Fig. 12.10. Pure binary counter

Again using the same convention, the operation of the counter, represented in Fig. 12.10 (with the help of the gates $\oplus$ and $\cdot$), can be expressed:

$$\begin{cases} Q_1^{n+1} = Q_1^n \oplus E^n \\ Q_2^{n+1} = Q_2^n \oplus E^n Z_1^n \end{cases}$$

$$\begin{cases} Z_1^n = Q_1^n \\ Z_2^n = Q_2^n \end{cases} \qquad (12.11)$$

Checking next that the two first equations are those defining the addition of the number OE to the number $Z_2 Z_1$

$$\begin{array}{c} Z_2 Z_1 \\ + \; O E \\ \hline = Z_2 \oplus Z_1 E, \; Z_1 \oplus E \end{array}$$

— *Eccles-Jordan flip-flop counter (Gray code)*

The memory unit is comprised of flip-flops.

By taking $X = Q_X$, $Y = Q_Y$, the equations (12.7) can be re-written in the form:

$$\begin{cases} Q_X^{n+1} = Q_X^n \oplus E^n \overline{(Q_X^n \oplus Q_Y^n)} = \overline{E^n} Q_X^n + E Q_Y^n \\ Q_Y^{n+1} = Q_Y^n \oplus E^n (Q_X^n \oplus Q_Y^n) = \overline{E^n} Q_Y^n + E^n Q_X^n \end{cases} \quad (12.12)$$

$$\begin{cases} Z_X^n = Q_X^n \\ Z_Y^n = Q_Y^n \end{cases}$$

— *Calibrated delay*

The circuit shown in Fig. 12.11 delays by one clock unit the pulse applied at E. Indeed, we have:

$$Z_{n+1} = Z_n \oplus T_n = Z_n \oplus (Z_n \oplus E_n) = E_n \qquad (12.13)$$

or again:

$$\begin{cases} Q_{n+1} = E_n \\ Z_n = Q_n \end{cases} \qquad (12.14)$$

— *Binary serial adder*

Two binary words:

$$X = (x_n, \ldots, x_j, \ldots, x_0)$$

12.2 Binary Sequential Networks. General Model

and

$$Y = (y_n, \ldots, y_j, \ldots, y_0)$$

are presented in series at the input to a serial adder (Fig. 12.12). At the instant j the adder calculates the digit s_j of the sum:

$$S = (s_n, \ldots, s_j, \ldots, s_0) \quad (S = X + Y)$$

and calculates the 'carry' to be used at the instant $j+1$ which will be known as r_{j+1}. We have:

$$\begin{cases} r_{j+1} = x_j y_j + r_j(x_j + y_j) & (12.15) \\ s_j = x_j \oplus y_j \oplus r_j & (12.16) \end{cases}$$

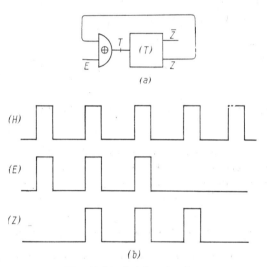

Fig. 12.11. A delay circuit

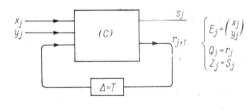

$(t = 0,1,2\ldots n)$

Fig. 12.12. Serial adder

Putting:
$$\begin{cases} E_j = (x_j y_j) \\ Q_j = r_j, \quad Z_j = s_j \end{cases}$$

provides equations conforming with the general model.

Fig. 12.12 depicts a symbolic delay of 1 time unit. The circuit C corresponds with the equations (12.15) and (12.16).

In practice, there will be one flip-flop. Using, for example, the flip-flop delay of the previous example produces Fig. 12.13.

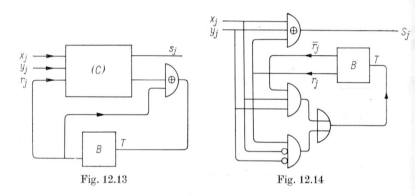

Fig. 12.13   Fig. 12.14

The logic circuit external to the flip-flop B can be simplified to take the form of the network (C') of Fig. 12.14. It should be noted that direct use of the flip-flop output $\bar{r}_j$ is made where there is need for $\bar{r}_j$ and that the same circumstances can arise for $x_j y_j$ in certain cases.

— *Pure binary — reflected binary conversions and vice versa*

Two binary representations are available for a number X:

In pure binary x is expressed $B = (B_n B_{n-1}, \ldots, B_j, \ldots B_0)$.

In reflected binary x is expressed $R = (R_n R_{n-1}, \ldots, R_j, \ldots R_0)$

The conversion formulae (Chap. 3) are:

$$(R \to B): B_p = \bigoplus \sum_{j=p}^{j=n} R_j$$

$$(B \to R): R_p = B_p \oplus B_{p+1}, \quad R_n = B_n$$

(12.17)

(the conversions are to be made with the most significant digit leading (subcript n))

12.2 Binary Sequential Networks. General Model

Under these conditions:

$$\begin{cases} Z_p = B_{n-p} \\ Q_p = B_{n-p+1} \quad (p \geq 1) \quad Q_0 = 0 \\ E_p = R_{n-p} \end{cases} \quad (12.18)$$

we have:

$$\begin{cases} Q_{p+1} = Q_p \oplus E_p \\ Z_p = Q_p \oplus E_p \end{cases} \quad (12.19)$$

(reflected → binary conversion)

From this the diagram of Fig. 12.15a is easily derived and, by the same procedure as for the preceding example, those of Figs. 12.15b and 12.15c, which make use of a type T flip-flop.

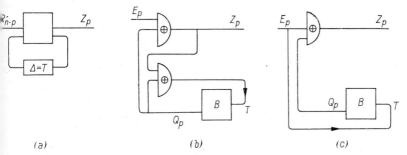

Fig. 12.15. Serial reflected binary/pure binary converter

For the pure binary to reflected binary conversion, by putting:

$$\begin{cases} Q_k = B_{n-k+1} \\ E_k = B_{n-k} \\ Z_k = R_{n-k} \end{cases} \quad (12.20)$$

is obtained:

$$\begin{cases} Q_{k+1} = E_k \\ Z_k = Q_k \oplus E_k \end{cases} \quad \text{with } Q_0 = 0. \quad (12.21)$$

The task of establishing the corresponding logical network is left to the reader.

— *Mechanical V-scan encoder*

The read-out from such an encoder (the principles of which have already been described in Chap. 9) can be made in a series or in a parallel mode.

Consider the case of the series mode. The read-out of the binary figure of rank n (which necessitates a knowledge of the figure of rank $n-1$), is made at an instant in time n linked to the instant $n-1$ by the following recurrence relation:

$$Z_n = Z_{n-1} B_n^r + \bar{Z}_{n-1} B_n^v \qquad (12.22)$$

This relation translates the sequential decision procedure on which the read-out is based: $(B_n^r B_n^v)$ comprises the input, i.e. the pair of values available at the two brushes, aft and forward respectively, on the track n. By putting:

$$\begin{cases} E_n = B_n^r B_n^v \\ Q_{n+1} = Z_n \quad (Q_1 = B_0) \end{cases} \qquad (12.23)$$

(12.22) can be re-written in the form:

$$\begin{cases} Q_{n+1} = Q_n B_n' + \bar{Q}_n B_n^v \\ Z_n = Q_n B_n' + \bar{Q}_n B_n^v \end{cases} (n \geq 1) \quad \begin{cases} Q_1 = B_0 \\ Z_0 = B_0 \end{cases} (n = 0) \qquad (12.24)$$

as illustrated in Fig. 12.16.

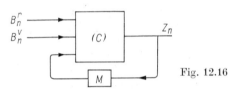

Fig. 12.16

— *Shift register*

An electronic shift register is made up from n memory units, shown here as $(B_1, B_2, \ldots, B_n)$ and assumed to be flip-flops.

It may contain a word $X = (x_n, \ldots x_i, \ldots x_1)$, which is conveniently interpreted as a number x written in pure binary. At every time interval the register admits an input E (0 or 1) and, during this same time interval, displaces the word X one 'bit' position to the right. For the version shown in Fig. 12.17a the figure contained in B_1 is lost during the shift. In the version illustrated in Fig. 12.17b, it can be re-introduced into B_n by a coupling of B_1 with B_n (*re-circulating register*).

The functioning of the register is a follows:

$$\begin{cases} \text{non-recirculating shift register:} \quad x^{(k+1)} = \left[\dfrac{x^{(l)}}{2}\right] + 2^n \cdot E^{(k)} \\ \text{re-circulating shift register:} \quad x^{(k+1)} = \left[\dfrac{x^{(k)}}{2}\right] + 2^n \, |x^{(k)}|_2 \cdot (1 - Q) \\ \qquad\qquad\qquad\qquad\qquad\qquad\qquad\qquad + 2^n \, E^{(k)} \cdot Q \end{cases}$$

12.2 Binary Sequential Networks. General Model

The variable x is represented by the internal (vector) state X. The input is E (presented in series). The outputs (R) and (R') are those of the flip-flops B_i.

Fig. 12.17 does not show the shift logic, which might have been expected, in accordance with the model shown in Fig. 12.6. In fact, for the majority of cases, this logic does not exist in the form of an explicit gate assembly. The memory functions, shifting and synchronisation are tightly integrated through purely electronic techniques.

In the majority of cases, it is convenient to consider the shift register as an indivisible unit (like that of the flip-flop).

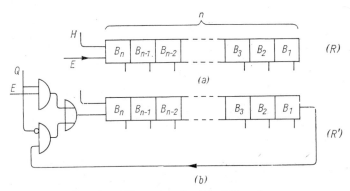

Fig. 12.17. Two examples of a shift register

12.2.7.2 Asynchronous Sequential Networks

— *Memory*

The passage from one phase to the next is effected by use of the built-in circuit delay units. Essentially, these delays are those of the gates of the combinatorial network (C) plus the delays brought about by the inter-gate connections and the delays Δ_l included in a certain number of *feedback loops*, by which the outputs r $(r \geq 1)$ of (C) are brought back to the inputs of the same network. Fig. 12.18 presents a diagrammatic representation. Δ is, in general, a function of the number l of the reaction loop $(\Delta = \Delta_l)$.

Among the variables which serve to describe the temporal evolution of the network can be taken, together with the input states and the logical states of the feedback lines, the output (C). It will later be shown that these latter variables can be used to describe the stable states of the network and to indicate to what extent they suffice to describe the operation of the system in the other configurations.

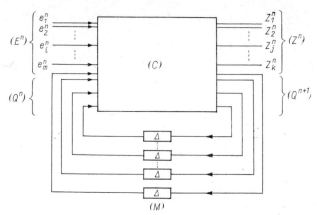

Fig. 12.18. General representation — asynchronous sequential network

— *Logic network*

The equations (12.1) to (12.4) which relate the binary vectors E_n, Q_n, S_n and Q_{n+1} are represented here by $k + r$ Boolean equations of the following form:

$$(Q) \begin{cases} q_1^{n+1} = f_1(e_1^n, e_2^n, \ldots, e_i^n, \ldots, e_m^n, q_1^n, q_2^n, \ldots, q_l^n, \ldots, q_r^n) \\ q_2^{n+1} = f_2(e_1^n, e_2^n, \ldots, e_i^n, \ldots, e_m^n, q_1^n, q_2^n, \ldots, q_l^n, \ldots, q_r^n) \\ \vdots \\ q_l^{n+1} = f_l(e_1^n, e_2^n, \ldots, e_i^n, \ldots, e_m^n, q_1^n, q_2^n, \ldots, q_l^n, \ldots, q_r^n) \\ \vdots \\ q_r^{n+1} = f_r(e_1^n, e_2^n, \ldots, e_i^n, \ldots, e_m^n, q_1^n, q_2^n, \ldots, q_l^n, \ldots, q_r^n) \end{cases}$$

and (12.25)

$$(Z) \begin{cases} Z_1^n = g_1(e_1^n, e_2^n, \ldots, e_i^n, \ldots, e_m^n, q_1^n, q_2^n, \ldots, q_l^n, \ldots q_r^n) \\ Z_2^n = g_2(e_1^n, e_2^n, \ldots, e_i^n, \ldots, e_m^n, q_1^n, q_2^n, \ldots, q_l^n, \ldots, q_r^n) \\ \vdots \\ Z_j^n = g_j(e_1^n, e_2^n, \ldots, e_i^n, \ldots, e_m^n, q_1^n, q_2^n, \ldots, q_l^n, \ldots, q_r^n) \\ \vdots \\ Z_k^n = g_k(e_1^n, e_2^n, \ldots, e_i^n \ldots, e_m^n, q_1^n, q_2^n, \ldots, q_l^n, \ldots, q_r^n) \end{cases}$$

The transition $Q_n \to Q_{n+1}$ is thus represented by Boolean r functions having $m + r$ variables, the determination of Z_n by Boolean k functions having $m + r$ variables, and these $r + k$ functions of $m + r$ variables are obtained from a combinational network C of type $[m + r,\ r + k]$.

12.2 Binary Sequential Networks. General Model

Example 1 (Fig. 12.19):

Consider the network of Fig. 12.19 where the delays Δ_l have been represented in the form of blocks of length proportional to Δ_l; here it is assumed that the internal delays of the combinational network are zero or negligible in relation to those of the feedback loops.

Firstly, how is the initial state to be defined?

It should be noted that for the circuit under consideration, the voltages E, X, Y can take the values 0 or 1 which play a similar rôle and for which neither is privileged: neither one nor the other can be considered as being characteristic of the state of rest of the system.

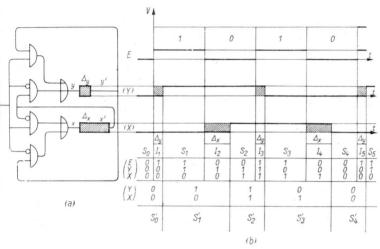

Fig. 12.19. Asynchronous network and its operation with time

If, as is the case here, the power supply is implicit, then the application of power is equally implicit (and it is this alone which puts the circuit into operation); the system of Fig. 12.19a must be considered to be *in action*. But, in order to define the phases (quantification of time) it is necessary to define, in particular, a phase 0, that of the initial state.

If at an instant t, $X'Y' = 00$ and $E = 0$, the diagram shows that $XY = 00$. Consequently, an output $\neq 0$, even after waiting a time $\Delta_x (\Delta_x > \Delta_y)$, will not appear at $EX'Y'$. If E has retained the value 0 the same conditions as before apply and XY still $= 0$.

Thus, when starting from a total state $EXY = 000$, the total state remains at 000 if E remains equal to 0: the state 000 is stable. With $E = 0$ consideration *can* be given to an initial state commencing at an instant t_o where $EXY = 0$ and finishing when E passes from 0 to 1 (Phase S_t of Fig. 12.19).

23*

It should be noted that in spite of an initial stable state in this present example, it is not necessarily so, from a purely logical point of view, for all networks. However, for *practical* reasons, it remains true that the majority of asynchronous networks effectively possess a stable state (after initiation of the power supply) which may be taken as the initial state. Later in this work an examination will be made of what occurs in the absence of such a state.

Phases

Starting from the phase S_0, Fig. 12.19 shows how the operation proceeds in a certain number of phases of unequal duration, denoted $I_1, S_1, I_2, S_2, \ldots I_5 S_5$ etc. During each of these phases S_j or I_k the total state EXY remains constant. It will, however, be noted that the phases I_k are of duration Δ_x of Δ_y (delay values of the feedback loops): they are in a way *transient* states, and their duration is *determined* by the internal delays of the circuit. On the other hand, the phases S_j have a duration *as long as may be desired* depending only on the behaviour of the input during which EXY is in a *permanent* state. It can still be said that during the phases I_k the states are *unstable* and that during the phases S_j they are *stable*.

Example 2:

In the above example the internal states XY follow a Gray code counting sequence 00, 01, 11, 10. Consider now the case of a combinatorial network (C) with transitions as in Table 12.20.

Table 12.20

	X_{n+1}		Y_{n+1}	
$X_n\ Y_n$	$E=0$		$E=1$	
0 0	0	0	0	1
0 1	1	0	0	1
1 0	1	0	1	1
1 1	0	0	1	1

Utilising the same symbol convention as above, Fig. 12.21 represents the three types of behaviour arising from the output E in the following three cases:

$$\Delta_X = \Delta_Y = \Delta, \quad \Delta_X > \Delta_Y, \quad \Delta_X < \Delta_Y.$$

12.2 Binary Sequential Networks. General Model

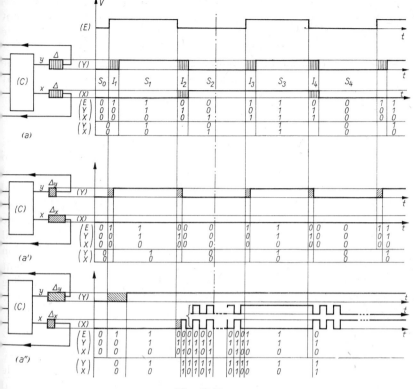

Fig. 12.21

In the first instance, XY takes the values 00, 01, 10, 11 etc. In the second instance the sequence is only 00, 01, 00, 01 etc. Finally, the third instance is characterised by the periodic eruption of an oscillatory regime of period Δ_x for X, while Y remains fixed at the value 1. This oscillatory regime can be terminated by any one of the two values 0 and 1, according to the instant at which the rising front E occurs relative to the castellation of oscillation X.

The example given above shows how, in general, the operation of an asynchronous sequential network is not independent of the values Δ_i of the delays which occur in the feedback loops. This could, however, be the case as it was in the first example considered.

The phases of operation are defined physically in the two examples given above, by the instants (initiation and extinction) recruited at the following times:

— the times of the fronts of the *input* variables,

— the times of the fronts of the internal state variables, i.e. by the total state fronts. These examples suffice to show that for an asynchronous sequential network, the definition of the phases is not only of external origin, as in the synchronous (time) case but is, at the same time, both external *and* internal.

12.3 Sequential Networks and Associated Representations

12.3.1 Logic Diagrams

These are the idealised representations which were introduced above. Compared with the electrical or electronic diagrams they are at a higher level of abstraction: while the latter are associated directly with mathematical representations having continuous variables (differential equations or, for very high frequencies, Maxwell equations), the former schemes shown above are concerned with discrete variables: binary variables, discrete time.

The representations of Figs. 12.4 and 12.5 are those of the general model for the equations (12.1) and (12.2) or for (12.3) and (12.4).

The division of the sequential networks into two specialised types, synchronous and asynchronous, results in slightly different diagrams (Fig. 12.6 and 12.18). They can differ by the number of inputs and outputs of the network C, by the presence of flip-flops in the synchronous networks, by indications (as necessary) of clock-time in the synchronous systems. However, in the case of the synchronous network, if the time base is not shown and if the memory unit comprises calibrated delays, to the exclusion of any flip-flop, the corresponding diagram (apart from any supplementary convention) cannot be distinguished from that of an asynchronous network having equal delays Δ_i.

12.3.2 Equations and Truth Tables

The equations (12.1) to (12.5) are, generally, applicable whether the system be synchronous or asynchronous. They can be associated with the diagrams of Figs. 12.4 and 12.5.

If they are to be translated into the form of Boolean equations, it will be necessary to introduce the distinction synchronous/asynchronous in order to obtain equations of the type (12.7) for example, or (12.25). The Boolean functions appearing in these relations can, in their turn, be translated by all the usual means (algebraic expressions, truth tables, Veitch-Karnaugh maps and other planar representations of the n-dimensional cube.

These relations thus translate the more detailed diagrams given in Figs. 12.6 and 12.8.

Notation

Various notations are utilised to distinguish the present and future states. Table 12.22 gives some examples:

Table 12.22

Present X	Future X_+
X_-	X
X_n	X_{n+1}
k	K
x	X

The pair X_n, X_{n+1} (numerical subscripts) are used, generally, with the synchronous systems while xX and kK are reserved for the asynchronous systems. In the text which follows the notation of type X_n, X_{n+1} is used indifferently for both types, the notation xX here being used only exceptionally and for the asynchronous circuits only.

12.3.3 Structural Representations

12.3.3.1 Alphabets

The set $\mathscr{E}$ of possible inputs E is known as the *input alphabet*.
The set $\mathscr{E}$ of possible outputs S is known as the *output alphabet*.
The set of internal states is denoted by $\mathscr{Q}$.

12.3.3.2 Words, Sequences, Sequence Lengths

A sequence of letters (of their input or output letters respectively) is known as a *sequence* or *word* (input or output word). The number n of letters in the sequence is its *length*: e.g. $s = 1232$ has length 4.

12.3.3.3 Transitions. Transition and Output Functions

The equations (12.1) and (12.2) represent:
— a mapping of the set $\mathscr{E} \times \mathscr{Q}$ into $\mathscr{Q}$
— a mapping of the set $\mathscr{E} \times \mathscr{Q}$ into $\mathscr{S}$ (Table 12.23)

Similarly, for each corresponding input there corresponds:

— a mapping of F_i of $\mathscr{Q}$ into itself
— a mapping of G_i of $\mathscr{Q}$ to $\mathscr{S}$ (Table 12.24)

Which may be expressed as:

$$\begin{cases} F(Q, E_i) = F_i(Q) \\ G(Q, E_i) = G_i(Q) \end{cases} \tag{12.26}$$

Particular case: Moore network:

All G_i mappings are identical (Table 12.25)

— the functions F_i are called "transition functions by E_i" or "succession functions by E_i";

— the functions G_i are the "output functions by E_i"

12.3.3.4 Successors. Direct Successors. Indirect Successors

For a given internal state Q, the state $F_i(Q) = F(Q, E_i)$ is known as the "successor of Q by E_i" (sometimes designated QE_i).

The application of a sequence S to this same initial state Q gives a state Q' designated by $F(Q, S)$, $F_s(Q)$ or QS. If the length n of S is greater than 1 it is said to be the indirect successor. If $(S = E_i)$ it is said that Q' is the direct (immediate) successor of Q by S, or more briefly, "the successor of Q by E_i".

It is to be noted that this concept of direct or indirect succession, like that of the length of a sequence, is linked to the breakdown of time into phases.

Thus there are 2 cases:

1. *Synchronous networks*

There is one phase per clock cycle. Consequently, the length n of S is, at the same time, equal to the number of clock cycles i (at which the inputs E_i are applied), to the number of transitions $Q \rightarrow F_i(Q)$ and to the number of outputs $S = G_i(Q)$ in the output sequence.

2. *Asynchronous networks*

The number of phases is no longer defined solely by the input phases (castellations) but also by those of the internal state.

12.3 Sequential Networks and Associated Representations

Thus, in the case of Fig. 12.21 a, between the origin and the vertical chain-dotted line:

— the input takes the value   0 1 0
— the total state takes the value   (E)   0 1 1 0 0
    (Y)   0 0 1 1 0
    (X)   0 0 0 0 1

thus, between these instants $n = 5$ phases for the total state,
   $n = 3$ phases for the input considered in isolation,
   $n = 3$ phases for the output considered in isolation.

In the case of Fig. 12.21 b there are also $n = 5$ total state phases and $n = 3$ for the output.

On the other hand, in the case of Fig. 12.21 a there are: $n = 9$ for the total state and $n = 7$ for the internal state.

Thus the length S must be considered as being equal to: 5 for the networks of Figs. 12.21 a and 12.21 b, and 9 for that of Fig. 12.21 a, this for the same input profile.

12.3.3.5 Transition Table — Table of Outputs

— *Transition table*

Each function $F_i(Q)$ can be represented by a table listing the pairs Q, $F_i(Q)$ (function variables). If these are grouped for all possible inputs E_i, they will provide a *transition table* or *flow table*.

Table 12.23 provides an example for the case of a network having 8 internal states ($Q_1, Q_2, ..., Q_8$) and two inputs (E_1, E_2).

— Each frame contains coordinates of the form E_i, Q_j ($i = 1, 2$; $j = 1, 2, ... 8$), and is thus associated with an element $E_i Q_j$ of the set $E \times Q$. The frame being arranged so that its content is $F(E_i, Q_j)$, successor of Q_j (line) by E_i (column).

— Associating a column (E_i) with the left hand column results in the transition table of functions F_i.

— *Output table*

Similarly, the function $G(E, Q)$ can be represented by a table for which the inputs are the same as for those above but where, this time, a frame E_i, Q_j contains the output $G(E_i, Q_j)$. Each column E_i defines a function $G_i(Q)$. Table 12.24 illustrates the procedure.

Particular case: Moore network

All the columns are identical and the output table can be reduced a to 2-column table as, for example, that of Table 12.25.

Table 12.23. *Transition table*
($\mathscr{E}$)

(Q)

Q \ E	E_1	E_2
Q_1	Q_2	Q_4
Q_2	Q_5	Q_6
Q_3	Q_3	Q_7
Q_4	Q_1	Q_2
Q_5	Q_5	Q_6
Q_6	Q_5	Q_4
Q_7	Q_3	Q_7
Q_8	Q_8	Q_8

Table 12.24. *Output table*
($\mathscr{E}$)

(Q)

Q \ E	E_1	E_2
Q_1	Z_1	Z_2
Q_2	Z_2	Z_1
Q_3	Z_2	Z_1
Q_4	Z_1	Z_2
Q_5	Z_1	Z_1
Q_6	Z_1	Z_2
Q_7	Z_2	Z_2
Q_8	Z_2	Z_2

Table 12.25

Q	Z
Q_1	Z_1
Q_2	Z_3
Q_3	Z_1
Q_4	Z_2

These two representations — transition table and output table, serve only to translate the relations (12.1) and (12.2). They are general and therefore make no distinction as between synchronous and asynchronous circuits. Taking into account that F and G cover the same range of variation, these two separate tables can be combined into one composite table in which the frames contain the pairs $F(Q_j, E_i)$, $G(Q_j, E_i)$. In practice it suffices to insert only the indices of the variables, their nature being obvious from the left hand column and from the top line, defined by the fixed order F, G adopted for the other frames.

This, for the example under consideration, leads to Table 12.26.

This table completely defines the sequential network, taking into account the conventions used. In theory it applies indifferently to a Mealy or to a Moore network. In the latter case it leads to a repetition of the output column $Z(Q)$.

12.3 Sequential Networks and Associated Representations

Table 12.26. *Composite transition and output table*

E \ Q	1	2
1	2, 1	4, 2
2	5, 2	6, 1
3	3, 2	7, 1
4	1, 1	2, 2
5	5, 1	6, 1
6	5, 1	4, 2
7	3, 2	7, 2
8	8, 2	8, 2

12.3.3.6 Graphical Representation of a Sequential Network (State Diagram, Transition Diagram)

A sequential network can also be represented in the form of a *graph*. Such a graph comprises:
— vertices or nodes, represented by circles (or points),
— directed arcs, connecting the pairs of nodes.

In the representations (Fig. 12.27) of sequential networks, 2 cases can be distinguished:

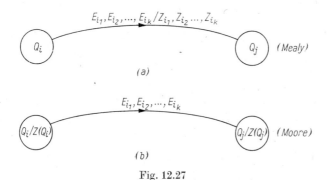

Fig. 12.27

1. *Mealy network*

Each circle (node) contains an indication of a state Q_i. There are as many nodes as there are states. For a given pair of states $Q_i Q_j$, if there are one or more transitions $Q_i \to Q_j$ for the inputs E_{i_1}, E_{i_2} ... E_{i_k}, the

346   12. Sequential Networks. Definitions and Representations

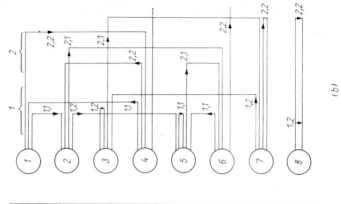

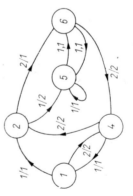

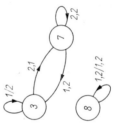

Fig. 12.28

12.3 Sequential Networks and Associated Representations

graph will possess a directional arc travelling from Q_i to Q_j. This arc first indicates the inputs E_{i_1}, E_{i_2}, ..., E_{i_k}, and then the outputs:

$$Z_i = Z(Q_i, E_{i_1}), \quad Z_{i_2} = Z(Q_i, E_{i_2}) \cdots Z_{i_k} = Z(Q_i, E_{i_k})$$

in the same order (Fig. 12.27a).

2. Moore network

Each circle contains the indication $Q_i/Z(Q_i)$. In the event of a transition from Q_i to Q_j only the inputs initiating this transition are indicated above the directed arrow (Fig. 12.27b).

The graph of a sequential network thus comprises exactly the same information as the transition and output tables. The reader will convince himself by Fig. 12.28 that it actually is the table (Fig. 12.28a, b), but for which the geometry is arbitrary (Fig. 12.28b, c).

Fig. 12.29 gives the graphs of several of the networks defined above (series adder, converter, V-scan encoder).

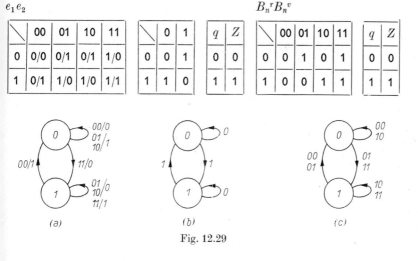

Fig. 12.29

It is also possible to draw up one diagram per input E_i by arranging that the different diagrams are superposable. This can prove convenient in certain cases.

Note: when all the outputs associated with an arc are identical, they will be written once only.

12.3.3.7 Connection Matrix

This consists of another transcription of the table of transitions (*Aufenkamp* and *Hohn*). From a given network having p internal states is constructed a square table having p lines and p columns, each line or column being associated with an internal state. The intersection of line i with column j thus defines an ordered pair of states (Q_i, Q_j). At this location the symbol $\varnothing$ is inscribed in the event that there is no input causing a transition from Q_i to Q_j. Otherwise the set of inputs which provoke the transition $Q_i \to Q_j$ is recorded.

Example: Table 12.30 shows the connection matrix for the conditions of Table 12.26.

Table 12.30. *Connection matrix*

	1	2	3	4	5	6	7	8
1	0	1	0	2	0	0	0	0
2	0	0	0	0	1	2	0	0
3	0	0	1	0	0	0	2	0
4	1	2	0	0	0	0	0	0
5	0	0	0	0	1	2	0	0
6	0	0	0	2	1	0	0	0
7	0	0	1	0	0	0	2	0
8	0	0	0	0	0	0	0	$1 \cup 2$

Clearly the connection matrix is equivalent to the table of transitions. It will be noted that each input symbol appears once, and once only, in each line of the matrix.

It is also possible to define a *composite matrix* of connections and outputs, by inscribing in each instance, not only the inputs which provoke the transitions but also the outputs (arranged in the same order) produced during these same transitions (Table 12.31):

Table 12.31. *Composite matrix of connections and outputs*

	1	2	3	4	5	6	7	8
1	0	1/1	0	2/2	0	0	0	0
2	0	0	0	0	1/2	2/1	0	0
3	0	0	1/2	0	0	0	2/1	0
4	1/1	2/2	0	0	0	0	0	0
5	0	0	0	0	1/1	2/1	0	0
6	0	0	0	2/2	1/1	0	0	0
7	0	0	1/2	0	0	0	2/2	0
8	0	0	0	0	0	0	0	$1 \cup 2/2$

12.3.4 Representations as a Function of Time

There is a need for this when representing, not the abstract mechanism which characterises the network (structural representation) but the behaviour of the latter during *particular* sequences. This representation is generally that of the correspondence between input sequences, internal and output states.

12.3.4.1 Literal Representation

The sequences can always be represented by concatenated symbols ('words'), with an indication of the order in which the letters are used relative to the passage of time.

Example: $E = E_1 E_2 E_3 E_4$ (time increasing from left to right).

12.3.4.2 Sequence Diagram (Chronogram)

This shows in graphical form the correspondence between the phases and the circuit variables.[1] Fig. 12.32 gives two variants of this ($a-b$, and $c-d$). The thin vertical lines divide the time into phases for which the number is indicated.

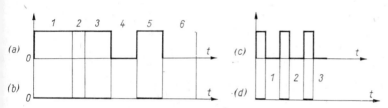

Fig. 12.32. Sequence diagram

This type of diagram can be utilised in the case of a finite number of sequences of finite length. Additionally, the specifications of operation for a sequential circuit can often conveniently be given in terms of such a diagram, as for example, Fig. 12.33.

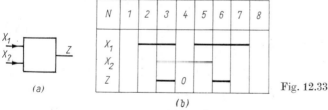

Fig. 12.33

For this reason the sequence diagram is sometimes known as the "phase diagram" (in the study of asynchronous networks).

12. Sequential Networks. Definitions and Representations

A diagram of this type can effectively be used in the study of the operation of sequential networks, and also for their synthesis.

Other representations

Instead of denoting the values of the variables, the *transitions* and their initial values may be indicated. To identify these transitions use can be made, either of special symbols $[(0 \to 1,\ 1 \to 0),\ (0/1, 1/0),\ (+, -)$ etc.], or of confining the representation to the first symbol of each block of identical values (Gavrilov).

N	1	2	3	4	5	6	7	8
X_1	0	1		0	1			0
X_2	0		1			0		
Z			1	0		1	0	

Fig. 12.34

N	1	2	3	4	5	6	7	8
X_1	0	1		1	1			1
X_2	0		1			1		
Z	0		1	1		1	1	

Fig. 12.35

Fig. 12.33 thus takes on the shape of Fig. 12.34 (the initial values are thus indicated automatically). Another method consists of the indication by an unique symbol (e.g. 1) for each transition, irrespective of it sense. The first column contains the initial values. Fig. 12.35 is thus derived from the preceding example.

12.4 Study of the Operation of Sequential Networks

Question:

What can be achieved from a given sequential network, set of logic circuits and of memory units? In particular, for a given input sequence what is the corresponding output sequence?

The study can be broken down as follows:

12.4 Study of the Operation of Sequential Networks

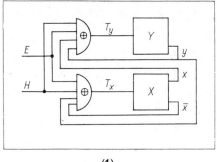

(1)

$$\begin{cases} Q_X^{n+1} = E_n Q_Y^n + \overline{E^n} Q_X^n \\ Q_Y^{n+1} = \overline{E^n} Q_Y^n + E^n \overline{Q_X^n} \end{cases}$$
$$\begin{cases} Z_X^n = Q_X^n \\ Z_Y^n = Q_Y^n \end{cases}$$

(2)

E	0	1
Q		
0	0, 0	1, 0
1	1, 1	2, 1
2	2, 2	3, 2
3	3, 3	0, 3

(4)

$(EQ_X Q_Y)_n$	$(Q_X Q_Y)_{n+1}$	$(Z_X Z_Y)_n$
0 0 0	0 0	0 0
0 0 1	0 1	0 1
0 1 0	1 0	1 0
0 1 1	1 1	1 1
1 0 0	0 1	0 0
1 0 1	1 1	0 1
1 1 0	0 0	1 1
1 1 1	1 0	1 0

(3)

(5) State diagram with states 0/0, 1/1, 2/2, 3/3 and transitions on 0 and 1.

Q (code)	
0 0	0
0 1	1
1 1	2
1 0	3

	0	1	2	3
0	0/0	1/0	∅	∅
1	∅	0/1	1/1	∅
2	∅	∅	0/2	1/2
3	1/3	∅	∅	0/3

(6)

Fig. 12.36

12.4.1 Synchronous Circuits

1. Starting from an electrical scheme, establish a logic sequential network showing the memory units and associated combinatorial network.
2. Determine the functions F and G. This may be effected in a number of ways:
    2.1 Determine a system of Boolean equations, or
    2.2 Produce truth tables representing the operation of the network C.
    Then, (as necessary)
3. Structural representations.
    3.1 Establish a table of transitions and outputs, or
    3.2 Establish a graphical representation of the network, or
    3.3 Establish a matrix of connections and outputs.
    At this stage the coding of inputs, internal states and outputs can be replaced by a more compact coding system, (e.g. decimal).
4. Study the correspondence between input, output and internal state sequences.

For any input sequence applied to a given initial state, it will then be possible to make a study of the corresponding sequences of internal states and outputs. Study, in particular, the correspondence between the sequences of input and output which constitutes the final stage.

Example: By way of an example, a return will be made to one of the networks examined earlier. The various stages of the study are summarised in Fig. 12.36 (for a Gray synchronous counter having 2 Eccles-Jordan flip-flops where the internal states are coded in decimal. Generally speaking it is often convenient to code the inputs and outputs in a similar manner.

12.4.2 Asynchronous Networks

In the general case, where the natural delays of the gates or of the connections of the combinational network (C) (Fig. 12.18), cannot be considered to be negligible (e.g. where all the delays, both those of C as well as those of the feedback loops are natural delays), the states of the feedback loops and of the corresponding binary variables are not always sufficient to take account of what occurs. In fact, given that an actual combinational network always requires a certain time in which to become stabilised after an input changes, it is not obvious, a priori, that the operation of the network of Fig. 12.8 may be divided into two unrelated operations: a decision by the network (C) followed by the return of the information. It will, on the contrary, appear — and the reader can pursue examples — that in the absence of a particular hypothesis for the relative

12.4 Study of the Operation of Sequential Networks

lengths of stabilisation times of (C) and the delays in the loops, that these two operations can actually be combined, in the sense that the output can react at the input to (C) even before (C) has become stabilised.

Keeping then to the state variables q_i of the feedback loops it will be necessary to examine a certain number of characteristic phenomena of the asynchronous networks, related at the same time to the delays of the logic elements, to the absence of a clock and to the structure of the network. To the extent that equations of the type (12.25) do not, of themselves, take account of these phenomena, and where it is necessary to cause the circuit delays to intervene, these latter may be considered parasitic. Where they result in erroneous outputs of a transient type, they may be considered acceptable and this is the hypothesis adopted here. If, however, they lead to errors of a permanent character ways must be found of eliminating them.

The study of an asynchronous network, in addition to a number of steps analogous to those defined for the synchronous networks, will necessitate an examination of a number of peculiarities as follows:

1. The stable states.
2. The constant input cycles.
3. The races between internal state variables.
4. The hazards.

Fig. 12.37 I, II, III summarises the different stages of the study with points 1 to 3 above.

From the flow diagram thus established, the following observations can be made:

12.4.2.1 Sequential Network Diagram

It is not, at this stage, necessary to show the delays in the diagram.

12.4.2.2 Equations

This time the phases n and $n+1$ are separated by a delay which is the common value of all the delays if they are equal, or equal to one of the delays Δ_i if they are unequal. This does not affect the writing of the Boolean equations which are expressed as though all the delays were equal.

12.4.2.3 Truth Tables

It is *customary*:

— to group all of the truth tables together,
— to separate the internal state variable combinations (listed vertically) and the input variables (listed horizontally).

12.4.2.4 Differential Truth Table (Fig. 12.37b)

This is obtained by inscribing in the "future state" frame the vectorial sum (which is also the difference):

$$Q^n \oplus Q^{n+1} = [(q_l{}^n) \oplus (q_l{}^{n+1})]$$

It reveals the simple or multiple character of the transitions dependent upon whether the vector represented has the weight 1 or more respectively. In Fig. 12.37 the multiple transition values are encircled.

12.4.2.5 Stable States

When the equations are available it is, in principle, possible to determine the stable states of the network. An internal state Q is *stable* if, for a given input E_i, $F(Q, E_i) = F_i(Q) = (Q)$. Otherwise it is *unstable*. If the state is stable (by hypothesis), as E_i is then constant, it can be considered that the *total state* (E_iQ) is itself *stable*.

Using the equations to determine the stable states necessitates the resolution of Boolean algebraic equations.

In this present work, with rare exceptions, only the truth table will be used. Its use in the resolution of the equations $F(E_i, Q) = Q$ then consists in allocating to the variables all possible values and finding those values which satisfy the equation. It will be seen that:

— for a state (total, internal) to be stable, it is necessary and sufficient that in the truth table, the binary code which appears at the intersection of the line Q_i with the column E_j be identical with that for Q_i, encircled in the truth table (Fig. 12.37a).

It will be noted that the stability of states is a property which influences only the characteristics of the state diagram, to the exclusion of the delay values.

12.4.2.6 Table of Transitions and Outputs. Graph

As for the truth tables, the stable states may be encircled. On the graph, the stable states are shown by a loop having as origin and extremity the same state value (Fig. 12.37f, h).

12.4.2.7 Cycles with Constant Input

A cycle is a progression of unstable internal states produced from a constant input. The progression of associated total states is thus itself comprised of unstable states. By reason of the instability of each state and of the finite number of such states, the circuit periodically recirculates through

12.4 Study of the Operation of Sequential Networks 355

these same states in the form of a closed cycle. Contrarily to the stability of states, the existence of cycles is not only a function of the properties of the states diagram, but of their codings, and of the delays. The cycles shown geometrically by the states diagram are not necessarily produced in this form. On the other hand, it is possible to have cycles other than those indicated by the diagram. For a given diagram and state coding, the appearance of these cycles is a function of the two characteristics of operation given below, the hazards and the races.

12.4.2.8 Hazards

In a general way it is usual to designate as hazards certain phenomena relating to the use of sequential circuits of 'real' logic circuits, possessing positive delays and able to cause parasitic operations. In this sense, hazards can manifest themselves both in synchronous and in asynchronous circuits. In the first case, however, the study of these phenomena and, as necessary, the elimination of the corresponding parasitic operations, necessitates an electrical study of the detailed character of the physical elements (times of ascent, descent, impedances, capacitances, etc.) which, in this context, do not themselves result from asynchronous network theory but, at any rate in part, from electrical circuit techniques. On the other hand, in the second case the properties of the circuit depend directly and entirely on the inherent delays in the circuit, in such a way that these delays which, in some cases, play only this rôle (e.g. output delays, lengths of cycles), can have a decisive effect on the input-output relationships.

It is convenient to refer to a hazard as a *possibility* of parasitic operation. When, with this terminology, a circuit "presents a hazard" there exists in theory a distribution of delays, which can provoke an operation but, for a particular implementation of the network, it could well be that the actual values of the delays, either by design or by chance, do not give rise to the parasitic operation in question.

The following hazards are usually considered:

1. Hazards of combinational networks

These hazards can only lead to parasitic transients: after stabilisation of the circuit the output functions take on their correct values. However, if there exists any possibility of storing these transient values — as is the case when the combinatorial circuit is inserted into a sequential network as in Fig. 12.18 — they can be transformed into permanent errors.

It can be said that a combinatorial network (C) presents a hazard if, for certain input transitions and for certain distributions of the delays, there exists the possibility of erroneous transient output in relation to that which would have been obtained from an ideal circuit (without delays). In the case of transitions applying to a single input variable, it will be seen that there exist two types of hazard as follows:

Static hazard. For a given transition $E_1 \to E_2$ such that[1]

$$D[E_1, E_2] = 1, \qquad Z(E_1) = Z(E_2).$$

the circuit presents a static hazard if possibly $Z \neq Z(E_1), Z(E)_2$.

Dynamic hazard. For a given transition $E_1 \to E_2$ such that

$$D[E_1, E_2] = 1 \qquad Z(E_1) = \bar{Z}(E_2),$$

the circuit presents a dynamic hazard if after a transition $Z(E_1) \to \bar{Z}(E_1)$ it possesses a value of Z equal to $Z(E_1)$.

A circuit possessing neither a static, nor a dynamic hazard, is known as a *hazard-free* circuit.

This leads to the introduction of the following definition:

Validation group. This is the name given to a minimum combination of input variables

$$G_1 = \{x_{i_k}^{a_{i_k}}\}, \qquad (a_{i_k} = 0 \text{ or } 1)$$

such that

$$\prod_{x_i \in G_1} x_i^{a_i} = 1$$

implies that $Z = 1$.

It can then be said that a combination $E = (e_1, e_2, e_i \ldots, e_n)$ is covered by a validation group if, for every subscript i_k such that $x_{i_k}^{a_{i_k}} E\, G_1$, $e_{i_k} = a_{i_k}$.

It can also be said that a transition $E_1 \to E_2$ is covered by G_1 if E_1 and E_2 are both covered by G_1, (the transition $E_2 \to E_1$, is then also covered). Next consider the circuits formed of AND or OR gates.

It follows (cf. for example [97]) that in order to ensure that a 2-level circuit be hazard-free, it is necessary and sufficient that each transition be covered by a validation group.

From this it can be deduced, in particular, that it *suffices* to include in the expression of the function all the prime terms thereof.

2. Essential hazards

This type of hazard, put forward by Unger, is linked to the diagram of states. It can be said that such a diagram presents an essential hazard if, from a basic state $Q(QE = Q)$, there exists a sequence $EE'E$ (with $D[E, E'] = 1$) such that

$$QEE'E = Q^* \neq Q.$$

[1] $D(E, E')$: Hamming distance of E and E'.

12.4 Study of the Operation of Sequential Networks

Whether or not this peculiarity of the structure of the diagram leads to a poor operation depends on the associated combinatorial circuit (C) and on the distribution of the delays therein. Assuming that, after the input transition $E \to E'$, one of the output variables had changed and that this change had been returned to the input to circuit (C) before the other outputs had changed. These outputs would be revealed as the former input (E) but in combination with the new state $Q' = QE'$.

Thus if the state Q' has as successor $Q'E = Q''$, and if $Q''E = Q^* \neq Q$ the resultant state Q^* will be erroneous.

The essential hazards can be brought to light by an examination of the transition table or of the phase diagram (§ 12.4.2.10).

Under these conditions, if the states diagram presents an essential hazard, and if the delays in the feedback loops are sufficiently large such that the circuit (C) becomes stabilised before the new states of the variables appear at the input to (C), there can be no parasitic operation as above. Conversely, it can be shown that if there is an essential hazard it is necessary, for the avoidance of permanent parasitic conditions, that the delays in the loops be greater than the time required for stabilisation of the circuit (C). It can be shown that the insertion of artificial delays can, at any rate in theory, be achieved by the use of a single delay, from which the delays of the other loops can be simulated (*Unger*).

Conversely, it can be shown that where the diagram does not include an essential hazard, this can be realised by the use of an asynchronous circuit without the necessity for introducing special delays to the loops (cf. [106]).

12.4.2.9 Races

It has already been seen that, for a given input sequence, the sequences of internal states (or of outputs) can differ and that they depend in some cases on the relative values of the delays. If all the delays are equal it is evident that only one type of transition is possible and that the phases are all of length Δ, if it is assumed that the circuit (C) is devoid of delays. In these circumstances, then, if they are seen to be unequal, there are two possible situations:

First case: the transition $Q \to F_i(Q)$ necessitates the changing (complementation) of one variable q_l only: i.e. $D[Q, F_i(Q)] = 1$, where D is the Hamming distance. It is then clear that the transition takes place after a delay Δ_i. The termination of this transition marks the start of a new phase. The other variables do not change, their reaction times play no part no matter what their values relative to those of Δ_i. The transition $Q \to F_i(Q)$ is thus made in a manner uniquely determined by the combinatorial logic circuit (C) independently of the delays. These latter only influence the required transition time (Δ_i if it is the variable internal state Q_i which switches).

Second case: the transition is multiple, i.e. it requires the change of several variables $q_{i_1}, \ldots q_{i_k}$ ($D[Q, F_i(Q)] = k > 1$). With these variables correspond the delays Δ_{i_l} which may be arranged in increasing order:

$$\Delta_{j_1}, \Delta_{j_2}, \ldots, \Delta_{j_k}.$$

After a delay Δ_{j_1}, q_{j_1} changes. The new internal state $(q_1, \ldots q_r)$ is then taken into account by the logic circuit (C) independently of the other transitions. This new state has, itself, a successor (by single or multiple transition), determined from the truth table for the circuit, which, of necessity, is to be found in one or other of the two situations described above.

This process may or may not continue, but will always take one of the two following forms (for a given distribution of delays).

— after passage through one or more unstable (total) states, arriving and finishing in a stable state.
— after passage through 0, 1 or several unstable states, arriving in an unstable state forming part of a cycle.

From a review of the different circuits associated with the different delay distributions can be obtained, for a given race, 0, 1 or several distinct stable states, 0, 1 or several distinct cycles in general.

As will be seen in the examples which follow, whether or not a race is critical generally depends on:

— the output function of the circuit,
— the definition of the time intervals, since the comparison of output sequences obtained from a unique sequence is, of necessity, made phase-by-phase (in the absence of any other time reference). The choice of *phases* can be considered from several viewpoints:

1. The output may be referenced in relation to physical time. It is rare, it is true, that the actual positions of the castellations is of interest. This could, however, be the case where it is desired, for example, to initiate a given sequence after a certain determined time interval. In certain cases, ignoring the instants of the fronts, importance may be given to the lengths of the castellations. Although this viewpoint is somewhat contrary to the general philosophy of asynchronous circuits, it is admissible in the following case:

It is required, in response to given sequences, to place the system in one or more states intended to activate a signal unit (e.g. audio, visual). With a single output there can be, apart from the continuous states 1 and 0, one (or even several) oscillatory regimes, identified by their time

12.4 Study of the Operation of Sequential Networks

periods (slow, normal, fast). In both cases the viewpoint adopted amounts to a breakdown of time into phases of equal length Δ_t, where Δ_t is considered as a quantum, a unit of measurement for the input castellations, generally small in relation to the length of the latter.

2. The output can be referenced in relation to the input/internal state phases.

3. The output can be referenced only in relation to its own phases.

4. The output can be referenced in relation to the input-output phases, taking no account of the internal state.

5. Account may be taken only of the stable states of the system and of a standard oscillatory state (without consideration of the length of the cycle or of the frequency).

6. Finally, account may be taken of the continuous stable states only, to the exclusion of all cycles. This is the case most often used in practice.

$q_1 q_2 q_3$ \ $e_1 e_2$	0 0	0 1	1 0	1 1
0 0 0	(0 0 0)	0 0 1	0 1 0	0 1 1
0 0 1	(0 0 1)	0 1 1	0 1 1	0 0 0
0 1 1	0 1 0	1 1 0	1 1 0	0 0 1
0 1 0	0 1 1	1 1 0	(0 1 0)	0 1 1
1 1 0	1 0 0	(1 1 0)	(1 1 0)	1 1 1
1 0 0	1 0 1	1 0 1	(1 0 0)	1 0 1
1 0 1	1 1 1	1 0 0	1 0 0	(1 0 1)
1 1 1	1 1 0	1 1 0	1 1 0	0 1 1

a)

$d_1 d_2 d_3$ \ $e_1 e_2$	0 0	0 1	1 0	1 1
0 0 0	0 0 0	0 0 1	0 1 0	(0 1 1)
0 0 1	0 0 0	0 1 0	0 1 0	0 0 1
0 1 1	0 0 1	(1 0 1)	(1 0 1)	0 1 0
0 1 0	0 0 1	1 0 0	0 0 0	0 0 1
1 1 0	0 1 0	0 0 0	0 0 0	0 0 1
1 0 0	0 0 1	0 0 1	0 0 0	0 0 1
1 0 1	0 1 0	0 0 1	0 0 1	0 0 0
1 1 1	0 0 1	0 0 1	0 0 1	1 0 0

b)

I

Fig. 12.37

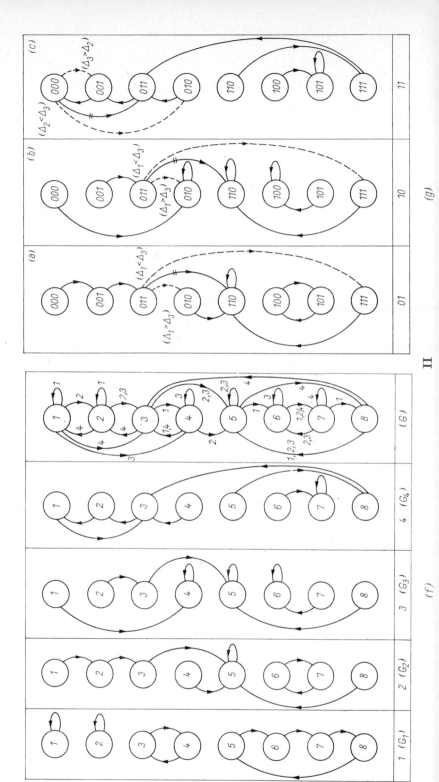

12.4 Study of the Operation of Sequential Networks

E \ Q	1	2	3	4
1	①	2	4	3
2	②	3	3	1
3	4	5	5	2
4	3	5	④	3
5	6	⑤	⑤	8
6	7	7	⑥	7
7	8	6	6	⑦
8	5	5	5	3

(h)

1	2	3	4
①₁	5₂	4₃	3₄
②₁	5₂	5₃	1₄
4₁	5₂	5₃	2₄
3₁	5₂	④₃	3₄
6₁	⑤₂	⑤₃	8₄
7₁	7₂	⑥₃	7₄
8₁	6₂	6₃	⑦₄
5₁	5₂	5₃	3₄

(i)

1	2	3	4
①₁	5₂	4₃	—
②₁	5₂	5₃	—
—	—	—	—
—	5₂	④₃	—
—	⑤₂	5₃	—
—	—	⑥₃	7₄
—	—	6₃	⑦₄
—	—	—	—

(j)

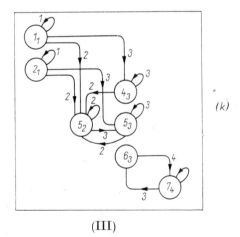

(k)

(III)

Fig. 12.37.

$Q \diagdown E$	1	2	3	4		Q	Z
1_1	1_1	5_2	4_3	—		1_1	$Z\,(1_1)$
2_1	2_1	5_2	5_3	—		2_1	$Z\,(2_1)$
5_2	—	5_2	5_3	—	(1)	5_2	$Z\,(5_2)$ (m)
4_3	—	5_2	4_3	—		4_3	$Z\,(4_3)$
5_3	—	5_2	5_3	—		5_3	$Z\,(5_3)$
6_3	—	—	6_3	7_4		6_3	$Z\,(6_3)$
7_4	—	—	6_3	7_4		7_4	$Z\,(7_4)$

IV

Fig. 12.37 (*Conclusion*)

By way of example, Fig. 12.39 a, b, c presents a number of chronogrammes from which have been extracted Tables 12.40 I and II. Table 12.40 I relates to the case of a single output $Z = q_2$, Table 12.40 II to that of a two-wire output $Z_1 = q_1$, $Z_2 = q_1 \oplus q_2$. The numbers 1 to 4 correspond to the phases 1 to 4 defined above. The sequences represented correspond to what appears to the left of the limiting line X. In certain cases is inscribed (in parentheses) the symbol which would immediately follow after the line. The input sequences used for each type of race a, b, c are not necessarily constant. In conformance with the definition, it is sufficient to find *one* input sequence and a pair of circuits producing divergent outputs to prove that a race is critical. Alternatively, to prove that a race is not critical, it is necessary that no input sequence of this type be present. This does not, however, necessitate the verification of an infinite number of input sequences, as might have been feared. In the case (4a) of Table 12.40 the output has the value 1 for phase N⁰ 1, and for the following phases, no matter what the value of input E, the same output sequence x will result. Thus the race is not critical, evidently due to the fact that the same internal state is attained with the 3 transitions from which the evolution of the system will be identical in the 3 cases.

It will be shown in the following chapter that the test for the equivalence of 2 states of 2 different circuits, or of the same circuit, is always effected in a finite number of operations despite the infinite character of the set of input sequences.

Fig. 12.37 g illustrates the different cases produced for the preceding example. For each 'race', the transition can assume 3 forms, if the direct transition corresponding to the simultaneous changes of the 2 variables is included.

12.4 Study of the Operation of Sequential Networks

Table 12.38

Input	Delay configurations	Transition	Duration of transition
0 1	$\Delta_1 = \Delta_3 = \Delta$ $\Delta_1 < \Delta_3$ $\Delta_1 > \Delta_3$	011 → 110 011 → 111 → 110 011 → 010 → 110	Δ $\Delta_1 + \Delta_3$ $\Delta_1 + \Delta_3$
1 0	$\Delta_1 = \Delta_3 = \Delta$ $\Delta_1 < \Delta_3$ $\Delta_1 > \Delta_3$	011 → 110 011 → 111 → 110 011 → 010	Δ $\Delta_1 + \Delta_3$ Δ_3
1 1	$\Delta_2 = \Delta_3 = \Delta$ $\Delta_2 < \Delta_3$ $\Delta_2 > \Delta_3$	000 → 011 → 001 → 000 000 → 010 → 011 → 001 → 000 000 → 001 → 000	3Δ $2(\Delta_2 + \Delta_3)$ $2\Delta_3$ (period)

For the input 01 there exists only one final state, 110, for the three transitions which can be initiated. The duration of these transitions is, however, not equal.

— For the input 10 there exist 2 stable final states.
— For the input 11 there exist 3 cycles.

'Race' condition: It can be said that a 'race' condition exists when between the two states Q, $F_i(Q)$, taking due account of the binary coding of these states and of the Boolean transition equations, there is a multiple transition.

Critical 'race': The race is said to be critical if there exist two delay distributions $(\Delta_1, ..., \Delta_r)$ and $(\Delta_1', ..., \Delta_r')$ and one input sequence commencing with E_i, such that, Q considered to be the initial state, the output sequences corresponding with each distribution are different.

Non-critical 'race': The race is said to be non-critical if, for every distribution of delays and for every input sequence commencing with E_i, with Q as the initial state, the output sequence remains the same.

In practice, the delays are natural delays (or time constants) which are quite small and which it is desirable to maintain as such. They are difficult to calculate: they depend on a large number of factors such as statistical scatter of the characteristics of the components, the loading conditions associated with the diverse configurations (and thus variable in general with time), and fluctuations in power supply.

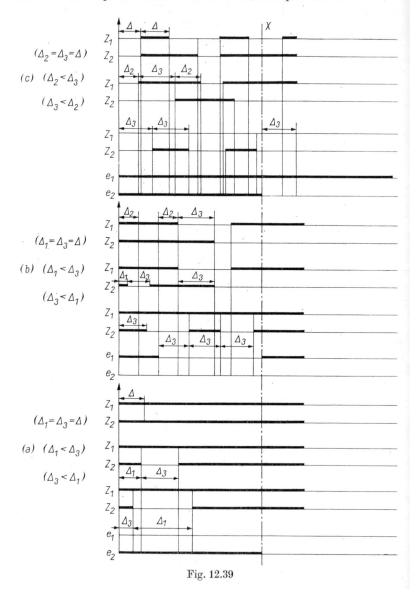

Fig. 12.39

The delays Δ_i therefore vary as functions of:

— the memory loop i.
— the particular example of the logic circuit (for a given concept of the latter).
— time (in the short term: phases, and in the long term: ageing).

12.4 Study of the Operation of Sequential Networks

Table 12.40 I

Type of race \ Race	(a)	(b)	(c)
(1)	C	C	C
(2)	1 1 x 1 1 1 x 1 1 1 x C	1 1 1 0 0 1 1 1 1 1 0 0 1 1 1 1 1 1 1 C	0 1 0 0 1 0 0 1 1 0 1 1 0 0 0 0 0 C
(3)	0 1 0 1 NC	0 1 0 1 NC	0 1 0 1 NC
(4)	1 x 1 x 1 x NC	1 1 0 1 1 1 0 1 1 1 (1) C	0 1 0 1 0 0 0 1 0 1 (1) 0 (0) C

Table 12.40 II. (C: critical; NC: non-critical)

Phase type \ Race	(a)	(b)	(c)
(1)	C	C	C
(2)	1 1 x 1 1 x 1 1 1 x 1 0 1 x 1 1 1 x 1 0 1 x C	1 1 1 0 0 1 1 1 1 1 0 0 1 1 1 1 0 0 1 1 0 1 1 1 0 0 1 1 1 1 1 1 1 0 0 1 0 1 C	0 1 0 0 1 0 0 1 1 0 1 1 0 1 1 0 1 1 0 0 1 1 1 0 0 0 0 0 0 0 1 0 1 0 C
(3)	1 x 1 x 1 1 1 x 1 0 1 x 1 1 1 x 1 0 1 x C	1 0 0 1 1 1 0 0 1 1 1 0 0 1 1 0 1 1 0 0 1 1 1 1 1 1 0 1 0 1 C	0 1 0 0 1 0 0 1 1 0 1 1 0 1 1 0 1 1 0 0 1 1 1 0 0 0 0 0 0 0 1 0 1 0 C
(4)	1 x 1 x 1 1 1 x 1 0 1 x 1 1 1 x 1 0 1 x C	1 1 0 0 1 1 1 1 0 0 1 1 1 1 0 0 1 1 0 1 1 1 0 0 1 1 1 1 1 1 1 0 0 1 0 1 C	0 1 0 0 1 0 0 0 1 0 1 1 0 1 1 0 1 1 0 0 0 1 1 0 0 0 0 0 0 0 0 0 1 0 C

It follows then that if their values are very close, their hierarchy itself may change with the passage of time. To escape the consequences of these random fluctuations in delays the circuits used in practice are designed to satisfy one of the following three policies:

1. The use of single transitions, or
2. The use of non-critical multiple transitions, or
3. The use of critical multiple transitions, but following a pre-determined distribution of delays (handicap technique). By thus inserting known delays into the feedback loops, so as permanently to isolate a particular distribution of delays which, at the same time, selects a particular circuit and automatically 'freezes' a unique trajectory among those capable of being produced.

It should be noted that this voluntary staggering of the delays can only be achieved by increasing the mean values of certain among them (by which they are 'handicapped'). The associated internal states variables are thus automatically "slowed down", quite apart from the 'race' (which motivated their choice) for all the other transitions which bring them into play.

This technique can thus become a delicate task. It is interesting to note that, to the extent that it depends on the acceptance of fixed delays augmenting the natural delays, it wanders a little from the basic principle, which consists of minimising delays in the loops and in more closely approaching the philosophy of the synchronous systems where the use of a constant delay (clock cycle) greater than that of the reaction time of the flip-flops is a method of simplification (at the expense of a general slowing-down of the operation).

In the first two cases the ambiguities are eliminated by purely logical means (modification to the binary code associated with the states, or to the transition graph (E), or to both).

The third method can be considered as using both logical and physical means; since, in addition to the above considerations, delay lengths must also be taken into account.

12.4.2.10 Phase Diagram (Huffmann)

There are two points of view to be considered in the study of the operation of an asynchronous network, that of the designer who has to give attention to the smallest details of the internal working of the system, and a second, common both to designer and user where the only interest lies in the operation as seen by the latter.

It is normal to apply certain restrictions dependent on the circumstances in which the equipment is to be used. They are as follows:

12.4 Study of the Operation of Sequential Networks

1. Only the stable states are of interest.
2. Between two stable states there is only one input change[1]

In other words the system proceeds from stable state to stable state in the following manner: commencing from a given stable state, the input changes. If the input initiates a transition trajectory, the completion of the transition and stabilisation of the output state are awaited before initiating a further input change.

This viewpoint is illustrated in Figs. 12.37 i, j, k.

Table 12.37 *j* represents the *phase diagram*. This is derived from the transition table 12.37 *h* where the stable states are as follows:

— in the place of any "future" unstable state is inscribed the stable state to be attained with constant input, so long as there is such a terminal stable state, and dashes are inscribed in place of the future state values in the contrary case.

There remains, therefore, a table of stable total states (encircled) and the directions of the transitions between them. It will be noted that the left hand column of present states, as in Fig. 12.37 *h*, is no longer needed and may be deleted.

The phase diagram is thus a notation special to asynchronous networks, and indicates:

— the stable states,
— the transitions between stable states.

Conversely, it does not show the intermediate unstable states. It therefore constitutes the end of the study of a sequential circuit, for which it describes the overall viewpoint, that of the user. For the same reason it can serve as the starting point for the synthesis of such a circuit based on user requirements. As far as the unstable states are concerned they are revealed by a 'detailed' study constituting an intermediate step in the study of operation as in the synthesis.

The phase diagram may be transcribed in the form of a *graph of stable states*, or as a table of transitions for stable states. The outputs associated with the former total states are, in turn, associated only with the internal states (Moore network) (Fig. 12.37 k, l, m).

It will be noted that in the particular case in question both the graph and the table are incompletely specified. The dashes have their origin, as was shown above, in the fact that certain input sequences have been rejected (those leading to cycles).

This is a particular case of a table often encountered (incompletely specified transition tables) to which further reference will be made later.

---

[1] This mode of operation is often called the fundamental mode.

12.5 Adaptation of the Theoretical Model to the Physical Circuit

This problem arises at the commencement of a study of the operation of a network, or at the end of its synthesis. The model of a sequential network which is to be associated with a given physical network (assembly of gates and memory units), is not unique; it depends to a large extent on the viewpoint adopted, i.e. essentially:

— on the choice of input variables, internal and output states,
— on the breakdown of time into phases.

Examples:

1. Consider the accumulator of a computer. It can, for example, consists of n flip-flops and of addition logic. Assume, for the purpose of the exercise, that it has 4 stages and that the logic unit consists of 1

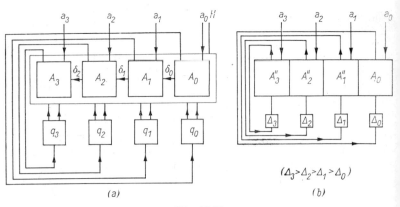

Fig. 12.41

adder having 2 inputs (for the least significant stage) and of 3 adders having 3 parallel-connected inputs. In a synchronous computer having a typical time base unit, the operation could be as follows: the number A is presented in parallel at the input during the pulse n and the resultant $A + Q$ (where Q is the content of the accumulator) could be read out at a time $n + 1$. Thus giving $Q_{n+1} = A_n + Q_n$ which may be completed by taking $Z_n = Q_n$.

This operation is *synchronous*. But is only so in an overall way. If an analysis be made of what occurs during the addition, it will be seen that the flip-flops do not change simultaneously, but in several steps linked

12.5 Adaptation of the Theoretical Model

to the delays between stages, and according to the circumstances, unequal. This operation is thus *asynchronous* and occurs during the application of the clock pulse. During this pulse the operation is that of the asynchronous circuit of Fig. 12.41 b. For example, when adding $A = 0011$ and $Q = 0111$, Q passes through the states $0111 \to 0110 \to 0110 \to 0010 \to 1010$. It is precisely this asynchronous operation which imposes a severe upper limit on the duration of the clock pulse.

If this pulse lasted too long the accumulator would continue to add the input to its content entailing a second addition $A + Q$ during the pulse.

Thus, in this example, according to the choice of phases and the phenomena observed, the resultant model may be either synchronous or asynchronous.

The pulse counters represent a particular case (often encountered) of an accumulator and give rise to analogous phenomena.

(2) The conditions earlier studied on the width and spacing of the clock pulses are intended to ensure that for the synchronous circuit there is one, and only one, transition per pulse, and that this pulse is correct. Otherwise it would be possible to produce other transitions during the same clock pulse. The reader can easily check that in these circumstances there can be a *local* operation of the *asynchronous* type and which here presents (as in the example above) the character of a parasitic operation.

(3) Reverting to the Gray asynchronous counter of § 11.4.3 and 12.2.7.2. Its operation does not depend on the lengths of the castellations. This circumstance authorises the adoption, in a completely legitimate manner, of the following viewpoint: since there is a single input it could be said that this is '*seen*' by the circuit as a regular succession of 0 and 1, in the absence of any other time reference. But, from a slightly different viewpoint, it is possible, instead of an irregular castellation (in relation to an external time reference), to make use of a periodic castellation (of form factor $1/2$) and the operation of the system is then of the synchronous type. Reverting to the case of 11.4.2, more generally, for a given counter with n internal state variables, synchronisation is always possible with a periodic castellation having phases (input-internal state) of equal duration. It is not even necessary to have the time factor scale $1/2$; it suffices, where the latter differs from $1/2$, to equip the timer with a circuit of the counter type described above (where the variable either x or y re-establishes the factor $1/2$).

This type of system synchronisation for which the internal operation is asynchronous is widely used. It is often used in the case of integrated circuits, the flip-flops of the shift register often being designed as asynchronous flip-flops, thus avoiding the use of capacitors which are always technically difficult to implement into integrated circuits.

370   12. Sequential Networks. Definitions and Representations

(4) In the study summarized in Table 12.37 it has been shown that, dependent on whether the interest centers on 'fine' details of operation or on stable states, the resultant graphs 12.37f or 12.37k respectively differ:

— by the nature of the internal states,
— by the number of states.

Moreover, the first is completely specified, the second is not. Finally, if the first is of the Mealy type, the second is of the Moore type. Here again the definition of the phases plays an important rôle and leads to two different models.

12.6 Some Supplementary Physical Considerations

12.6.1 Synchronous Networks

Synchronous circuits are controlled by a clock which provides a periodic waveform. It is customary to distinguish between operation by pulses and at continuous levels. From the point of view of the *clock* there is only a difference of degree (form factor) between the two. It is only from the viewpoint of *internal states* and *outputs* that there is a difference which is basically a matter of technology; giving rise to the following cases:

1. The memory unit is made up from calibrated delays. It can be considered that the information re-circulates within the feedback loops at the frequency of the clock. The state of each feedback loop is characterised by the presence or absence of a re-circulating pulse within the loop. For this reason it is often known as the *dynamic* mode of circuit utilisation.

2. The memory unit is comprised of flip-flops (e.g. Eccles-Jordan type). These have outputs in the form of constant voltages which are available during most of the time and, in particular, are available even outside the clock pulse times. The information thus being available during these steady states is said to be in *static* form. If consideration be given only to input-output relations the circuit may be considered (generally as a function of the use of the outputs) as either operating in *pulse-pulse* mode (if there is sampling of the outputs during the time cycles), or as operating in the mixed *impulse-continuous voltage* mode.

12.6.2 Asynchronous Networks

The circuits we have defined function with continuous voltages to represent the internal states. It is the state of tension (of transient or prolonged duration) which characterises the internal state of the system. The output or input fronts can equally be transformed into relatively brief impulses (very small form factor) by differentiation and rectification.

12.6.3 Physical Duration of States

Whether or not the system be synchronous, the states are only used during part of the time (sampling period), i.e.;
— duration of time pulses (synchronous circuits),
— intervals of time other than those of the rising and descending fronts so that the states (stable or unstable) may be considered as being on a 'plateau' (asynchronous circuits).

This duration is therefore constant for the synchronous case. For the asynchronous case the unstable states have a (brief) duration of the order of the delays and time constants of the feedback loops, while ths stable states, once attained, continue so long as the input remaine unchanged.

12.6.4 Delay Elements

The delays of asynchronous circuits are, in principle, the natural delays of the feedback loops, or of the time constants. In exceptional cases (critical races) they may be extended by synthetic delays. These latter which may be used in synchronous circuits can consist of portions of the delay lines, (passive element) or, in certain instances, of a monostable flip-flop (active element) which produces rectangular waves of adjustable length on receipt of the input impulse.

The bi-stable (e.g. Eccles-Jordan) flip-flop provides a delay whose duration may be programmed.

12.6.5 Synchronisation

Synchronisation provides:

— standardisation of the duration of states,
— systematic stabilisation of all states,
— a 'buffer' effect between the excitation of the memory units and their read-out for all the feedback loops. In a rather sketchy way the time period is thus divided into two phases:

— the first phase where the outputs are active, where the future state has been determined, but where the flip-flops (or feedback loops) receive only their operating instructions (as appropriate) for starting the transitions, without having, effectively, sufficient time to switch completely; — a second phase where the logic of the system is inactive, where the outputs are therefore ignored, and during which the transitions launched earlier are produced in an autonomous manner between the time pulses.

This separation between the transition "command" (the circuit being active) and the "hidden" realisation (the circuit being inactive) of the transition itself often constitutes a delicate problem of circuit design.

If it can be solved, the following benefits are derived:

— suppression of the influence of the delays (of flip-flops or feedback loops) on the operation. In particular, the elimination of 'races': multiple transitions are possible even though several internal state variables may switch in a staggered manner, since the corresponding intermediate states occur when the circuit is inactive (second phase of the time cycle), and the circuit does not return to the active condition until the staggered transitions are completed.

— often, a relaxation of the requirements relating to certain parameters (e.g. delays) and a simplification of design.

Synchronisation therefore presents advantages in the design of circuits and also in their utilisation. The presence of a clock in a system provides a structure and simplifies some repair and maintenance work. But to the extent to which it increases the natural delays of the circuit (loops, flip-flops) by an additional safety margin, synchronisation, from the speed point of view, is a worst case policy: the time quantum is the clock period. Conversely, with asynchronous circuits, the object is to make the best possible use of the intrinsic speed of the circuit at the price of greater complexity (generally) of the logic design.

Although, in general, this last concept leads to an overall gain in speed in relation to the synchronous solution, this is not always the case, locally: a staggered transition involving several variables can take longer than a simultaneous transition with the same variables in the synchronous mode, depending on the delay values on the one hand, and on the clock period on the other. Experience of asynchronous systems which have been built and compared with synchronous systems of the same type (e.g. digital computers) seems to confirm these two points: gain in average speed and increased complexity in the asynchronous case.

One consequence of this is that, for the synchronous system, the time of completion of a sequence of n states is constant, and is equal to n time periods. For an asynchronous system, even with constant input, this

12.6 Some Supplementary Physical Considerations

duration for a sequence of n distinct states varies as a function of all the dispersion factors described earlier. It will thus vary from one realisation of the network to another, and for a given example, with time. In certain applications of asynchronous circuits, this could be a disadvantage.

In the asynchronous case, despite the absence of a clock, there still remain reference points in the temporal evolution of such a circuit: these are the *beginnings of phases* linked with the fronts of the inputs and the internal states and, in a certain sense, can still be considered as a lax, degraded form of clock peculiar to asynchronous circuits. By reason of the internal origin of certain of these signals, it is sometimes said that the system is, in part, (sometimes wholly) *self-synchronised.*

In both cases, synchronous or asynchronous, the delays in the circuit depend upon the characteristics of the components, of the power supply values, and parameters describing the ambient conditions, which are not fixed but distributed statistically around the average values. Thus they are, themselves, of a random character. To a certain extent this can be avoided in synchronous circuits by the use of a time period which increases these fluctuating delays, i.e. essentially *electronic* means. Conversely, the basic idea of asynchronous circuits is to deal with them by purely *logical* means.

12.6.6 State Stability

Since for asynchronous networks it is possible to distinguish two kinds of state, stable and unstable, one is led, for synchronous circuits, to consider that all the states have the same character of stability. Indeed, the rhythm of operation remains the same whether or not the future state Q_{n+1} be identical with the present state Q_n. When, for example, a given input is applied in a repetitive manner, from a given state, it is possible to change that input at any later time, without the least ambiguity, whether this results in a fixed state or in a chain of different states, closed or not. It will therefore be convenient to say that for a synchronous circuit, *all states are stable.*

12.6.7 Memory

A system possesses a memory if it can accept an item (bit) of information at a time t (inscription) and reproduce it at a later time $t' > t$ (read-out).

In the sequential circuit, the memory function appears in several forms:

— in the (rudimentary) form of delays in the loops (asynchronous circuits and, as necessary, synchronous), where the memorised

output appears at a fixed time after inscription and cannot be re-interrogated.
— in the form of the states of the feedback loops which, on the contrary, can be re-interrogated at more or less arbitrary instants (arbitrary in the asynchronous case, co-incident with the time base in the synchronous case).
— in the form of flip-flops (which a close analysis would show to be analogous to the memory type).

12.6.8 Synchronous and Asynchronous Viewpoints

The above examples are of systems which may be studied as synchronous or as asynchronous circuits depending on the viewpoint adopted. In the case of the accumulator, the *overall* concept was that of a *synchronous* set intended to be integrated with the computer, itself synchronous. However, adhering to the internal functional *details* leads to consideration of the phenomenon of 'carry' propagation which is entirely *asynchronous*.

This is generally the case. It must be remembered that the operation of actual electronic systems (electronic, mechanical) is continuous, and the systems intended to be synchronous are only so from a relatively overall viewpoint. In general, the study of a so-called "synchronous" system may be divided into two parts: an overall study of the macroscopic type, and an asynchronous system of the microscopic type, the latter being directed to the internal details of operation. The boundary between the two, a function of the degree of the organisation studied and more or less from the point of view of the synthesis adopted, can be described in several ways although, usually, with respect to the characteristics of the clock.

This detailed asynchronous study is itself only a stage between the physical study of the system (differential equations, etc.) and the study as a synchronous system.

The remainder of this work is devoted to synchronous sequential circuits. The methods of synthesis which will be discussed in Chapter 15 have no part, themselves, in the asynchronous type phenomena which appear within a clock period or clock pulse. These phenomena cannot, however, be ignored at the logical design stage: it has been shown that they can intervene, among other things, in the internal functioning of the logic blocks employed in the process of synchronisation and in preventive treatment for all parasitic asynchronous functions. The logical designer charged with the design of a synchronous circuit must remember that there may remain a certain number of 'fine' asynchronous phenomena capable of forming the subject of a logical study and which must not be

thrown purely and simply onto the electronic circuit design specialist. There is a risk of imposing very severe restrictions on the latter, from the electronic viewpoint, vis-à-vis the performance of the units at his disposal or which may be achieved within a given state of technology. The study techniques for asynchronous circuits developed in this chapter will suffice for the asynchronous study:
— of the internal functioning (as necessary),
— of the synchronisation,
— of the parasitic phenomena.

a study, although limited, remains nevertheless indispensible in the majority of cases.

12.7 Incompletely Specified Circuits

The circuits defined at the beginning of the chapter were completely specified: for every pair $E_n Q_n$ (present instant), the future state Q_{n+1} and the output S_n (present) are indicated and all frames in the transition and output tables are filled. However, it can happen that there is a need to deal with incompletely specified circuits. For such a circuit there exist pairs $E_n Q_n$ such that either the future state, or perhaps the output, or even neither are indicated. It is normal to show these pairs in the transition or output tables by the use of a dash (—) in the place of each internal state or output not specified. The transition Table 12.37 is an example of such a table. The frames containing the dashes resulted from the non-availability of certain inputs based on stable states. Generally speaking, these incompletely specified tables are presented, during the synthesis, by reason of the restrictions imposed on the sequences of input or of output as, for example, the non-acceptability of certain sequences, the inadmissibility of certain inputs based on certain states, the inability to examine the output during certain phases (sampling).

The signification which may be given to the unspecified frames is as follows:

1. Unspecified future state:
   Commencing with this state, the evolution of the circuit is of no importance and can, therefore, be arbitrary.
2. Unspecified output:
   The output encountered during a sequence at the point represented by the dash may be arbitrary. In accordance with the hypothesis of Paull and Unger, it is assumed that not only can any value whatsoever replace the dash in an output sequence, but that this value can still vary for different appearances of the same dash (associated with the same pair $E_n Q_n$) with the passage of time.

12.8 Sequential Networks and Combinatorial Networks

A combination of gates without the inclusion of feedback loops, results in a combinatorial network. It does not possess a storage (memory) unit and an application of the same input at diverse times, always produces the same output. This property is true for an ideal circuit into which the gates introduce no delays, and remains so for an actual circuit containing delays, assuming that when an input is applied it is of sufficient duration to permit the propagation of electrical signals and for the output to attain a permanent state. A subsequent change to the input cannot intervene until the transitions due to the original input have been extinguished. Moreover, it is certain that the output will be stable by reason of the absence of feedback loops. Without the above restrictions, i.e. if the inputs and outputs are not limited to permanent states, there will be no univocal correspondence between input and output, in certain cases, during the transitions.

Conversely, a sequential circuit does possess a storage (memory) unit which may be considered as the state of the feedback loops. Consequently, in general, the output depends not only on the input but on its 'history', by means of changes of internal states.

That this 'history' intervenes in the elaboration of the output, results, in effect, from the relations:

$$\begin{cases} Q_{n+1} = F(Q_n, E_n) \\ Q_n = F(Q_{n-1}, E_{n-1}) \\ \cdots\cdots\cdots\cdots\cdots \\ Q_1 = F(Q_0, E_0) \end{cases} \quad (12.27)$$

which gives:

$$Q_{n+1} = H(Q_0, E_0, E_1, \ldots, E_{n-1}, E_n). \quad (12.28)$$

Examples:

1. Serial adder (§ 12.2.7.1).

It was shown in Chapter 9 that the equations for the adder can be put in explicit form with the formula:

$$S_k = x_k \oplus y_k \oplus \left[\sum_{j=0}^{j=k-1} \prod_{i=j+1}^{i=k-1} (x_i + y_i) x_j y_j \right] \quad (k = 1, 2, \ldots) \quad (12.29)$$

2. Reflected binary-pure binary conversion.

We have:

$$B_k = \bigoplus \sum_{j=k}^{j=n} R_j \quad (12.30)$$

or with the notation adopted:

$$Z_k = \bigoplus_{j=0}^{j=k} E_j \quad \begin{array}{l}(E_j = R_{n-j}) \\ (k = 1, 2, \ldots)\end{array} \quad (12.31)$$

This then is the general situation. Whilst a combinatorial circuit establishes an input-output correspondence *letter by letter*, a sequential circuit produces a correspondence between input and output *sequences*. To take a further example: a combinatorial padlock is a combinatorial system: it is sufficient to compose a certain combination of figures, e.g. 2317 (input) to permit the padlock to be opened. The manner in which this number is actually produced plays no part[1].

Conversely, the locks fitted to safes are of the sequential type, it is the *sequence* of manoeuvres of the various controls which leads to the internal state which corresponds to the opening of the door. In particular, in the case of error of operation, it is generally necessary to recommence the operation after returning to the initial condition.

It should be noted that it is possible to imagine circuits conforming to the sequential models given at the beginning of the chapter, but where the output would depend only on the input, to the exclusion of the internal state. Such a circuit, of no practical interest, would behave like a combinatorial circuit. In fact, such a peculiarity would, in any case, be revealed with the aid of the minimisation techniques to be described in the following chapter, which would restore the number of internal states to 1. But, the minimum number of states having meaning is two, without which a knowledge of the state carries no information. Thus a circuit having one state can be considered to be a combinatorial circuit and vice versa.

Finally, the combinatorial circuits are subsets of sequential circuits for which they control the memory units. Conversely, in a certain sense, the memory unit stores, in coded form, the history of the inputs of a combinatorial circuit.

---

[1] There are, however, sequential models, although less common.

Chapter 13

Regular Expressions and Regular Events

13.1 Events

13.1.1 Definition

Consider a finite automaton having an input alphabet of k symbols and 1 binary output.

The set of *finite* sequences which can be formed from these symbols is infinite; it will be denoted by I. In the alphabet {0, 1}, I is thus the set of binary sequences (without limitation of length).

A subset E of the set I of all possible sequences is known as an *event*. It is said that an event E *occurs at an instant* t when the input sequence formed by the symbols received between the instant 1 and the instant t form part of an event (or set) E. It can thus be seen that, at a given instant t, the question "will an event occur?" can have only two possible answers — *yes* or *no* — and that the set of possible sequences is thus divided into two disjoint sets; the sequences for which the event is produced at the instant t and those for which the event does not occur. It is said that the automaton represents the event E if an output 1 is produced each time that E occurs.

Examples of events:

a) In using the alphabet {0, 1}, consider the following set of sequences:

$$E = \{011, 100\} \tag{13.1}$$

This is a finite event (with 2 sequences); at an instant $t \neq 3$ it would certainly not occur. At an instant $t = 3$, it would occur if the input sequence $a_{i_1} a_{i_2} a_{i_3}$ received at the instants $t = 1$, $t = 2$ and $t = 3$ is identical to one of the two sequences of E.

b) In using the alphabet {0, 1, 2, 3}, consider the set of sequences commencing with 23 or 14. This event cannot occur before a time $t = 2$.

At a time $t \geq 2$, it will occur if the sequence has commenced with 23 or 14 no matter what the remainder of the sequence up to the instant t. Obviously this event comprises an infinity of sequences. The events to be described below (§ c, d, e) are also of the infinite type.

c) With the alphabet {0, 1} the event E is the set of sequences containing an even number of 1.

d) With the alphabet {0, 1} the event E is comprised of the sequences for which the last 4 symbols are: 0101.

e) With the alphabet {0, 1, 2, 3} the event E is the set of sequences containing the symbol 2 three times.

More events will be described later. Before so doing it is necessary to introduce two distinguished events which will be of service in what now follows.

13.1.2 Empty Sequence λ

It will be seen to be convenient to introduce a particular "sequence" known as an *empty* sequence (devoid of symbols) and an event (set) made up from this unique sequence. Although, strictly speaking, a set containing a unique element is a mathematical entity differing in respect of the element it contains, it will be convenient to represent by the symbol λ, both the empty sequence and the event composed of this empty sequence, in order to simplify the notation. The length of λ is zero. A return to the properties of λ will be made later.

13.1.3 Empty Set of Sequences $\emptyset$

As in the general theory, the empty set, by definition, contains no elements, i.e. it contains no sequences.

Note that, in this respect, the set reduced to the empty sequence λ is not empty.

13.2 Regular Expressions. Regular Events

The events are sets of sequences. Different operations known as *regular operations* will be defined for these.

13.2.1 Regular Operations

13.2.1.1 *Union of A and B ($A \cup B$)*

For two given events A and B, the union of A and B, denoted by $A \cup B$ (read as "A union B" or "A plus B") is the name given to the event comprising the sequences belonging either to A, or to B, or to both. This is merely the union of 2 sets in the normal sense, the sequences being the "elements".

Example: consider the 2 events A and B defined by the alphabet $\{0, 1, 2, 3\}$ such that

$$A = \{0, 13, 21\} \qquad B = \{00311, 223\} \qquad (13.2)$$

this gives:

$$A \cup B = \{0, 13, 21, 00311, 223\}. \qquad (13.3)$$

13.2.1.2 Product (Concatenation) of A and B ($A \cdot B$)

The *product* of A and B, or again, the *concatenation of A and B*, is the name given to the set denoted as $A \cdot B$ (read as AB), or, more briefly, AB, of sequences formed by a sequence of A followed by a sequence of B.

Giving, for the events A and B considered below:

$$A \cdot B = \{000311, 0223, 1300311, 13223, 2100311, 21223\} \qquad (13.4)$$

It can easily be verified, with the help of the events A and B above, that in a general way the product $A \cdot B$ is not commutative, i.e. that $A \cdot B \neq B \cdot A$.

Particular case: an event of form $AA, AAA, \ldots$ we write by convention:

$$\begin{cases} A \cdot A & = AA & = A^2 \\ A \cdot A \cdot A & = AAA & = A^3 \\ \underbrace{A \cdot A \cdots A}_{n \text{ factors}} & = AA \cdots A & = A^n \end{cases} \qquad (13.5)$$

In conformance with the definition given for $\cdot$, it will be seen that A^n is the set of the sequences formed by the juxtaposition of n sequences belonging to A. It is important to remember that these n sequences are not necessarily identical, in the case where A contains more than 1 sequence.

Example:

$A = \{0, 13\}$

$A^2 = \{00, 013, 130, 1313\} \qquad (13.6)$

$A^3 = \{000, 0013, 0130, 01313, 1300, 13013, 131313\}$, etc.

It could be said that:

$$A^2 = \{0 \cup 13\} \{0 \cup 13\} = \{0\}^2 \cup (013) \cup (130) \cup \{13\}^2 \qquad (13.7)$$
$$= \{00 \cup 013 \cup 130 \cup 1313\}.$$

13.2.1.3 Iteration of an Event A (A^*)

The iteration of A, denoted by A^* (read as A star) is the event:

$$A^* = \lambda \cup A \cup A^2 \cup A^3 \cup \cdots = \lambda \cup \bigcup_{1}^{\infty} A^n \qquad (13.8)$$

The *definite iteration* of A, denoted by A^+ (read as A plus) is defined by the expression:

$$A^+ = A \cup A^2 \cup A^3 \cup \cdots = \bigcup_{1}^{\infty} A^n \qquad (13.9)$$

It will be seen that when A does not include λ, neither does A^+, although A^* does contain λ by definition.

In a general manner this gives:

$$A^* = \lambda \cup A^+ \qquad (13.10)$$

Example: if i designates the event formed by all sequences of length 1, i.e. those formed from a single letter, this gives with k different symbols:

$$i = \{0 \cup 1 \cup 2 \cdots \cup (k-1)\}. \qquad (13.11)$$

It will be seen that the event i^* defines the set of all possible sequences including the sequence of length zero; here it is the universal set:

$$I = i^* \qquad (13.12)$$

13.2.2 Function $\delta(A)$

It will later be shown that it is often necessary to know whether an event A contains λ. For this purpose a function designated as $\delta(A)$ is defined, such that:

$$\left. \begin{array}{l} \delta(A) = \lambda \quad \text{if } \lambda \in A \\ \delta(A) = \varnothing \quad \text{if } \lambda \notin A \end{array} \right\} \qquad (13.13)$$

which possesses the following properties:

$$\left. \begin{array}{l} \delta(\overline{A}) = \lambda \quad \text{if } \delta(A) = \varnothing \\ \phantom{\delta(\overline{A})} = \varnothing \quad \text{if } \delta(A) = \lambda \end{array} \right\} \qquad (13.14)$$

$$\delta(a_i) = \varnothing \qquad (13.15)$$

$$\delta(\lambda) = \lambda \qquad (13.16)$$

$$\delta(\varnothing) = \varnothing \qquad (13.17)$$

$$\delta(A^*) = \lambda \qquad (13.18)$$
$$\delta(A \cup B) = \delta(A) \cup \delta(B) \qquad (13.19)$$
$$\delta(A \cap B) = \delta(A) \cap \delta(B) \qquad (13.20)$$
$$\delta(A \cdot B) = \delta(A) \cap \delta(B) \qquad (13.21)$$

13.2.3 Properties of Regular Operations

The following properties are easily verified:

$$A \cup B = B \cup A \qquad \text{(Commutativity of } \cup\text{)} \qquad (13.22)$$
$$(A \cup B) \cup C = A \cup (B \cup C) \qquad \text{(Associativity of } \cup\text{)} \qquad (13.23)$$
$$(AB)C = A(BC) \qquad \text{(Associativity of } \cdot\text{)} \qquad (13.24)$$

(It should be remembered that, in general, the product here is *not* commutative.)

$$\begin{cases} A(B \cup C) = AB \cup AC & \text{(distributivity to the left and} \\ (B \cup C)A = BA \cup CA & \text{right of } \cdot \text{ with respect to } \cup\text{)} \end{cases} \qquad (13.25)$$

$$A \cup A = A \qquad \text{(idempotence)} \qquad (13.26)$$

— *Properties of $\varnothing$ and of λ.*

$\varnothing$ being the symbol for the empty set of sequences gives:

$$A \cup \varnothing = \varnothing \cup A = A \qquad (13.27)$$

Also it is natural to write:

$$A \cdot \varnothing = \varnothing \cdot A = \varnothing \qquad (13.28)$$

(i.e. there can be no sequence of type $A \varnothing$ or of type $\varnothing A$ if there is no sequence in $\varnothing$).

Under these conditions, $\varnothing$ plays the part of a *additive zero* (operation $\cup$) and of a *multiplicative* (operation $\cdot$).

Then, for λ:

$$A\lambda = \lambda A = A \qquad (13.29)$$

(λ plays the role of a multiplicative '1').

— *Properties involving iteration*

$$\lambda^* = \lambda \qquad (13.30)$$
$$\varnothing^* = \lambda \qquad (13.31)$$

Other useful properties:

$$(A \cup B)^* = (A^* B^*)^* \tag{13.32}$$

$$(A \cup B)^* = (A^* B)^* A^* \tag{13.33}$$

$$(AB)^* A = A(BA)^* \tag{13.34}$$

(( )* may be moved in the above expressions)

$$(A^*)^* = A^* \quad \text{(idempotence of iteration)} \tag{13.35}$$

$$A^* = \lambda \cup AA^* = \lambda \cup A^*A \tag{13.36}$$

and more generally:

$$A^* = \bigcup_{k=0}^{k=n-1} A^k \cup A^n A^* \tag{13.37}$$

$$AA^* = A^*A = A^+ \tag{13.38}$$

$A^* A^* = A^*$ (law of absorption for iteration)   (13.39)

$A^k \cup A^* = A^*$ (law of disjunctive absorption for iteration)   (13.40)

If $A = A_1 \cup \delta(A)$, with $\delta(A_1) = \varnothing$, then

$$A^* = (A_1)^*. \tag{13.41}$$

Indeed:

$$A^* = [A_1 \cup \delta(A)]^* = [(A_1)^* (\delta(A))^*]^* \quad \text{(from 13.32)} \tag{13.42}$$

but:

$$[\delta(A)]^* = \begin{cases} \lambda \text{ if } \delta(A) = \varnothing \\ \lambda^* = \lambda \text{ if } \delta(A) = \lambda \end{cases} \tag{13.43}$$

$$A^* = [(A_1)^* \lambda]^* = [(A_1)^*]^* = (A_1)^* \tag{13.44}$$

13.2.4 Generalised Regular Operations

The three operations ($\cup$, $\cdot$, *) defined above suffice, as will be shown later, to represent all the events which are of interest for a study of finite automata. Thus, in a certain sense, they constitute a complete set of operations, for the class of events under consideration, which play a role analogous to that of the complete set of operators used to represent the functions in Boolean algebra. However, by reason of the small number of operations brought into play, it is sometimes difficult, in practice, to make use of them in the translation into ordinary language of the events described. Moreover, in order to give greater flexibility to this formalism, other operations, also revealed as being convenient, are sometimes defined, such as:

13.2.4.1 Intersection of 2 Events A and B $(A \cap B)$

This is the intersection in the sense of the theory of sets and is denoted as $A \cap B$.

Example:
$$A = \{01, 23\} \qquad B = \{02, 23, 03\} \qquad (13.45)$$
giving:
$$A \cap B = \{23\} \qquad (13.46)$$

13.2.4.2 Complement of an Event $A(\bar{A})$

The complement of the event A, denoted by $\bar{A}$ (read A bar), is the set of those sequences which do not belong to A, i.e. it is the set $I - A$, where I denotes the set of all possible sequences.

13.2.4.3 Boolean Expression $E(A, B)$

Given a Boolean expression $E(A, B)$ obtained by the application of a finite number of operations $\cap, \cup, ^-$, the event $E(A, B)$ is the set of sequences thus obtained from A and B.

13.2.4.4 Event $E_1 = (t)_{t \leq p} E$

Given an event E, $E_1 = (t)_{t \leq p} E$ is the set of sequences of E, for which all the initial segments of length $\leq p$ also belong to E, the initial segments of sequence S being obtained by truncating S after l symbols ($l = 1, 2, ..., p - 1$). It is easily seen that:

$$E_1 \subset E. \qquad (13.47)$$

The signification of E_1 is as follows: when a sequence E_1 of length p has been applied to an input, the event E is produced at every instant $1, 2, ...$

13.2.4.5 Event $E_2 = (Et)_{t \leq p} E$

This is the set of sequences such that at least one of their initial segments belongs to E. Thus, when a sequence E_2 of length p has been applied to the automaton at the instants $t = 1, ..., p$, the event E is produced at least once during the period $[1, p]$. Thus, evidently

$$E_2 \supset E \tag{13.48}$$

There are also the following relations for which verification is immediate:

$$\begin{cases} (t)_{t \leq p} E = \overline{\left(E t_{t \leq p} \overline{\overline{E}}\right)} & (13.49) \\ (E t)_{t \leq p} E = \overline{\left((t)_{t \leq p} \overline{\overline{E}}\right)} & (13.50) \end{cases}$$

13.2.5 Regular Expressions

The events are represented by symbols. Manipulation of events is necessarily effected through the intermediary of operations on the corresponding symbols, thanks to which the operations can be described. Essentially 3 types of symbols are utilised:

1. Symbols representing sets of sequences.
2. Symbols representing operations.
3. Auxiliary signs such as parentheses, commas, etc.

It is convenient, as has been done for λ, to represent by a unique symbol and both a sequence and the event containing the sequence. This being so, operations on the symbols permit the definition of a class of algebraic expressions to be associated with certain events.

These are the so-called regular expressions which are defined in a recurrent manner as follows:

1. A symbol representing a letter of the alphabet, the symbol λ and the symbol $\emptyset$ are regular expressions.
2. If A and B are regular expressions, $A \cup B$, AB and A^* are regular expressions.
3. Only the expressions derived by the application of 1 and 2 a finite number of times, may be called regular expressions.

In other words an expression is not regular unless it is obtained from a succession of operations $\cup, \cdot, *$, applied in finite number, to the letters of the alphabet, to $\emptyset$ and to λ.

The regular expressions defined above which bring in to play only the operations $\cup, \cdot,$ and $*$ are known as *regular expressions in the restricted sense*, or *restricted regular expressions*.

If, within § 2 of the definition, these operations are not restricted and other regular operations are also used, the expressions so obtained are known as "generalised regular expressions". Where, in the text which follows, there is reference to an expression without other definition, by convention this will refer to a *restricted* expression.

13.2.6 Regular Events

By definition it is said that an event is regular if it can be represented by a regular expression. Evidently such a definition makes sense if there exist events which are not regular. Such events, known as *irregular* events, do exist. By way of example, reference is made to the following event, quoted by Kleene:

With the alphabet {0, 1} consider a set of sequences of lengths equal to a perfect square and for which only the last symbol is 1, all preceding symbols having, therefore, the symbol 0. It can be shown (cf. [64] or [85]) that this event cannot be represented by a regular expression.

Examples of Regular Events

1) *Finite sets*

Every *finite* set of sequences is regular. It is, in fact, expressed as a union of a finite number of sets, each containing one sequence and which, in their turn, are expressed as the product of a finite number of terms.

2) *Sets A^k*

If A is regular, A^k is also regular.

3) With the alphabet {0, 1} consider an event E formed from sequences having an even number of 1. It consists of an infinity of sequences. However, it can be seen to be capable of being represented by a finite expression and thus it is regular.

In fact, a sequence including the figure 1 once, and once only, is of the form $S_1 = 0^*10^*$. This relationship indeed shows that there is one figure 1 preceded and followed by two sequences of 0 of any length, finally null, which takes account of the cases where 1 is to be found at the start, at the end, or alone.

A sequence S_2 containing the symbol 1 exactly twice is obtained by the juxtaposition of 2 sequences of the preceding type. Giving:

$$S_2 = (0^*10^*)^2 = (0^*10^*)(0^*10^*)$$
$$= (0^*1)(0^*0^*)(1\ 0^*) \tag{13.51}$$

Since $0^*0^* = 0^*$, it follows that:

$$S_2 = (0^*1)(0^*)(1\ 0^*) = 0^*1\ 0^*1\ 0^* \tag{13.52}$$

a sequence containing $2p$ symbols 1 is, in its turn, formed by the juxtaposition of p sequences S_2

$$S_3 = (0^*1\ 0^*1\ 0^*)^p. \tag{13.53}$$

13.2 Regular Expressions. Regular Events

The event is the set of all possible sequences S_3 for all p

$$E = S_2 \cup \{S_2\}^2 \cup \{S_2\}^3 \cdots$$
$$E = (S_2)^+ \tag{13.54}$$
$$E = (0^*1\ 0^*1\ 0^*)^+$$

If it is convenient to include the sequence of length 0, the resultant is:

$$E_1 = \lambda + (S_2)^+ = (S_2)^*$$

whence:

$$E_1 = (0^*\ 1\ 0^*\ 1\ 0^*)^*. \tag{13.55}$$

E could be re-written differently. For example it could be said that:

$$(0^*\ 1\ 0^*\ 1\ 0^*)^p = (0^*\ 1\ 0^*\ 1)^p\ 0^* \tag{13.56}$$

which then gives:

$$E = (0^*\ 1\ 0^*\ 1)\ 0^* \cup (0^*\ 1\ 0^*\ 1)^2\ 0^* \cup \cdots$$
$$E = (0^*\ 1\ 0^*\ 1)^+\ 0^* = [(0^*\ 1)^2]^+\ 0^* \tag{13.57}$$

4) The event E is formed by all the sequences including the symbol 1 at least three times. It is an infinite set, which can be represented thus:

$$E = i^*\ 1\ i^*\ 1\ i^*\ 1\ i^*$$
$$E = (i^*\ 1)^3\ i^*. \tag{13.58}$$

It should be noted that the event formed of sequences containing exactly three 1's is written, in the alphabet $\{0, 1\}$:

$$E_1 = (0^*\ 1)^3\ 0^*. \tag{13.59}$$

5) With the alphabet $\{0, 1, 2, 3\}$ the event E contains all the sequences where one of the last two symbols is either 1 or 2. This infinite event can be represented by:

$$\left.\begin{aligned}
E &= i^*\ 1\ i \cup i^*\ 2\ i \cup i^*\ 1 \cup i^*\ 2 \\
&= i^*\ (1\ i \cup 2\ i \cup 1 \cup 2) \\
&= i^*\ [(1 \cup 2)\ i \cup (1 \cup 2)] \\
&= i^*\ [(1 \cup 2)\ (i \cup \lambda)] \\
&= i^*\ (1 \cup 2)\ (i \cup \lambda)
\end{aligned}\right\} \tag{13.60}$$

6) The event comprises (with the alphabet {0, 1}):

1) All the sequences of 0's of whatever length terminating in 1.
2) All the sequences of the preceding type, followed by a sequence $i*$ terminating in 0.

$$E = 0^* 1 \cup (0^* 1)(i^* 0)$$
$$E = 0^* 1 (\lambda \cup i^* 0) \tag{13.61}$$

The reader can easily verify that the automaton which presents this event can do nothing more than determine the true complement of a number presented in series at its input, the low 'weights' leading.

7) The event, based on the alphabet {0, 1, 2, 3}, comprises all the sequences formed only of 0 or 3.
Thus:

$$E = (0 \cup 3)^* \tag{13.62}$$

The identity (13.32) permits this also to be written:

$$E = (0 \cup 3)^* = (0^* 3^*)^* \tag{13.63}$$

for which the interpretation is obvious: E is composed of blocks 0 and 3 formed from a chain of 0's, of any length, followed by a chain of 3's, also of any length, the blocks themselves being repeated any number of times in order to obtain a sequence of the event.

8) With the alphabet {0, 1, 2, 3}, the event E comprises:

1) All the sequences ending in 2
2) All the sequences ending with a block 3, of any length, itself preceded by a 2.
Thus:

$$E = i^* 2 \cup i^* 2\, 3^*$$
$$E = i^* 2 (\lambda \cup 3^*) \tag{13.64}$$
$$E = i^* 2\, 3^*$$

9) The event E, with the alphabet {0, 1, 2, 3}, comprises:

1) All the sequences commencing with a sequence of 2, followed by 1, followed by any sequence of 0 or of 3.

2) All sequences commencing with 1 and followed by any sequence of 0 or 3.

3) All sequences commencing with 1, followed by a 2, followed in turn by any sequence of 0 or 3.

4) All sequences commencing with 2, followed by any sequence of 0 and 3.

Thus:

$$E = 2^* 1 (0 \cup 3)^* \cup 1 (0 \cup 3)^* \cup 1^* 2 (0 \cup 3)^* \cup 2 (0 \cup 3)^*$$
$$E = (2^* 1 \cup 1 \cup 1^* 2 \cup 2) (0 \cup 3)^*$$
$$E = [(2^* 1 \cup 1) \cup (1^* 2 \cup 2)] (0 \cup 3)^* \quad (13.65)$$
$$E = [(2^* \cup \lambda) 1 \cup (1^* \cup \lambda) 2] (0 \cup 3)^*$$

10) Definite events

It is said that an event is definite if it can be described by a regular expression of the form:

$$R = E \cup i^* F, \quad E = \bigcup_i e_i, \quad F = \bigcup_j f_j \quad (13.66)$$

where E and F are finite sets of finite sequences

R is said to be *initial* if $R = E$

non-initial if $R = i^* F$

otherwise it is *composite*.

It can easily be verified that in the case of a non-initial event, every sequence f_j terminating in a sequence $f_k (f_j, f_k \in F)$ can be withdrawn from the set F without in any way modifying the event $i^* F$ [1]. In the text which follows, except for explicit mention to the contrary, it will be assumed that this withdrawal has been made for the expression F [2].

13.2.7 Set Derived from a Regular Event. Derivation of a Regular Expression

For a given *regular event* E, represented, therefore, by a *regular expression* R_E, the set of sequences x, such that Sx belongs to E, is known as the *derived set* of E or, more briefly, the *derivative* of E with respect to S. It will be shown how this new event can be represented by an algebraic expression which we shall call the *derivative* of R_E with

---

[1] If R is composite, every e_i terminating in f_j can be eliminated (this will be assumed to have been done, except for explicit mention to the contrary). For a non-initial event R, the form (13.66) obtained is thus unique upto the order of factors [19].

[2] When R is composite, it is necessary and sufficient that every term E or F terminating in e_j is such that $i^* e_j$ belonging to R be withdrawn, in order that the form obtained be unique (always upto the order of factors). E and $i^* F$ are then disjoint. This withdrawal is also assumed to have been made. The resulting forms, if an order has been specified for the e_i and f_j terms, are callad canonical. [19]

respect to S, written as $D_S(R_E)$ or $D_S R_E$ thus denoting the derivative set. A the same time it will be shown that $D_S R_E$ is regular. The determination of the derivative of a regular expression is made in accordance with the following rules, to be justified below in *each case*. First it must be noted that:

$$D_\lambda R = R \tag{13.67}$$

13.2.7.1 Derivatives with Respect to a Sequence of Length 1

$$R = \{a\}$$

— Clearly:
$$D_a(a) = \lambda \tag{13.68}$$

So with $b \neq a$
$$D_b(a) = \varnothing \tag{13.69}$$

Moreover:
$$\begin{cases} D_a(\lambda) = \varnothing & (13.70) \\ D_a(\varnothing) = \varnothing & (13.71) \end{cases}$$

— *Derivative of a sum:* $D_a(A \cup B)$

The set of sequences of the form ax such that x belongs to A, or B or to both, is equal to the union of of the sequences x', where ax' belongs to A and to the sequences x'' such that ax'' belongs to B. In other words:

$$D_a(A \cup B) = D_a A \cup D_a B \tag{13.72}$$

— *Derivative of a complement* $D_a(\bar{A})$

$$D_a(\bar{A}) = \overline{D_a(A)} \tag{13.73}$$

It is indeed clear that:
$$D_a(A) \cup D_a(\bar{A}) = i^* \tag{13.74}$$
$$D_a(A) \cap D_a(\bar{A}) = \varnothing \tag{13.75}$$

hence the identity (13.73).

— *Derivative of an arbitrary Boolean expression.*

For any expression containing $\cap, \cup, ^-$, the derivative can be obtained with the aid of the rules given above, taking into account the relation:

$$A \cap B = \overline{\bar{A} \cup \bar{B}}$$

and this gives:

$$D_a F[A, B] = F[D_a A, D_a B]. \tag{13.76}$$

13.2 Regular Expressions. Regular Events

— *Derivation of a product $A \cdot B$:*

$$D_a(AB) = (D_a A)B \cup \delta(A) D_a B. \qquad (13.77)$$

It can, in fact, be said, as in § 13.2.3, that $A = A_1 \cup \delta(A)$, where A_1 represents the set of sequences A of length ≥ 1, (i.e. that $\delta(A_1) = \varnothing$, and where $\delta(A)$ is λ if A contains λ in addition to the sequences of A_1. Thus giving:

$$D_a AB = D_a\{[A_1 \cup \delta(A)][B]\} = D_a[A_1 B \cup \delta(A)B] \qquad (13.78)$$
$$= D_a(A_1 B) \cup D_a[\delta(A)B] \quad \text{(from 13.72.)} \qquad (13.79)$$

There are 2 cases:

1) $\delta(A) = \lambda$; therefore:

$$\delta(A)B = \lambda B = B \qquad (13.80)$$
$$D_a[\delta(A)B] = D_a B \qquad (13.81)$$

2) $\delta(A) = \varnothing$; therefore:

$$\delta(A)B = \varnothing\, B = \varnothing \qquad (13.82)$$
$$D_a[\delta(A)B] = D_a \varnothing = \varnothing \quad \text{(from 13.70)} \qquad (13.83)$$

which can be summarised by the formula:

$$D_a[\delta(A)B] = \delta(A) D_a B. \qquad (13.84)$$

Proceeding to the term $A_1 B$ it will be seen that the derivation only affects A_1 (sequences of length ≥ 1) thus giving:

$$D_a A_1 B = (D_a A_1) B \qquad (13.85)$$

hence, finally the relation (13.77).

— *Derivative of an expression of the form A^**

It has been seen (13.41) that:

$$A^* = A_1^* \qquad (13.86)$$

A_1 being defined as in § 13.2.3
Whence

$$D_a(A^*) = D_a(A_1^*) = D_a[\lambda \cup A_1 A_1^*]$$
$$= \varnothing \cup D_a(A_1 A_1^*) = D_a(A_1 A_1^*)$$
$$= [D_a(A_1)] A_1^* \quad = (D_a A) A^* \qquad (13.87)$$
$$D_a A^* \quad = (D_a A) A^* \qquad (13.88)$$

Also $A^* = A^+ \cup \lambda$, whence

$$D_a A^* = D_a A^+ \cup D_a \lambda = D_a A^+ \cup \varnothing \qquad (13.89)$$

$$D_a(A^*) = D_a(A^+) \qquad (13.90)$$

13.2.7.2 Derivatives with Respect to a Sequence of Length >1.

By reference to the general definition of § 13.2.7 it will be seen that the determination of a derivative can be broken down and carried out by a step-by-step procedure, in the following manner:

$$D_{a_1 a_2} R = D_{a_2}(D_{a_1} R)$$

$$D_{a_1 a_2 a_3} R = D_{a_3}(D_{a_1 a_2} R), \text{ etc.} \qquad (13.91)$$

and, more generally:

$$D_{a_1 a_2 \cdots a_{k-1} a_k}(R) = D_{a_k}(D_{a_1 a_2 \cdots a_{k-1}} R) \qquad (13.92)$$

With the aid of the rules given above the derivative of any regular expression with respect to any sequence can be calculated.

13.2.8 Properties of Derivatives

— The derivative of a regular expression is, itself, a regular expression. The justification is as follows: every regular expression is constructed with the aid of a finite number of regular operations. But, for each regular operation, the method of calculation of the resultant derivative with respect to a sequence of length 1 is known. This requires only a *finite* number n of regular operations, again, on the operands appearing in the expression and on their derivatives. By step-by step progression it will be seen that the derivative, with respect to a sequence of length 1, is expressed with the aid of a *finite* number of regular operator units applied to the operators a_i, λ and $\varnothing$ and their derivatives (which are λ or $\varnothing$), and, therefore, themselves regular.

By taking derivatives in this manner k times for a sequence of length k, the number of operator units can only remain finite and the derivative with respect to any sequence is then regular. The reader may, himself, carry out a detailed proof to this effect, or turn to reference [23].

— In order that a sequence s shall belong to a regular event R, it is necessary and sufficient that $D_s R$ contain λ.

Indeed, $s = s \cdot \lambda$. If s belongs to R, then $s\lambda$ does too, which signifies, by definition, that λ belongs to the set denoted by $D_s R$.

Conversely, if λ belongs to $D_s R$ it means that the sequences of type $s\lambda$ belong to R, i.e. since $s\lambda = s$, then, quite simply, s belongs to R.

13.2 Regular Expressions. Regular Events

— Every regular expression has a *finite* number d_R of distinct derivatives (i.e. representing distinct sets).

This theorem is evident in the case of a regular expression representing a finite set. An indirect justification will later be given for the general case — Chapter 15. The reader should also turn to reference [23].

Assuming that it is necessary to determine the derivative $D_s R$ for sequences of increasing length 0, 1, 2 etc., and arranged in an increasing order of numbers $a_{i_1}, a_{i_2}, a_{i_3} \ldots$ For each length k of s, are obtained a certain number of derivatives which are to be compared with those previously found. If no new derivative be found, it is useless to continue: every subsequent derivation operation can only produce a ("characteristic") derivative already known. In fact, if for a length k there is no distinctly new derivative, then all the derivatives are equal to the derivatives relating to sequences of length $\leq k-1$, such that every supplementary derivative (corresponding to a sequence of length $k+1$) really only produces a derivative relating to a sequence of length k, causing a return to the derivatives relating to sequences of length $\leq k-1$, etc. Thus the derivatives related to sequences of length $k+1$ are identical to those for sequences of length $\leq k-1$ and, progressively step-by-step this remains true for derivations of the order $k+l \, (l \geq 0)$.

So that for $d(i)$, the number of new and distinct derivatives obtained for length i, and k the smallest value for which $d(i) = 0$, we have:

$$d_R = \sum_0^k d(i) \geq 1 + k - 1 \quad \text{(since } d(i) \geq 1 \quad \text{for } 1 \leq i < k$$
$$\text{and } d(0) = 1)$$

whence:

$$k \leq d_R. \tag{13.93}$$

Consequently the last number i such that there is at least one new derivative such that:

$$i \leq d_R - 1 \tag{13.94}$$

Moreover, for sequences of length $i \leq d_R - 1$, there exists at least one distinct derivative among the derivatives $D_s R$ relating to sequences of length i.

Each regular expression can be written, with the alphabet A, in the form:

$$R = \delta(R) \cup \bigcup_{a_i \in A} a_i D_{a_i} R. \tag{13.95}$$

Indeed, R takes the form:

$$R = R^1 \cup \delta(R)$$

where R^1 is the set of sequences of R of length ≥ 1. R^1 is, itself, composed of the set $R^1_{a_1}$ of sequences commencing with a_1, of the set $R^1_{a_2}$ the sequences commencing with a_2 etc., and, in a general manner, of the different sets $R^1_{a_k}$ of sequences commencing with a_k, of which some sets may, ultimately, prove to be empty.

Hence:
$$R = \delta(R) \cup \bigcup_{a_i \in A} a_i R^1_{a_i} \tag{13.96}$$

and, for every i
$$D_{a_i} R = D_{a_i}(a_i R^1_{a_i}) = \lambda R^1_{a_i} = R^1_{a_i} \tag{13.97}$$

(which, moreover, is evident from the definition of a derivative). Hence the desired identity:
$$R = \delta(R) \cup \bigcup_{a_i \in A} a_i D_{a_i} R \tag{13.98}$$

It will also be noted that the different terms $a_i D_{a_i} R$ and $\delta(R)$ represent disjoint sets.

— Each of the distinct derivatives d_R may be expressed:
$$D_s R = \delta(D_s R) \cup \bigcup_{a_i \in A} a_i D_{u_{a_i}} R \tag{13.99}$$

where $D_s R$ is a characteristic derivative and where each $D_{u_{a_i}} \cdot R$ is the characteristic derivative equal to $D_{sa_i} R$.

This property becomes evident when applying the relation (13.95) to $D_s R$ which, as has been shown, is itself a regular expression. From the foregoing, it will be seen that the derivatives in the second members form part of the derivatives appearing in the first members of the analogous equations d_R from which are obtained the required derivatives.

13.2.8.1 Examples of the Determination of Derivatives

a) $R = 0(12 \cup 33) \cup 11 \cup 32 = 012 \cup 033 \cup 11 \cup 32$

$A = \{0, 1, 2, 3\}$

$D_0 R = D_0(012) \cup D_0(033) \cup D_0(11) \cup D_0(32)$

$ = 12 \cup 33 \cup \emptyset \cup \emptyset$

$ = 12 \cup 33 \tag{13.100}$

Similarly
$$\begin{cases} D_1 R = 1 \\ D_2 R = \emptyset \\ D_3 R = 2 \end{cases} \tag{13.101}$$

13.2 Regular Expressions. Regular Events

Then
$$D_{01}R = D_1(D_0R) = D_1(12 \cup 33) = 2 \cup \varnothing = 2$$
$$D_{02}R = D_2(D_0R) = D_2(12 \cup 33) = \varnothing \qquad (13.102)$$
$$D_{03}R = D_3(12 \cup 33) = \varnothing \cup 3 = 3$$

etc.

b) $R = [(0^* 1)^2]^+ 0^* = [0^* 1 \, 0^* 1]^+ 0^* \qquad (13.103)$

Putting
$$A = 0^* 1 \, 0^* 1 \qquad (13.104)$$
$$D_0 R = (D_0 A^+) 0^* \cup (\delta A^+) D_0 0^* \qquad (13.105)$$

$\delta A^+ = 0$ since A^+ contains no sequences of length < 2: $(0^* 1 \, 0^* 1)$ is of length ≥ 2 and consequently does not contain λ which is of length 0. Whence:
$$D_0 R = (D_0 A^+) 0^*$$
$$D_0[A^+] = D_0[A^*] = (D_0 A) A^*$$
$$D_0 A = D_0[0^* 1 \, 0^* 1] = (D_0 0^*) 1 \, 0^* 1 \cup D_0(1 \, 0^* 1) \quad \text{since} \quad \delta(0^*) = \lambda$$
$$D_0 0^* = (D_0 0) 0^* = \lambda 0^* = 0^*$$
$$D_0(1 \, 0^* 1) = \varnothing \qquad (13.106)$$

whence:
$$D_0 A = 0^* 1 \, 0^* 1 \qquad (13.107)$$

and
$$D_0 A^* = A A^* = A^+ \qquad (13.108)$$

whence:
$$D_0 R = A^+ 0^* = R. \qquad (13.109)$$

In an analogous manner:
$$D_1 R = (D_1 A^+) 0^* = ((D_1 A) A^*) 0^*$$
$$= 0^* 1 \, (0^* 1 \, 0^* 1)^* 0^* \qquad (13.110)$$

After which it is useless to derive $D_0(R)$ since $D_0 R = R$ and no further derivatives can be obtained.

Proceeding at once to $D_1(R)$:
$$D_{10}(R) = D_0[0^* 1 \, (0^* 1 \, 0^* 1)^* 0^*]$$
$$= (D_0 0^* 1)(0^* 1 \, 0^* 1)^* 0^* \qquad (13.111)$$

since
$$\delta(0^* \, 1) = \varnothing$$
$$D_0 0^* \, 1 = D_0 0^* \, 1 \cup \lambda \, D_0 1 \qquad (13.112)$$
$$= D_0 0^* \, 1 = 0^* \, 1 \qquad (13.113)$$
whence:
$$D_{10}(R) = 0^* \, 1 \, (0^* \, 1 \, 0^* \, 1) \, 0^* = D_1(R) \qquad (13.114)$$

Proceeding to the determination of $D_{11}(R)$

$$D_{11}(R) \, [D_1(0^* \, 1)] \, (0^* \, 1 \, 0^* \, 1)^* \, 0^*$$
$$= \lambda \cdot (0^* \, 1 \, 0^* \, 1)^* \, 0^*$$
$$= (0^* \, 1 \, 0^* \, 1)^* \, 0^*$$
$$= [\lambda \cup (0^* \, 1 \, 0^* \, 1)^+] \, 0^* = \lambda 0^* \cup R$$
$$= 0^* \cup R. \qquad (13.115)$$

Finally we obtain:
$$D_{111}(R) = D_1(0^* \cup R)$$
$$= D_1 0^* \cup D_1 R$$
$$= \varnothing \cup D_1 R$$
$$= D_1 R \qquad (13.116)$$
and
$$D_{110}(R) = D_0(0^* \cup R)$$
$$= D_0 0^* \cup D_0 R$$
$$= 0^* \cup D_0 R$$
$$= 0^* \cup R$$
$$= D_{11} R \qquad (13.117)$$

Thus no new derivatives for the length $l = 3$ have been discovered and the process is terminated. Thus, there are 3 derivatives in all: R, $D_1 R$ and $D_{11} R$. The interpretation is immediate: $D_1 R$ is the set of sequences containing an *odd* number of 1, and $D_{11} R$ is the set of sequences containing an even number of 1.

c) *Derivatives of* $R(n) = (0^* \, 1)^n \, 0^*$ *with alphabet* $\{0, 1\}$

$R(n)$ is the set of sequences containing the symbol 1, n times.

$$D_x[R(n)] = D_x[(0^* \, 1)^n \, 0^*]$$
$$= \{D_x[(0^* \, 1)^n]\} \, 0^*$$
$$= [D_x(0^* \, 1)] \, (0^* \, 1)^{n-1} \, 0^*$$
$$= [(D_x 0^*) \, 1 \cup D_x 1] \, (0^* \, 1)^{n-1} \, 0^*$$
$$= [(D_x 0) \, 0^* \, 1 \cup D_x 1] \, (0^* \, 1)^{n-1} \, 0^* \qquad (13.118)$$

whence:
$$D_0[R(n)] = (\lambda 0^* 1 \cup \emptyset)(0^* 1)^{n-1} 0^*$$
$$= (0^* 1)(0^* 1)^{n-1} 0^*$$
$$= (0^* 1)^n 0^* = R(n) \tag{13.119}$$
$$D_1[R(n)] = (\emptyset\, 0^* 1 \cup \lambda)(0^* 1)^{n-1} 0^*$$
$$= (0^* 1)^{n-1} 0^* = R(n-1). \tag{13.120}$$

Thus only $D_1[R(n)]$ is distinct from $R(n)$ and it can be seen that only the derivatives of form $D_{11\ldots 1}[R(n)]$ are distinct. Consider now a sequence formed by the symbol 1 repeated x times:

$$\begin{cases} D_{11\ldots 1}[R(n)] = R(n-x) & (x < n) \tag{13.121} \\ D_{11\ldots 1}[R(n)] = 0^* & (x = n) \tag{13.122} \end{cases}$$

Consequently:
$$D_0 0^* = 0^*$$
$$D_1 0^* = \emptyset \tag{13.123}$$

and this ends the list which, including $R(n) = D_\lambda[R(n)]$ and $\emptyset$, gives a total of $n+2$ distinct derivatives. These are the sets of sequences respectively including the symbol 1 n times, $(n-1)$ times etc., once and zero times, and the empty set.

d) *Successive derivatives of a non-initial definite event*

It has been seen that this is the name given to any event which can be expressed in the form:
$$R = i^* F \tag{13.124}$$

where F is an event formed by a finite number of finite sequences.
— Derivatives with respect to a sequence of length 1.

$$D_a(R) = D_a(i^*) F \cup \delta(i^*) D_a F$$
$$= D_a(i^*) F \cup D_a F \quad \text{(since } \delta(i^*) = \lambda)$$
$$= \bigl(D_a(i)\bigr) i^* F \cup D_a F$$
$$= \lambda i^* F \cup D_a F$$
$$= i^* F \cup D_a F = R \cup D_a F$$

then:
$$D_a(R) = R \cup D_a F. \tag{13.125}$$

— Derivatives with respect to sequences of length ≥ 1.

$$D_a(R) = R \cup E_a \quad (E_a = D_a F) \tag{13.126}$$

subsequently giving:

$$D_{ab}(R) = D_b[R \cup D_a F] = D_b R \cup D_{ab} F$$

$$= R \cup D_b F \cup D_{ab} F. \tag{13.127}$$

Proceeding thus step-by-step it will be seen that $D_s(R) = R \cup E_s$ where E_s is a finite set of finite sequences. It can also be shown (Brzozowski [19]) that $E_s \cap R = \emptyset$.

Indeed assuming that $E \cap R \neq \emptyset$, i.e. that there exists a sequence e belonging to both R and E_s at the same time. Belonging to R, e thus terminates in a sequence f of F $(e = e'f)$; as a sequence of E_s, e is a derivative of a sequence g of F, i.e. a termination of sequence $g \in F$, giving:

$$g = e''e = e''e'f \tag{13.128}$$

which is a contradiction if is assumed that F satisfies the hypotheses of § 13.2.6. Thus:

$$E_s \cap R = \emptyset.$$

It is clear that in the particular case where $E_s = \emptyset$ the latter is verified automatically.

Supposing that it is necessary to compare 2 derivatives $D_s(R)$ and $D_t(R)$:

$$\begin{cases} D_s(R) = R \cup E_s & (R \cap E_s = \emptyset) \\ D_t(R) = R \cup E_t & (R \cap E_t = \emptyset) \end{cases} \tag{13.129}$$

whence:

$$D_s(R) = D_t(R) \text{ is equivalent to } E_s = E_t. \tag{13.130}$$

It will thus be seen that in the case of the non-initial definite events R, it is simple to verify the equality of 2 regular expressions representative of the derivatives of R. For this purpose it suffices to verify the equality of the finite sets E_s. It will later be shown that this property is particularly useful for the synthesis of states diagrams of automata signalling events of this type. By way of an exercise the reader may enlarge on this result for the case where R is composite.

13.3 Regular Expressions Associated with a States Diagram

Consider a diagram (Moore or Mealy) having m internal states, an input of k values and 1 binary output. If it be assumed that the initial state of the machine is q_i, the set of input sequences which will provoke an output $Z = 1$ constitutes an event E_i. An algebraic expression can be found to denote E_i and this expression will be *regular*, which, by definition, ensures that the event E_i is regular. For this the method used is due, for the essentials, to Arden and can be described as follows.

13.3.1 Moore Diagram

The events E_i associated with the states q_i are not independent, they are, in effect, linked by the following relations for which the interpretation is immediate:

$$E(q_i) = \quad a_1 E(q_i a_1) \cup a_2 E(q_i a_2) \cdots \cup a_j E(q_i a_j) \cup \cdots$$
$$\cup \, a_k E(q_i a_k) \cup \delta[E(q_i)] \quad (i = 1, \ldots, m) \qquad (13.131)$$

$E(q_i) = E_i$ is the event associated with q_i, $q_i a_j$ is the state into which the system passes when it is in q_i and receives a_j ("successor of q_i by a_j") and $E(q_i a_j)$ is the event associated with every state $q_i a_j$ thus obtained. Finally, $\delta[E(q_i)]$ takes account of the possible presence of λ in $E(q_i)$: if $\lambda \in E(q_i)$, $\delta[E(q_i)] = \lambda$ and conversely, if $\lambda \notin E(q_i)$, $\delta[E(q_i)] = \varnothing$. Thus there are m equations (as many as there are internal states) which, in the algebra of events, are *linear* with respect to the k input symbols, with a constant term whose role is played by $\delta[E(q_i)]$ for each event $E(q_i)$. The symbols a_j are the coefficients, the "unknowns" are the sets E_i.

Additionally, if a particular coefficient $a_j = a$ is considered, each equation takes the form:

$$E = aE \cup b. \qquad (13.132)$$

It can be demonstrated (*Arden, Bodnarcuk* [2.15]) that in a general manner an equation of the type $E = AE \cup B$, where A does not include λ, is equivalent to:

$$E = A^*B. \qquad (13.133)$$

In this instance a does not include λ and the method applies:

$$E = aE \cup b \quad \text{is equivalent to} \quad E = a^*b$$

Thus the system of equations (13.131) may be solved in the following manner:

An equation defining E_i is chosen and solved by formula (13.133). This value is carried over to the remaining equations. The operation has no effect on the value of the function $\delta(a)$ for the a coefficients of the other equations, so that these may also be solved with formula (13.133). This procedure is followed step-by-step so as to eliminate all the expressions E_i other than the particular expression E_k being sought.

Example: On the basis of the diagram of Fig. (13.1), the following equations can be written:

$$\begin{cases} E_1 = 0\,E_1 \cup 1\,E_2 \\ E_2 = 0\,E_2 \cup 1\,E_3 \\ E_3 = 0\,E_3 \cup 1\,E_2 \cup \lambda \end{cases} \quad (13.134)$$

whence:

$$E_3 = 0^*(1\,E_2 \cup \lambda) \quad (13.135)$$

$$E_2 = 0\,E_2 \cup 1\,0^*(1\,E_2 \cup \lambda)$$
$$= (0 \cup 1\,0^*\,1)\,E_2 \cup 1\,0^* \quad (13.136)$$
$$= (0 \cup 1\,0^*\,1)^*\,1\,0^*$$

Fig. 13.1   and finally:

$$E_1 = 0\,E_1 \cup 1\,E_2$$
$$E_1 = 0^*\,1\,(0 \cup 1\,0^*\,1)^*\,1\,0^*. \quad (13.137)$$

13.3.2 Mealy Diagram

This is a relation analogous to (13.131) with the difference that for every input a_j applied to the automaton in the state q_i, the event E_i will include 2 types of terms:

1. The events of form $a_j E(q_i a_j)$ where $E(q_i a_j)$ denotes the set of sequences which give a system output 1, once the system has attained $q_i a_j$.
2. The events represented by the inputs a_j such that $Z(q_i, a_j) = 1$, events associated with the appearence of 1 at the output during the passage from q_i to $q_i a_j$:

$$E(q_i) = a_1[E(q_i a_1) \cup \alpha_1 \lambda] \cup a_2[E(q_i a_2) \cup \alpha_2 \lambda] \cup \cdots$$
$$\cup a_j[E(q_i a_j) \cup \alpha_j \lambda] \cup \cdots a_k[E(q_i a_k) \cup \alpha_k \lambda] \quad (13.138)$$

with

$$\begin{cases} \alpha_j = \lambda & \text{if} \quad Z(q_i a_j) = 1 \\ \alpha_j = \varnothing & \text{if} \quad Z(q_i a_j) = 0 \end{cases}$$

13.3 Regular Expressions Associated with a States Diagram

Example: the diagram of Fig. (13.2) supplies the equations:

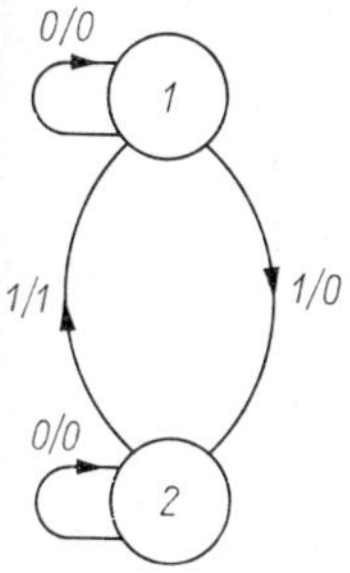

$$\begin{cases} E_1 = 0\,E_1 \cup 1\,E_2 \\ E_2 = 0\,E_2 \cup 1\,(E_1 \cup \lambda) \end{cases} \quad (13.139)$$

whence

$$E_2 = 0^*(1\,E_1 \cup \lambda) \quad (13.140)$$

$$E_1 = 0\,E_1 \cup 1\,0^*\,1\,E_1 \cup 1\,0^*$$

$$E_1 = (0 \cup 1\,0^*\,1)\,E_1 \cup 1\,0^*$$

$$E_1 = (0 \cup 1\,0^*\,1)^*\,1\,0^*. \quad (13.141)$$

Fig. 13.2

Thus Arden's method admits the association of every state of a state diagram (Moore or Mealy) with one (or several) expressions denoting the event that the machine *represents* while it is initially in the state q_i, in particular, in the initial state (the so-called q_λ) with which the expression E denoting the event represented corresponds. It will be seen that the expressions obtained in this way really are regular expressions.

It is clear that the elimination of variables, carried out in a *finite* number of steps, by means of a formula containing only 2 regular operators (* and ·) leads to an expression containing only alphabetic symbols, again combined through a finite number of regular operations.

Other methods exist, notably those of McNaughton and Yamada, Lee, Glouchkov [cf. e.g. 64.100] for the determination of the regular expressions associated with a diagram of states.

It should be noted that:

— the above method always produces regular expressions in the restricted sense, i.e. containing only $\cup$, · and * as operators.

— for a states diagram and a given state q_i, there is obviously only one associated *event* E_i. There may, however, be several expressions denoting this event but which differ in their algebraic forms. In the case of the Arden method, the expression will, in general, depend on the order in which the variables are eliminated.

Reverting, for example, to the equations of (13.134) which can be expressed:

$$E_2 = 0^*\,1\,E_3$$

$$E_3 = 0\,E_3 \cup 1\,0^*\,1\,E_3 \cup \lambda$$

$$= (0 \cup 1\,0^*\,1)\,E_3 \cup \lambda \quad (13.142)$$

whence:
$$E_3 = (0 \cup 1\ 0^*\ 1)^*\ \lambda$$
$$= (0 \cup 1\ 0^*\ 1)^*$$
$$E_1 = 0E_1 \cup 1\ 0^*\ 1(0 \cup 1\ 0^*\ 1)^*$$
$$= 0^*\ 1\ 0^*\ 1(0 \cup 1\ 0^*\ 1)^* \qquad (13.143)$$

E_1 represents the same event but possesses a different algebraic form.

13.3.3 Example: Series Adder

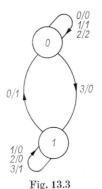

Fig. 13.3

Consider the 2-states diagram of Fig. (13.3) which represents a series adder, the inputs being coded decimally.

The following equations may be written:
$$\begin{cases} R_0 = (0 \cup 1 \cup 2)\ R_0 \cup 3R_1 \cup 1 \cup 2 & (13.144) \\ R_1 = (1 \cup 2 \cup 3)\ R_1 \cup 0R_0 \cup 0 \cup 3 & (13.145) \end{cases}$$

From the second equation is extracted:
$$R_1 = (1 \cup 2 \cup 3)^*\ (0R_0 \cup 0 \cup 3). \qquad (13.146)$$

By transferring the expression thus derived for R_1 to the equation (13.444) is obtained:

$$R_0 = (0 \cup 1 \cup 2)R_0 \cup 3(1 \cup 2 \cup 3)^*\ 0R_0 \cup 3(1 \cup 2 \cup 3)^*\ (0 \cup 3) \cup 1 \cup 2$$
$$= [(0 \cup 1 \cup 2) \cup 3(1 \cup 2 \cup 3)^*\ 0]\ R_0 \cup [3(1 \cup 2 \cup 3)^*\ (0 \cup 3) \cup 1 \cup 2]$$
$$(13.147)$$

where finally:
$$R_0 = [(0 \cup 1 \cup 2) \cup 3(1 \cup 2 \cup 3)^*\ 0]^*\ [3(1 \cup 2 \cup 3)^*\ (0 \cup 3)$$
$$\cup 1 \cup 2]. \qquad (13.148)$$

Other expressions may be derived for the first factor: for example:

$$[0 \cup 1 \cup 2 \cup 3\ (1 \cup 2 \cup 3)^*\ 0]^*$$
$$= [(0 \cup 1 \cup 2)^*\ 3(1 \cup 2 \cup 3)^*\ 0]^*\ (0 \cup 1 \cup 2)^* \quad \text{(from (13.33))}$$
$$= (0 \cup 1 \cup 2)^*\ [3(1 \cup 2 \cup 3)^*\ 0(0 \cup 1 \cup 2)^*]^* \quad \text{(from (13.34))} \qquad (13.149)$$

Next consider the Moore diagram obtained from the preceding example by deleting the output indications on the arrows and attributing to it,

13.3 Regular Expressions Associated with a States Diagram

at the state i ($i = 0$ or 1) the output i, which thus describes the behaviour of the carry bit. This gives the following equations;

$$\begin{cases} R_0 = (0 \cup 1 \cup 2) R_0 \cup 3 R_1 \\ R_1 = (1 \cup 2 \cup 3) R_1 \cup 0 R_0 \cup \lambda \end{cases} \quad (13.150)$$

This time, proceeding in an analogous manner,

$$R_1 = (1 \cup 2 \cup 3)^* (0 R_0 \cup \lambda) \quad (13.151)$$

then:

$$R_0 = (0 \cup 1 \cup 2) R_0 \cup 3(1 \cup 2 \cup 3)^* 0 R_0 \cup 3(1 \cup 2 \cup 3)^* \quad (13.152)$$
$$= [0 \cup 1 \cup 2 \cup 3(1 \cup 2 \cup 3)^* 0] R_0 \cup 3(1 \cup 2 \cup 3)^* \quad (13.153)$$

whence:

$$R_0 = [0 \cup 1 \cup 2 \cup 3 (1 \cup 2 \cup 3)^* 0]^* 3(1 \cup 2 \cup 3)^*. \quad (13.154)$$

At first sight this expression is relatively complex. Note, however, that the sequences formed of 0, 1, 2 or 3 are of the following two types:
— sequences including an initial sequence of any length (possibly null) of 0, 1 and 2 followed by an intermediary sequence of blocks commencing with 3 and finishing with 0 separated by blocks of 0, 1 and 2, — this intermediary sequence being, itself, terminated by a sequence of 0, 1 and 2.
— sequences of the preceding type terminating in a block commencing with 3 and not terminating in 0.
From which:

$$i^* = (0 \cup 1 \cup 2)^* [3(1 \cup 2 \cup 3)^* 0 (0 \cup 1 \cup 2)^*]^* [\lambda \cup 3(1 \cup 2 \cup 3)^*]$$
$$= A [\lambda \cup 3(1 \cup 2 \cup 3)^*] \quad (13.155)$$

and

$$i^* 3(1 \cup 2 \cup 3)^* = A [\lambda \cup 3(1 \cup 2 \cup 3)^*] [3(1 \cup 2 \cup 3)^*] \quad (13.156)$$
$$= A [3(1 \cup 2 \cup 3)^* \cup 3(1 \cup 2 \cup 3)^* 3(1 \cup 2 \cup 3)^*]$$
$$= AB \quad (13.157)$$

The content B of the square bracket reduces to $3(1 \cup 2 \cup 3)^*$ so that:

$$3(1 \cup 2 \cup 3)^* = 3(1 \cup 2 \cup 3)^* (1 \cup 2 \cup 3)^* (1 \cup 2 \cup 3)^*. \quad (13.158)$$

whence:

$$B = 3(1 \cup 2 \cup 3)^* [(1 \cup 2 \cup 3)^* (1 \cup 2 \cup 3)^* \cup 3(1 \cup 2 \cup 3)^*]$$
$$= 3(1 \cup 2 \cup 3)^* [(1 \cup 2 \cup 3)^* \cup 3](1 \cup 2 \cup 3)^*$$
$$= 3(1 \cup 2 \cup 3)^* (1 \cup 2 \cup 3)^* (1 \cup 2 \cup 3)^*$$
$$= 3(1 \cup 2 \cup 3)^* \quad (13.159)$$

from which:
$$R_0 = i^* \, 3 \, (1 \cup 2 \cup 3)^*. \tag{13.160}$$

It is easily verified that this relation expresses, in the language of regular expressions, the explicit value of the carry bit of a binary addition as established in Chapter 9, i.e.:

$$R_{k+1} = \sum_{j=0}^{j=k} \prod_{i=j+1}^{i=k} P_i G_j, \, (P_i = x_i + y_i, \, G_j = x_j y_j) \tag{13.161}$$

Indeed $G_j = 1$ corresponds to $x_j = y_j = 1$, $x_j y_j = 11 \, (= 3$ in the decimal code). $P_i = x_i y_i = 1$, (corresponds to $x_i y_i = 0 \, 1$ or $1 \, 0$ or $1 \, 1$, $= 1$ or 2 or 3 in the decimal code.)

Chapter 14

The Simplification of Sequential Networks and Minimisation of Transition Tables

14.1 Introduction

It has been seen that for combinatorial networks an important problem has been that of minimisation. This problem, still imperfectly resolved, and for which it is frequently necessary to accept an approximate solution can be formulated thus: it is desired to design a circuit $[m, n]$ having m inputs and n outputs, where each of the outputs is a Boolean function of m variables which may or may not have been completely specified. The input-output correspondence thus defined (with a certain arbitrary margin where the outputs are incompletely specified), constitutes the 'specification' for the "performance" expected of the logic network. Then there exists an infinity of circuits providing a solution to the problem: to minimise the circuit is to search out from among these solutions, those which not only satisfy the performance requirements, but also reduce to a minimum the "merit figure" (cost function) pertaining to the circuit. In particular, in the case of a single output circuit $[m, 1]$ the performance reduces to a Boolean function which can be represented by an infinity of algebraic expressions and which can be associated with just as many versions of the required circuit. Thus two circuits providing the same input-output relation can, none the less, have different internal layouts and unequal costs. The problem is posed in similar terms for the sequential networks and it will first be necessary to examine the sense to be given to the concepts of *performance* and of *cost function*.

14.1.1 Performance

The sequential network effects a mapping $\mathcal{A}$ of a set of *input sequences* into a set of *output sequences*. Although other definitions are admissible, it will be assumed here that this mapping $\mathcal{A}$ is the mathematical expression of the performance concept. It is therefore necessary to distinguish two cases:

14.1.1.1 Completely Specified Circuits

The application $\mathcal{A}$ is defined for every input sequence.

14.1.1.2 *Incompletely Specified Circuits*

The mapping is defined only for the input sequences belonging to a subset $\mathcal{E}'$ of the set $\mathcal{E}$ of possible sequences, to which the name *applicable sequences* has been given and for which a definition will be given later.

14.1.2 Cost Function

The sum of the costs of the logic circuits and of the memory circuits can be taken as the cost (economics) of a sequential network.

Two cases exists:

— The memory unit consists of flip-flops.

The design and production of a flip-flop are, in general, more complicated and more costly than those of a simple gate. This is certainly true for conventional circuits having discrete elements, where a ratio of 10 can be considered to be typical. This ratio is already considerably less for integrated circuits where, for example, the ratio can fall to 2.

— The memory unit consists of delays.

The cost of these is nil for the natural delays (time lags), low for the passive synthetic delays, much higher for the monostable type delays where it is comparable to that for the bistable flip-flops.

Even when adhering to a cost function of this sort which, as has already been seen (Chap. 10), can only be an extreme simplification of reality, it is scarcely necessary to affirm that the problem of the minimum circuit remains of considerable difficulty. In practice, it is customary to be content with an approximate simplification procedure which can be divided into two stages:

1. Minimisation of the number of internal states.
2. Minimisation of the associated logic.

The aim of the first step is the minimisation of the number of variables q_i of the memory unit. If there are p internal states then there must, in fact, be:

$$2^r \geq p \tag{14.1}$$

$$r \geq r_0 \quad \text{where} \quad 2^{r_0-1} < p \leq 2^{r_0} \tag{14.2}$$

(r_0 is thus the minimum number of binary variables with the aid of which the p states can be encoded).

The minimisation of the logic necessitates:

— the choice of a binary code to be attributed to the internal states.
— the minimisation of the combinatorial network (C).

The first step, *the binary coding of the internal states*, known as the "secondary assignment", is a problem for which there exists no general method. Some examples, for the particular case of counters, will be given in Chapter 16: they bring to light the influence of the choice of the number, and for r-fixed, of the binary code retained. In practice this 2-stage minimisation method gives relatively satisfactory results. Here too, despite the absence of a general method, it is often possible in a particular case to obtain a near-minimal solution and for an experienced designer to get a good 'feel' for the degree of minimisation approached. It is possible, however, to encounter cases where the network possesses the minimum number of states but, nevertheless, has a combinatorial section much more complicated than would be the case with a non-minimal number of internal states and resulting in a more costly sequential network. This is a basic weakness of the minimum number of states approach.

A reduction in the number of internal states, even though its influence on the total cost of the network be of slight account does, *in practice*, present the advantage of economy. It also presents two other advantages:

— the first, of an order both practical and theoretical, is that it enables a decision to be taken as to whether or not a race is critical, in the case of asynchronous circuits. This problem is linked with the equivalence (subject to various criteria) of two states. But the diagnosis of the equivalence of 2 states (whether for a single machine or for two different machines) is the very basis for the reduction of the number of internal states of a table.

— the second, of an essentially theoretical order, is to furnish a table which is simple to comprehend and which, among the infinity of possible tables, preserves the character of a reference table.

In the case of completely specified tables this is effectively so since, as will be seen, there can be only one minimum table (upto notation). For that of the incompletely specified tables there may well be several but their number always remains finite. Here the viewpoint is akin to that which, in the studies of Boolean functions and expressions, has led to the privileged position accorded to the two canonical forms.

14.2 Minimisation of the Number of States for a Completely Specified Table

14.2.1 Some Preliminary Considerations

Commencing from an initial state q, the response of the network is defined for every input sequence S with which correspond a sequence of internal states and one of outputs. The following notation will be adopted:

qS the successor of q by S

$Z(q)$ the output associated with q

$\overrightarrow{qS}$ a sequence of states obtained by applying S to q

$\vec{Z}(qS)$ an output sequence obtained by applying S to q

Consider the Table 14.1 which is that for a network comprising 7 internal states and 3 outputs.

Table 14.1

	E_1	E_2	E_3
1	2, 0	3, 0	2, 1
2	1, 0	4, 0	1, 1
3	2, 1	1, 0	1, 1
4	2, 1	1, 0	1, 1
5	3, 0	1, 0	1, 1
6	4, 1	3, 0	6, 1
7	4, 1	4, 0	7, 1

This gives for example

$$\begin{cases} Z(4E_1) = 1 \\ Z(5E_1) = 0 \end{cases} \qquad (14.3)$$

Assume the network to be known to be in one of the states 4 or 5, but without further precision as to which of the two. The application of E_1 constitutes a test which will enable the states to be distinguished and, with a knowledge of the table, effectively determines in which state the network is to be found. In this sense the states 4 and 5 can be *distinguished*. It can be said that they are "distinguishable" and that they are not equivalent.

Consider next the states 1 and 2. Can they be distinguished? With the pairs (1, 2) correspond the pairs of immediate successors (2, 1), (3, 4) and (2, 1) and the output pairs (0, 0), (0, 0) and (1, 1). Since these are formed from identical elements, it will be seen that the "test of length 1" which comprises the application of a sequence E_1 or E_2 or E_3 reduced to one letter, does not enable it to be said whether or not 1 or 2 can be distinguished. After a test of length 1, therefore, judgement must be reserved. However, there remains the possibility of examining the response of the network to sequences of length 2 applied to 1 and 2. For this it is sufficient to examine the response to sequences of length 1 applied to the successors of 1 and 2 to sequences of length 1 which form the pairs indicated above.

14.2 Minimisation of the Number of States

Among these pairs of successors, the pair (1, 2) appears again twice. No account will be taken of this pair since this is precisely the pair for which it is required to determine whether or not it is composed of elements which can be distinguished. The reply to the question above can only be given by an examination of the pair (3, 4). Not only are the outputs obtained from the application of an identical input for 3 and 4 themselves identical but, in addition, 3 and 4 have, here, a unique successor for every input.

Consequently it is useless to continue with the experiment by recourse to sequences of length > 2; it is now clear that the output sequences $\vec{Z}(1; S)$ and $\vec{Z}(2, S)$ will be identical for every input sequence S. If as before, the process be confined to the observation of inputs and outputs alone, it will not be possible to distinguish between the two states: no input "stimulus" will cause any difference to appear. It will be said that the states 1 and 2 are equivalent. It will, here, be noted that it has been necessary to have recourse to test sequences of length 2 to arrive at this conclusion. In a similar manner the case could occur where, after an input sequence S of length >1, the initial state of the pair $(q_1 q_2)$ became transformed into a pair $(q_1 S, q_2 S)$ which had previously been found to be distinguishable. This is, therefore, the case for the pair $(q_1 q_2)$. It will thus be seen that, in order to determine whether or not two states can be distinguishable, an examination of the states and outputs obtained from the different inputs does not always suffice and that, in general, it is necessary to examine the pairs (more or less remote) of indirect successors. The situation to which this type of test leads can be stated as follows:

1. In order that two states q_i and q_j be distinguishable, it is necessary and sufficient that there be a sequence S which leads to successors $q_i S$ and $q_j S$ which can be distinguished.

Particular case: $q_i S$ and $q_j S$ are distinguishable when there exists an input x such that $Z(q_i S x) \neq Z(q_j S x)$.

2. In order that two states be equivalent, it is necessary and sufficient that for every input sequence S, the successors $q_i S$ and $q_j S$ are, themselves, equivalent.

Example: The pair (6, 7) of Table 14.1 has as successors the pairs (4, 4), (3, 4) and (6, 7) and as outputs pairs: (1, 1), (0, 0) and (1, 1). It suffices to consider the sequences commencing with E_1 and E_2; it will then be seen that the only pair likely to give rise to a difference in output is (3, 4) which has been shown to be a pair of equivalent states. The states 6 and 7 thus enjoy the same property.

The above considerations, for the cases which can be encountered, lead to the presentation of the following definitions which permit the formulation and resolution of the problem in a rigorous manner.

14. The Simplification of Sequential Networks

14.2.2 Equivalent States

14.2.2.1 States in the Same Table T

Two states q_i and q_j in a single table are said to be *equivalent* (indistinguishable) if, when the same input sequence S is applied to one or other of these two states, the two output sequences obtained are identical, and this for every input sequence. In other words, q_i and q_j are equivalent, if for every sequence S, $\vec{Z}(q_i S) = \vec{Z}(q_j S)$. Written as $q_i \sim q_j$.

The reader can verify without difficulty that the relation for the set of states is a binary relation which is:

— symmetric
— reflexive
— transitive

in other words, that this is a genuine equivalence relationship in the sense of Chapter 1. It thus defines a partition of the set of internal states into equivalence classes such that:

— two states belonging to the same class are equivalent
— two states belonging to two distinct classes are not equivalent.

The interest of this partition can be seen at once. Since nothing permits the states of the same class to be distinguished, all can be given the same name, that of the class in question, and repeating the operation for every class, will thus define a new table T' in which the states are associated with these classes and which is the equivalent of the former table. This leads to the following definition:

14.2.2.2 State Belonging to 2 Tables T and T'

The states q and q' respectively belonging to two tables T and T', having the same input and output symbols, are said to be *equivalent* if, for every input sequence S; $\vec{Z}(qS) = \vec{Z}'(q'S)$, where Z and Z' are the output functions for T and T' respectively.

14.2.3 Covering of Table T by T'

It is said that T' covers T if for every state q of T, there exists an equivalent state in T'.

14.2.4 Successor to a Set of States

The successors to a state have been defined in § 12.3.3.4. This concept can be generalised for a set of states. The successor to a set of states (by a sequence S of length $k \geq 1$) is the set of successors of states through S. It will be noted that the successor to a set of states could possess fewer states than that of the initial set. The same distinction between direct successor ($k = 1$) and indirect successors ($k > 1$) remains valid.

14.2.5 Reduction of a Table. Successive Partitions P^k

The above definition of the equivalence of two states q_i and q_j of a table requires that the output sequences generated by q_i or q_j be identical for all input sequences. This brings into play, therefore, an infinity of sequences of length 1 then of length 2, 3, etc., l ... However, as was stated in § 12.4.2.9 with respect to races, it is unnecessary, for determining whether or not q_i and q_j are equivalent, to investigate an infinity of sequences. This circumstance without which there would, in practice, be no solution to the problem, has its origin in the finite character of the table, as will now be shown.

For this it will be necessary to consider a relationship of equivalence between states to be known as *k-equivalence* or "equivalence of order k".

Two states q_i and q_j are k-equivalent if $Z(q_i S) = Z(q_j S)$ for every sequence of length k.

It is clear that two k-equivalent states are i-equivalent with $i \leq k$. It will be noted that two equivalent states are necessarily k-equivalent (the converse is not necessarily true).

The reader will easily verify that the k-equivalence is also an equivalence relation (in the sense of Chapter 1). It will be denoted k. It thus defines a partition P^k of the set $Q = (q_1, ..., q_p)$ of internal states into subsets

$$Q_i^k \left(\bigcup_{Q_1}^k = Q, Q_i^k \cap Q_j^k = \varnothing \right).$$

Thus with the different "test lengths" 1, 2, 3, ..., k, ... are associated the partitions of Q, $P^1, P^2, \ldots P^k$, with $P^k = \{Q_i^k\}$. This then gives the following property: two states q_i and q_j which belong to two different (and thus disjoint) sets Q_l^k and Q_m^k, are to be found again in two disjoint sets of the partition $P^{k+1} = \{Q_i^{k+1}\}$. Otherwise, they would in fact be equivalents of order $k + 1$ and thus (a fortiori) of order k, which would be a contradiction. In other words, the partition P^{k+1} can only be obtained from the partition P^k by breaking down (partitioning) certain sets Q_i^k of P^k. As a result the rank of the partition P^{k+1} is higher than or equal to that of P^k: the rank of the partitions P^k is a non-decreasing function of k.

The same reasoning applies, in particular, to the partition of P into equivalent states. It can always be considered that this is obtained from a partition P^k by breaking down the latter, since two states which are not k-equivalent are, a fortiori, non-equivalent.

This gives the following property:

1. If in each equivalence class Q_i^k of a partition P^k the states of Q_i^k are all pairwise equivalent, then the partition P^{k+r} is identical with that for P^k for every $r \geq 1$.

Indeed, if two states of a partition are equivalent they are, in particular r-equivalent. Consequently the state of the corresponding class are $(k + r)$-equivalent for every $r \geq 1$.

2. Conversely, if there exists a class Q_i^k containing two states q_i and q_j which are not equivalent then, of necessity, there exist two successors $q_i S$ and $q_j S$, where S is of length $\geq k$, which are also not 1-equivalent. By putting $S = S'S''$, with S'' of length k, it will be seen that $q_i S'$ and $q_j S'$ are equivalents of order k and no longer of order $k + 1$. They are thus to be found in distinct classes of the partition P^{k+1}, which in turn differs from P^k. In particular, the rank of P^{k+1} is higher than that of P^k.

The consequence, therefore, is that in order to determine the partition P, it is necessary and sufficient to determine the partitions of P^k until such time as is found an index i such that $P^{i+1} = P^i$. It is then certain that the partition P^i is the required partition of P.

In addition, as for $k \leq i$, the rank of P^k increases strictly with k, but is always less than the number of internal states p, thus: $i \leq p - 1$. It will be seen that, as stated, it suffices to test the input sequences of length $\geq p - 1$. There are cases for which it is necessary to go thus far (there is then only one state per equivalence class and consequently no reduction). At each step of the test, characterised by k, the rank P^k represents in a way the minimum number of states to which it may be expected to reduce the table. If this number remains less than the number of states p for the index i (such that $P^{i+1} = P^i$) there is, effectively, some reduction.

14.2.6 Algorithm for the Minimum Number of Internal States and its Practical Application

Algorithm

1. Q_0 being the set of internal states p,

$$Q_0 = (q_1, q_2, \ldots, q_p)$$

determine the set $E_0 = (Q_0 \times Q_0)$ of the pairs (q_i, q_j).
2. Determine the set E_1 of the pairs (q_i, q_j) for which $Z(q_i E) = Z(q_j E)$ for every input E.
3. Determine the set E_2 of the pairs (q_i, q_j) for which the immediate successor is a pair (q_e, q_m) belonging to E_1.

14.2 Minimisation of the Number of States

4. Repeat the procedure step-by-step: from the set E_n, determine a set E_{n+1} of the pairs having an immediate successor in E_n, and continue the procedure until, for an index i, $E_{i+1} = E_i$. The process is then complete: the set E_i defines the partition P of Q_0 in equivalence classes.
5. Take one element in each equivalence class. Replace all other elements of the corresponding class by this element wherever they appear in the transition table.
6. Delete the identical lines.

14.2.7 Example: Minimisation of Table 14.2

Table 14.2. *Table to be minimised*

Q\E	E_1	E_2	E_3	E_4
1	2, 0	6, 1	4, 1	5, 0
2	3, 0	6, 1	5, 1	4, 0
3	1, 0	6, 1	5, 1	5, 0
4	6, 0	2, 0	6, 1	4, 1
5	6, 0	1, 0	6, 1	5, 1
6	3, 1	6, 1	4, 0	2, 0

This table possesses six internal states, four inputs and one binary output.

1. $Q_0 = (1, 2, 3, 4, 5, 6)$
2. Set $E_0 = Q_0 \times Q_0$:

The order of the elements of a pair is of no importance (on account of the symmetry of an equivalence relation), work will be done only on the pair (i, j) where $i < j$. The table of pairs of states, limited in this way, possesses $C_n^2 = n(n-1)/2$ elements and takes the triangular form of Table 14.3 below.

Table 14.3

$$E_0 = \begin{Bmatrix} 1,2 & & & & \\ 1,3 & 2,3 & & & \\ 1,4 & 2,4 & 3,4 & & \\ 1,5 & 2,5 & 3,5 & 4,5 & \\ 1,6 & 2,6 & 3,6 & 4,6 & 5,6 \end{Bmatrix}$$

E_1 is now determined, by taking the above pairs successively and by examining the outputs associated with each. If for an input the outputs diverge, then the pair does not belong to E_1.

In practice, it suffices to delete the non-equivalent pairs directly on the table for E_0, whence the set E_1 (Table 14.4):

Table 14.4

$$E_1 = \begin{Bmatrix} 1,2 & & & & \\ 1,3 & 2,3 & & & \\ - & - & - & & \\ - & - & - & 4,5 & \\ - & - & - & - & - \end{Bmatrix}$$

E_2 is now determined by taking the pairs belonging to E_1 and which have as successor a pair belonging to E_1 or formed from two identical states (Table 14.5)

Table 14.5

$$E_2 = \begin{Bmatrix} 1,2 & & & & \\ 1,3 & 2,3 & & & \\ - & - & - & & \\ - & - & - & 4,5 & \\ - & - & - & - & - \end{Bmatrix}$$

Having $E_2 = E_1$, repetition of the procedure will produce no further change to E_2, and as has been seen this terminates the test.

The various equivalence classes: (1, 2, 3), (4, 5) and (6) can now be determined.

Whence the of Q_0:

$$Q_0 = (1, 2, 3) \cup (4, 5) \cup (6)$$

Next, by coding the equivalence classes with the aid of new symbols (Table 14.6)

Table 14.6

Class	New state
(1, 2, 3)	A
(4, 5)	B
(6)	C

is obtained a new transition-output (Table 14.7)

14.2 Minimisation of the Number of States

Table 14.7 *Minimal table*

Q\E	E_1	E_2	E_3	E_4
A	A, 0	C, 1	B, 1	B, 0
B	C, 0	A, 0	C, 1	B, 1
C	A, 1	C, 1	B, 0	A, 0

which is the simplified form of Table 14.2. It possesses only 3 internal states in lieu of 6.

In practice, the following can be adopted for the determination of the equivalences:

— Draw up a triangular array comprising as many elementary frames as there are distinct pairs, in accordance with the layout of Table 14.8.

The pair corresponding to a frame is defined by the column number and by the line number. The table thus reproduces that which was drawn earlier for $E_0 = Q_0 \times Q_0$.

— In every frame corresponding to a non-equivalent pair, inscribe a cross ($\times$).

— In every frame which is not found to be non equivalent at a given stage of the test, inscribe the pairs of states formed by immediate successors of the associated pair of the frame.

— Take each of the frames containing a $\times$ one after another; e.g. the the frame i, j. Then inscribe a cross in every frame where the pair of states i, j is to be found.

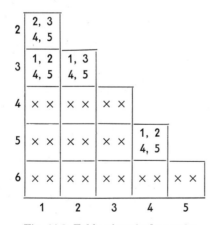

Fig. 14.8. Table of equivalent pairs.

At the same time inscribe a second cross in the frame i, j to indicate that this frame will no longer be taken into account.

After a finite number of steps a table is obtained for which the frames contain 0 or 2 crosses. The cases which do not contain crosses correspond with the equivalent pairs.

The remaining different pairs permit the definition of the equivalence classes for the table to be reduced, and thus the states of the reduced table.

14.3 Minimisation of the Number of Internal States for an Incompletely Specified Table

As has been seen from the definition for sequential networks, such a table possesses pairs (future state, output) which are incompletely specified, i.e. the future state or the output or both are not shown. The unspecified state or unspecified output are then denoted by a dash —. When neither the future state, nor the output are specified, a single (common) dash is inscribed. These dashes can be given various (non-equivalent) interpretations.

14.3.1 Interpretation of Unspecified Variables

— *Future state unspecified* (output specified or unspecified).

It is assumed that from the moment when the input drives the system into an unspecified (future) state, the subsequent evolution of the system can be of any type (from the viewpoint of states and outputs) in accordance with the requirements of the circuit designer.

— *Unspecified output*.

It is assumed that when, for a given present state, the input provides an unspecified output, this output is permitted to be of any type.

This is to say that, at a given *instant*, not only can the dash take one or other value at will but, in addition, that it is not, a priori, required that the dash be always replaced by the same constant value. In other words, if it is possible to arrange that at 2 different *instants* the dash is replaced by 2 different values, this will remain compatible from the 'user' point of view as chosen above, and the corresponding network is not excluded.

14.3.2 Applicable Input Sequences

An input sequence $E_{i_1}, E_{i_2}, \ldots, E_{i_n}$ of length n is said to be *'applicable'* to a circuit assumed to be in the state q_i if the future state is defined by the table for all the letters $E_{i_l}(l=1, \ldots, n)$ which form it with the possible exception of the last.

The applicable sequences are thus the only input sequences to be used. With each 'departure' state q_i there corresponds a set $\mathcal{E}(q_i)$ of applicable input sequences.

For a given set of states $q_{i_1}, \ldots, q_{i_k}$ the set of sequences applicable simultaneously to all the states is:

$$\mathcal{E}(q_{i_1}) \cap \mathcal{E}(q_{i_2}) \cap \cdots \mathcal{E}(q_{i_k}).$$

14.3.3 Compatibility of 2 Partially Specified Output Sequences

Two partially (incompletely) specified output sequences are said to be *compatible* if, for every rank l, where Z_{i_l} and Z'_{i_l} are both defined ($\neq -$) then:

$$Z_{i_l} = Z'_{i_l}.$$

This will be expressed:

$$Z \# Z'.$$

Example:
$$Z = (1\ 3 - 5\ 4 - - 4\ 6 -)$$

and
$$Z' = (1 - 4 - 4\ 3\ 1\ 4\ 6\ 2)$$

are compatible.

The relation of compatibility is a binary relationship between sequences but which, contrary to the equality of sequences, *is not transitive:* consider a sequence Z'' defined by:

$$Z'' = (1\ 5\ 4 - 4\ 3 - - 6\ 2);$$

which gives:
$$Z \# Z'$$
$$Z \# Z''$$

but does not give $Z \# Z''$ since Z and Z'' differ in the second symbol (3 and 5 respectively).

This can be generalised to n sequences: n output sequences will be said to be compatible if at all points at which at least one letter is specified, the outputs are equal.

14.3.4 Compatibility of 2 States q_i and q_j

Two states q_i and q_j are said to be *compatible* if for every input sequence S applicable (to q_i and q_j at the same time)

$$\vec{Z}(q_i S) \# \vec{Z}(q_j S) \qquad (14.4)$$

28*

it will also be said that q_i and q_j comprise a compatible pair, *a compatible set of the order* 2 or, more briefly, *a 'compatible'* (of order 2).

Two states q_i and q_j are thus *incompatible* if there exists a input sequence S applicable at q_i and q_j such that $\vec{Z}(q_i S)$ and $\vec{Z}(q_j S)$ are not compatible, e.g. that they possess at least one position where both are specified, but different.

14.3.5 Compatibility of a Set of n States ($n > 2$)

The compatibility of 2 states is a binary relation (with 2 terms). In general, it can be said that n states $q_{i_1}, \ldots, q_{i_n}$ comprise a compatible of the order n if, for every sequence applicable at the same time to $q_{i_1}, \ldots, q_{i_n}$, the output sequences $\vec{Z}(q_{i_l} S)$ ($l = 1, \ldots, n$) are compatible. It can be shown (Paull and Unger [96]) that in order for a group of n states to form a compatible of the order n, it is necessary and sufficient that the $C_n^2 = n(n-1)/2$ pairs of states that can be formed from it be compatible pairs (compatibles of the order 2). A compatible will, henceforth, be denoted by pairs of brackets (parentheses) containing the states of which it is composed. Conversely, a set of states which does not possess the above property will be called an "incompatible".

Example:

The compatibility of order 2 between states, like that between sequences, is symmetric, reflexive, but not transitive. Consider, for example Table 14.9 (with 3 states):

Table 14.9

Q \ E	0	1
1	2,	1, –
2	2, –	–, 1
3	3, 0	3, –

(1, 2) is a compatible pair. Indeed, the pairs of successor states are (2, 2) and (–, 1). The pair (–, 1) signifies that, from the state 2, there is only one "applicable" sequence commencing with 1, that precisely reduces to the single 1 written once ($S = 1$). For this sequence, the output sequences are – and 1, which are compatible. With respect to the other acceptable sequences, (those which, of necessity, commence with 0) it will be seen that from (1, 2) they produce the pair (2, 2). This ensures that for the sequences of form (0, x) the output sequences obtained from 1 and 2 will be compatible, if they are compatible for the sequence 0. This is, indeed, the case.

In the same way (23) is a compatible pair if (2, 3) and (−, 3) are compatible, i.e. if (−, 3) is compatible, which is the case. It will thus be seen that (12) and (23) are compatible pairs. It must, however, be verified that $Z(1, 0) = 1$ and that $Z(3, 0) = 0$, i.e. that (13) is not a compatible pair. As a result it will be seen that (123) is not a compatible of order 3. Thus, in general, compatibility between states is not transitive.

14.3.6 Compatibility of States and Simplification of Tables

As for the unspecified tables, the desire is to associate the state in the table T to be minimised, with those of the minimal table T''. Compatibility is a relation which permits the generalisation of the equivalence between states of a single table. In view of the need to replace T by T'' there is a need for the following definitions:

14.3.6.1 Covering of a Sequence Z by a Sequence Z'

It is said that a sequence

$$Z' = Z'_{i_1} Z'_{i_2} \cdots Z'_{i_l} \cdots Z'_{i_k}$$

covers a sequence

$$Z = Z_{i_1} Z_{i_2} \cdots Z_{i_l} \cdots Z_{i_k}$$

if for every specified position Z_{i_x} of Z, the corresponding position Z'_{i_x} is filled with the same symbol. We will write:

$$Z' \geq Z$$

Example:

$$Z' = 3\ 2\ 5\ 3$$
$$Z = -\ 2\ 5\ -$$

we have:

$$Z' \geq Z$$

Particular case: $Z' = Z$. Thus $Z' \geq Z$ and $Z' \leq Z$, and conversely, these last two conditions give $Z' = Z$. It will be noted that in the case of a completely specified table, the relation $Z' \geq Z$ when true always reduces to $Z' = Z$.

14.3.6.2 Covering of a State q_i of a Table T by a State q_i' of a Table T'' (same input-output alphabet)

It is said that q_i is covered by q_i' if for every input sequence S applicable to q_i and q_i':

$$\vec{Z}_T(q_i S) \leqq \vec{Z}_{T'}(q_i' S). \tag{14.5}$$

14.3.6.3 Covering of a Table T by a Table T''

It is said that T is covered by T'' if for every state q_i of T, there exists a state q_i' of T'' such that q_i is covered by q_i'. As an immediate result of the above definitions, in order that a set of states $(q_{i_1}, q_{i_2}, ..., q_{i_n})$ may be associated with a state in a new table T'', it is *necessary* that it be a compatible of order n.

In order that a table T'' may cover a table T it is necessary:

1. That there should exist a cover Q for the set states of T, of which the sets are associated with the states of T''.
2. That the sets of the cover be "compatibles".
3. That the cover Q, with the aid of the compatible defined in 1 and 2, be a *closed* cover. A cover Q of compatibles $C_p (\cup\limits_p C_p = Q)$ is said to be closed if, for every p and for every input E_i, the successor to C_p is included in a compatible C_k of this cover, i.e. if:

$$C_p E_i \subset C_k \tag{14.6}$$

In other words, in a closed cover the successor to one of the compatibles does not "break down" into subsets belonging to compatibles which are different. In order that T'' be minimal, it is necessary that:

4. The number of compatibles in the cover be minimal.

The conditions 1 to 4 are thus sufficient as can easily be verified.

14.3.7 Properties of Compatibles

— It is clear that every subset of a compatible is, itself, a compatible.
— For a given compatible $(q_1, ..., q_k)$ and an associated state q_l, in order that $(q_1, ..., q_k, q_l)$ be compatible, it is necessary and sufficient that q_l be separately compatible with $q_1, q_2, ... q_k$.
— Maximal compatible
 This is a compatible such that, if another external state be added to it, the new set is not a compatible no matter what that state may be. In other words, there is no state, distinct from those of the compatible, for which the above property is true. A maximal incompatible is defined in a similar way.

14.3 Minimisation of the Number of Internal States   421

— Upper bound on the number of states for a minimal table.
It will easily be seen that the set E of all the maximal compatibles is a closed set. Indeed, every successor to a compatible of E is a maximal compatible. Since E contains *all* of them it will be seen that for *every* compatible of E, the successors are included in the other compatibles of E, i.e. that by the same definition, the set E of all maximal compatibles really is a closed set. Thus it provides a cover for the initial table. In consequence, the number of maximal compatibles constitutes an upper bound on the number of states comprising a minimal table. This limit is of no great importance in practice except when it is effectively less than the number of states of the initial table T.
— Lower bound on the number of states for a minimal table.
Clearly the number of states of the largest maximal incompatible is *one* lower bound for the states of a minimal table.

14.3.8 Determination of Compatibles

14.3.8.1 Compatible Pairs

The procedure is similar to that used above for the completely specified tables. Firstly, all of the "direct" incompatibles are determined by successive examination, for a pair of states (i, j) of the output pairs associated with the different inputs. If a pair for which these outputs are different is found, it is useless to continue further with this pair (i, j). It is incompatible. Conversely, if no incompatibility be found (thus exhausting all the inputs), a second attempt is made, using the columns of the table for which the successors to those of (i, j) are both specified and the corresponding pairs are inscribed in the frame of the triangular table of coordinates (i, j) each time that they differ from the initial pair and are composed of different elements. If no pair of successors is completely specified the sign $\vee$ is inscribed in the frame (i, j). The frames containing this sign mark the compatibles. Conversely, the frames containing an indication of the pairs of successors are at this stage, only candidates for compatibility and a check is made to determine whether they do not indicate indirect incompatibilities, i.e. if they are compatibles, by proceeding as for the completely specified case. After a finite number of steps a final table emerges for which the frames can be of 3 types at most: they may contain two crosses, or none, or contain the symbol $\vee$. The latter two types of frame have the compatible pairs as coordinates. This procedure is illustrated by the example of Table 14.10. The search for the compatibles then leads on to Table 14.11. (This example will continue to be used in the determination of the other operations of minimisation.)

14. The Simplification of Sequential Networks

Table 14.10. *Table to be minimised*

Q \ E	1	2	3	4
1	2, –	1, 1	—	5, 0
2	—	3, –	—	—
3	3, 0	2, –	—	—
4	2, 0	4, –	4, –	3, 1
5	2, –	3, 1	4, 1	7, 1
6	3, –	4, 1	5, 1	2, 1
7	2, –	1, 0	4, 1	6, –

Determination of compatible pairs: in the table which follows the frames which do not contain crosses are associated (by their coordinates) with the compatible pairs:

Table 14.11. *Compatible pairs*

2	1, 3					
3	1, 2 2, 3 5, 6	2, 3				
4	× ×	3, 4	2, 3 2, 4			
5	× ×	γ	2, 3	3, 4 3, 7		
6	× ×	3, 4	2, 4	2, 3 4, 5	2, 3 2, 7 3, 4 4, 5	
7	× ×	γ	1, 2 2, 3	1, 4 ×3, 6×	× ×	× ×
	1	2	3	4	5	6

14.3.8.2 Compatibles of Order ≥ 2. Maximal Compatibles

As has been shown above, for the determination of a maximal compatible, it is necessary and sufficient to determine all the compatibles 2 by 2. There are a number of ways of doing this:

— *First method* (Paull and Unger)

This is a step-by-step procedure based on the final Table 14.11 in the following manner:

1. Commencing from the extreme right hand column a search is made for the first column containing one or more compatible pairs. These are then listed.
2. The column on the left is then examined. If it has frames which do not contain crosses on the lines for which the numbers are those of the two states of the preceding pair, this means that the state marking the column is compatible pairwise with each of the two states of this pair, and a new compatible can be found by the addition of this state to the pair in question.

At this stage the list of compatibles comprises:

the closed compatibles of order 2 of the preceding rank which it has not been possible to combine with the present rank to give compatibles of the order 3;
— the compatibles of order 2 determined for the present rank;
— the compatibles of order 3 formed from those of the present rank and of the compatibles previously obtained.

Remark: The compatible pairs which have given rise to the compatibles of order 3, are the subsets of the latter and thus do not form part of the list of maximal compatibles.

Continue step-by-step progressively to the left and examine each new column to determine whether the pairs contained therein (as necessary) will, with the aid of the compatibles (of different orders) already formed, form new pairs. In the general case (in passing from column k to column $k + 1$), the procedure set out above can be formulated as follows;

— take the first of the compatibles so far obtained ('old' compatibles — i.e. C_1, comprising m elements $(m \geq 2)$). Examine the m frames of the present column $(k + 1)$ for which the line numbers are those of C_1. If none contains the sign $\times$ the column number may be added to that of C_1 to form a new compatible. This compatible is added to the present list and C_1 deleted. If one or more lines carry an $\times$ this indicates that it is not possible to form a compatible with *all* the states of C_1. But, the set of states C_1, which corresponds with the

lines of frames having no ×, is a compatible and can, therefore, be added to the present column to form a new compatible. The latter is added to the list C_1 and retained.
- take the second of the 'old' compatibles, i.e. C_2 and carry out the same operation. If this leads to the consideration of a subset of C_2 already used to form a compatible, the latter already appears in the list and it is pointless to write it in again. The process is continued — the 'old' compatibles being taken in turn until all have been exhausted.
- add to the list of compatibles which existed before the examination of the column $k + 1$, the newly determined compatibles (as necessary) if they have not already been included in the list.
- continue until the table has been completely explored.

Example: The following sequences of compatibles have been derived from Table 14.11 (only the new compatibles have been indicated for each step):

1. ~~(56)~~

2. ~~(456)~~

3. ~~(3456)~~   ~~(37)~~ (14.7)

4. (23456)   (237)

5. (123)

The 'cancelled' terms are included in a compatible of a higher order. The maximal compatibles are the terms which, at the end, have not been cancelled. Maximal compatibles: (123), (23456), (237).

— *Method II (Ch. Ioanin)*

Explanation of the method.

The compatible pairs are grouped as in Table I. (This can be the triangular table used for the determination of compatible pairs). Reverting to the above example gives Table 14.12.I. The pair 1, 2 is compared with 1, 3. They differ with respect to the second elements (2 and 3). The second column is examined to determine whether 2, 3 is a compatible. This is the case and, therefore, 1, 2, 3 is a compatible. 1, 2, 3 is inscribed in a second Table 14.12.II. Crosses are inscribed with respect to the two pairs used in this way. Proceeding to the column of pairs commencing with 2 then 2, 3 is compared with all the following pairs, then 2, 4 etc. For each pair of elements (i, j) for which two pairs $(2, i)$ and $(2, j)$ differ, $(i > 2)$, a check is made to determine whether they appear in one of the following columns. If they do, the compatible $(2, i, j)$ is formed, and the terms

14.3 Minimisation of the Number of Internal States

Table 14.12

× 1, 2	× 2, 3	× 3, 4	× 4, 5	~~5, 6~~
× 1, 3	× 2, 4	× 3, 5	× 4, 6	
	× 2, 5	× 3, 6		
	× 2, 6	~~3, 7~~		
	× 2, 7			

I

1, 2, 3	× 2, 3, 4	× 3, 4, 5	~~4, 5, 6~~
	× 2, 3, 5	× 3, 4, 6	
	× 2, 3, 6	× 3, 5, 6	
	2, 3, 7		
	× 2, 4, 5		
	× 2, 4, 6		

II

× 2, 3, 4, 5	~~3, 4, 5, 6~~
× 2, 3, 4, 6	
× 2, 3, 5, 6	
× 2, 4, 5, 6	

III

2, 3, 4, 5, 6

IV

used are marked. If they do not, proceed to the next comparison. Table 14.12.II is formed by this step-by-step process.

The latter table contains compatibles of the order 3 while the non-cancelled terms not included in those of Table 14.12.II are the maximal compatibles. In the example in question, the only pairs are 3, 7 and 5, 6; they are included in the compatibles of 14.12.II and are thus deleted in their turn.

This continues in a similar way to obtain the compatibles of order k based on those of order $k-1$. All the pairs which differ in one position are retained for each column and a check is made to determine whether the variables i and j, by which they differ, form a compatible pair in Table 14.12.I. If two pairs differ in more than one position, they are disregarded. After establishing a table in this way, a check is made of the non-deleted terms to see whether any exist which are contained in the newly obtained compatibles. If they do, they are deleted. The terms neither marked nor deleted are maximal compatibles. In this way, Tables 14.12.II, III and IV are derived for the example in question.

It will be noted that this procedure is similar, from the formal point of view, to that of the Quine-McCluskey method, the maximal compatibles playing the rôle of the prime implicants.

14. The Simplification of Sequential Networks

— *Method III* (Marcus)

The method indicated by Marcus can be stated as follows:
1. For every incompatible pair $(i \cdot j)$ form the Boolean sum $i + j$.
2. Unite all the sums thus obtained by the sign $\cdot$ for the product.
3. Develop the product in the form of a sum of products.
4. Eliminate the redundant terms by applying the law of absorption.
5. For every monomial X of the sum thus obtained write down a new one X^* for which the factors are all of those not included in X.

Example: returning once more to the above example:

$$P = (7 + 4)(7 + 5)(7 + 6)(1 + 4)(1 + 5)(1 + 6)(1 + 7)$$

$$= (7 + 456)(1 + 4567)$$

$$= 17 + 1456 + 4567 \tag{14.8}$$

whence the set of compatibles maxima:

$$\{(23456)(237)(123)\}. \tag{14.9}$$

This procedure can be justified as follows: if the set of incompatible pairs is of the form (i, j) it will be seen that to obtain a maximal compatible, it will be *necessary* to remove at least one of the 2 elements of each incompatible pair, since a compatible must not contain a single incompatible pair.

It is clear that this *suffices* for the set obtained to be a compatible. Additionally it will be noted that by removing, not one, but both elements of one or more pairs, a compatible always results — but a compatible which is contained in one of the preceding ones. Consequently, the compatibles obtained by removing a single element from each pair are the largest possible, i.e. the maximal compatibles.

If

$$I = \{i_1 j_1, i_2 j_2, \cdots i_k j_k \ldots, i_m j_m\} \tag{14.10}$$

is the set of incompatible pairs, the set of the complements of the maximal compatibles is obtained by taking 1 element in each pair. There are 2^m ways of making this choice and, at most, 2^m corresponding distinct combinations of states. (Since the same state can exist and can be taken from several pairs.)

It can then be said that this amounts to the determination of the set of monomial terms of the product:

$$(i_1 + j_1)(i_2 + j_2) \cdots (i_k + j_k) \cdots (i_m + j_m) \tag{14.11}$$

14.3 Minimisation of the Number of Internal States

Each term of the product is, in fact, obtained by taking one of the two terms of each factor.

If the above product is considered as a Boolean product, this will have the effect, in the final reduction, of leaving only factors such that:

— any factor is not contained in any other (law of absorption);
— the letters which form a factor are not repeated (law of idempotence).

Remark: Applying the same procedure to the compatible pairs will result in the set of maximal incompatibles, as the reader will easily prove for himself.

Methods other than the three demonstrated above, can also be used for the determination of maximal compatibles (cf. for example [96]).

14.3.9 Determination of a Table T' Covering T, Based on a Closed Set of Compatibles

Having made the choice of a closed set of compatibles, the following procedure is adopted to form the new table;

Take the first compatible, call it C_1.

With the aid of the initial table form the "successor" compatibles obtained by applying E_k to this compatible. Form the output sets associated with the states of this compatible and with the input E_k.

— *States*

Replace each successor $(C_1 E_k)$ by the symbol of the compatible (of the closed set) in which it is contained. Given that certain *states* can belong to several compatibles, it is possible that certain "successor" compatibles may, themselves, be contained in several of the maximal compatibles of the cover (and which appear in the extreme left-hand column of the table under consideration). It is then possible either to take any one of them and inscribe it as the "successor state" under E_k, or to inscribe the set of possible symbols $S_1, S_2, \ldots$

— *Outputs*

The set of outputs associated with E_k and C_1 can have values of two types only: the specified value or the dash (—). When there is a common specified value, this value will be inscribed as the output value for E_k in the new table.

By completing this operation for all input values of E_k the first line of the new table is determined. A second covering compatible is then taken and the above procedure repeated.

14.3.10 Determination of a Minimal Table T'

This consists of the determination of a cover satisfying the conditions 1 to 4. In the general case there is no method other than the successive examination of covers by compatibles increasing in number from a lower bound. The cardinality i of the largest incompatible constitutes such a limit.

If it proves possible easily to determine the minimal number r of compatibles necessary to give a cover for all the states, the lower limit can be taken as the greater of the two numbers i and r.

This choice of a lower limit as large as possible, as well as of the process of successive enumeration of the diverse solutions, can be improved in certain cases, taking account of the peculiarities of the table to be minimised.

Example:

— *Determination of a covering for a minimal table.*

Two of the three maximal compatibles will not suffice to cover the set of states (1, 2, 3, 4, 5, 6, 7). Consequently, neither will any two compatibles (maximal or not) suffice. At least 3 are, therefore, required to cover the set of states. It does then suffice to take the 3 maximal compatibles since it is known that they comprise a closed cover. They can then be given new names with the aid of the following table (Table 14.13):

Table 14.13

(1 2 3)	A
(2 3 4 5 6)	B
(2 3 7)	C

— *Determination of a minimal table.*

Table 14.10 leads to the following table (Table 14.14):

Table 14.14

Q \ E	1	2	3	4
A	(A, B, C), 0	A, 1	—	B, 0
B	(A, B, C), 0	B, 1	B, 1	C, 1
C	(A, B, C), 0	A, 1	B, 1	B, –

14.3 Minimisation of the Number of Internal States

Table 14.15

(a)

E\Q	1	2	3	4
A	A, 0	A, 1	—	B, 0
B	B, 0	B, 1	B, 1	C, 1
C	C, –	A, 1	B, 1	B, –

```
E  = 1 2 4 3 3 2 4 1 4 3
Za = 0 1 0 1 1 1 1 – – –
Zb = 0 1 0 1 1 1 1 – – –
Zc = 0 – – 1 1 1 1 – – –
```
(applicable to each table at A)

(b)

E\Q	1	2	3	4
A	A, 0	A, 1	—	B, 0
B	A, 0	B, 1	B, 1	C, 1
C	A, –	A, 1	B, 1	B, –

(c)

E\Q	1	2	3	4
A	C, 0	A, 1	C, –	B, 0
B	C, 0	B, 1	B, 1	C, 1
C	C, –	B, 1	B, 1	B, –

(d)

E\Q	1	2	3	4
A	A, 0	A, 1	B, 1	B, 0
B	B, 0	B, 1	B, 1	C, 1
C	C, 0	A, 1	B, 1	B, 0

```
E  = 1 2 4 3 3 2 4 1 4 3
Zd = 0 1 0 1 1 1 1 0 0 1
Ze = 0 1 0 1 1 1 1 1 1 1
Zf = 0 1 1 1 1 1 1 1 0 0 1
```
(initial state A)

(e)

E\Q	1	2	3	4
A	A, 0	A, 1	A, 0	B, 0
B	A, 0	B, 1	B, 1	C, 1
C	A, 1	A, 1	B, 1	B, 1

(f)

E\Q	1	2	3	4
A	C, 0	A, 1	C, 1	B, 0
B	C, 0	B, 1	B, 1	C, 1
C	C, 0	A, 1	B, 1	B, 1

430   14. The Simplification of Sequential Networks

Thus there are the following possibilities:

1. The choice of state when there are several possible successors (e.g. A, B, C). Thus are obtained different transition tables.
2. The choice of successor when this is not specified (this is the case for $A3$).
3. The choice of outputs where they are not specified.

In each case this choice can be made in a manner which will simplify the combinational portion of the circuit. For this present example there are:

$$3 \times 3 \times 3 \times 2 \times 2 \times 2 \times 2 = 432$$

possibilities. Note that the output sequences (which will always be defined once the circuit has been designed) can, in general, be different for the same input sequence.

Tables 14.15 a–f illustrate this phenomena.

14.3.11 Another Example of Minimisation

Minimise the table 14.16:

Table 14.16

$Q \backslash E$	1	2	3	4
1	2, 3	—	5, 0	8, 1
2	—	2, 0	6, 0	9, 1
3	—	4, 0	—	—
4	5, –	5, 0	—	10, –
5	5, 3	—	7, –	—
6	—	7, 1	9, 3	—
7	7, 3	6, 1	8, 3	—
8	7, –	6, –	—	12, 0
9	—	—	9, –	13, 0
10	8, 1	9, 2	11, –	—
11	8, 1	8, 2	10, 1	—
12	—	12, 2	13, 1	1, –
13	—	14, –	15, 1	1, –
14	17, 3	—	—	2, –
15	—	—	16, 1	4, 2
16	—	14, 2	—	5, 2
17	17, 3	11, 2	—	1, 2

(1) *Determination of compatible pairs*

We obtain Table 14.17.

14.3 Minimisation of the Number of Internal States

Table 14.17. *Compatible pairs*

	1	2	3	4	5	6	7	8	9	10	11	12	13	14	15	16
2	5,6 8,9															
3	✓	2,4														
4	2,5 8,10	2,5 9,10	4,5													
5	2,5 5,7	6,7	✓	✓												
6	××	××	××	××	7,9											
7	××	××	××	××	5,7 7,8	6,7 8,9										
8	××	××	4,6	5,6 5,7 10,12 ××	5,7	6,7	✓									
9	××	××	✓	10,13 ××	7,9	✓	8,9	12,13								
10	××	××	××	××	××	××	6,9 7,8	9,11								
11	××	××	××	××	××	××	6,8 7,8	9,10	8,9 10,11							
12	××	××,	××	××	7,13 ××	××	××	1,12 6,12 ××	1,13 9,13 ××	9,12 11,13 ××	8,12 10,13 ××					
13	××	××	4,14	1,10 5,14 ××	7,15 ××	××	××	1,12 6,14 ××	1,13 9,15 ××	11,14 11,15 ××	8,14 10,15 ××	12,14 13,15				
14	2,8 2,17 ××	2,9 ××	✓	2,10 5,17 ××	5,17	✓	7,17 ××	2,12 7,17 ××	2,13 ××	××	××	1,2	1,2			
15	××	××	✓	4,10 ××	7,16 ××	××	××	××	××	11,16 ××	10,16 ××	1,4 13,16	1,4 15,16	2,4		
16	××	××	××	××	✓	××	××	××	××	9,14 ××	8,14 ××	1,5 12,14	1,5	2,5	4,5	
17	××	××	××	××	5,17	××	××	××	××	××	××	11,12 ××	11,14 ××	1,2	1,4	1,5 11,14 ××

(2) *Determination of maximal compatibles*

Proceeding as in method I, the following sequences of compatibles are obtained:

~~(15, 16)~~ ~~(15, 17)~~
~~(14, 15, 16)~~ (14, 15, 17)
~~(13, 14, 15, 16)~~
(12, 13, 14, 15, 16)

~~(10, 11)~~
~~(9, 10, 11)~~
(8, 9, 10, 11)
~~(7, 8, 9)~~
~~(6, 7, 8, 9)~~ ~~(6, 14)~~
(5, 6, 7, 8, 9) (5, 6, 14) (5, 14, 16) (5, 14, 17)

~~(4, 5)~~
~~(3, 4, 5)~~ (3, 5, 8, 9) (3, 5, 14) (3, 14, 15)
~~(2, 3, 4, 5)~~
(1, 2, 3, 4, 5)

The non-cancelled terms are the maximal compatibles. There are 11 of these which guarantees the existence of a table of 11 states at most — a number less than that (17) of the table to be minimised.

(3) *Study of minimal covering*

We have the following table (Table 14.18):

Table 14.18

```
1 2 3 4 5
    3  5  8 9
    3  5           14
    3              14 15
       5 6         14
       5           14    16
       5           14       17
       5 6 7 8 9
             8 9 10 11
                   12 13 14 15 16
                         14 15    17
```

The states 1, 2, 4 necessitate at least one compatible.
The states 10, 11 necessitate at least one compatible.
The states 12, 13 necessitate at least one compatible.
The state 17 also needs one compatible

14.3 Minimisation of the Number of Internal States

Thus at least 4 compatibles are necessary. But each of the 4 states or groups of states above belong to a single maximal compatible. Even by covering each of them with this maximal compatible (which of all the cover compatibles is the most extensive and, therefore, most likely to cover other states) the 17 states cannot be covered. Thus at least 5 compatibles are needed. It will be found that this will then cover the 17 states effectively. Taking for example:

    12345
        56789
            89 10 11
                12 13 14 15 16
                    14 15   17

It is easily seen that this collection is indeed *closed*. Renaming the constituent compatibles, in accordance with the following table (Table 14.19):

Table 14.19

1 2 3 4 5	A
5 6 7 8 9	B
8 9 10 11	C
12 13 14 15 16	D
14 15 17	E

Whence the table (Table 14.20)

Table 14.20

$Q \diagdown E$	1	2	3	4
A	$A, 3$	$A, 0$	$B, 0$	$C, 1$
B	$B, 3$	$B, 1$	$B, 3$	$D, 0$
C	$B, 1$	$B, 2$	$C, 1$	$D, 0$
D	$E, 3$	$D, 2$	$D, 1$	$A, 2$
E	$E, 3$	$C, 2$	$D, 1$	$A, 2$

Chapter 15

The Synthesis of Synchronous Sequential Networks

15.1 General

Consideration is to be given to the following problem: given a function which assigns to every sequence of symbols of a first set (input sequences) another sequence belonging to a second set (output sequences), design a sequential network to achieve this mapping.

It will be noted that in order to specify the above function it is necessary to have recourse to a language which, in line with the origin of the problem, is more or less mathematical (e.g. tables, regular expressions). The first problem which presents itself, and which from the theoretical viewpoint, cannot be resolved until such time as the language to be used is rid of all imprecision or ambiguity, is to determine whether or not a sequential network is effectively capable of providing a solution. This is the so-called decision problem. Taking account of the difficulties mentioned above, and linked to the language to be used, the study will be limited to two cases where the problem can effectively be solved.

— the case of the correspondence between two finite sets of sequences;
— the case where the specifications are expressed in the language of regular expressions (cf. § 15.4).

This problem, supposedly solved, the following two methods will be described at a later stage:

15.1.1 The Direct Method

This consists of directly combining certain logic elements (gates) and certain elementary sequential networks taken as basic constituent elements (flip-flops, shift registers, counters etc.) by reasoning applied to their logical and memory behaviour. In numerous cases (e.g. where the memory function to be implemented is obvious) this method will be found to be suitable but, in the general case, it will remain difficult to appreciate whether the resulting design will be economical. However, to the extent that microminiaturisation techniques permit the fabrication of relatively complex assemblies (shift registers, counters etc.) this technique is of interest. Some simple examples will be given at a later stage.

15.1.2 The State Diagram Method

This method can be broken down as follows:

a) the determination of a state diagram,
b) binary coding of the states,
c) determination of the control functions of the flip-flops and of the output functions.

In practice, however, the determination of the minimal network is attempted by observance of the following steps:

a') minimisation of the state diagram,
b') determination of a coding minimizing the combinatorial portions of the network.

As might a priori be expected, and as experience confirms, this break-down of the optimization problem into separate steps aimed at "local" optimization may not result in the absolute minimum. Nevertheless, experience seems to indicate that in many practical cases the limited simplification thus obtained will often closely approach the true minimum and that a more 'global' method, not separating the problem into distinct steps as above, would often produce a refinement of interest but not a radical change from the first solution so far as overall complexity is concerned. Consequently there is a measure of continuity and stability in the solutions which, for lack of precise statistics, is confirmed by experience and thus confers a non-negligible practical value on the above procedure. Moreover, the resolution of the problem, even within this limited scope, remains extremely instructive.

15.2 Direct Method. Examples

(1) *Series circuit for parity check digit.*

A binary word is presented in series at the input to the network.

It is assumed that at every time pulse the network calculates the parity check corresponding to the symbols already passed, without seeking to signal the end of the word, assumed to have been obtained externally.

Evidently a type T flip-flop (elementary counter) will suffice here, it being set back to zero at the start of the word and calculating the residue modulo 2 of the sum of digits received. An initial setting to zero provides a check on imparity.

Hence for (Fig. 15.1):

$$S_k = Q_0 \oplus \sum_{i=1}^{i=k-1} x_i \tag{15.1}$$

(Q_0 being the initial state).

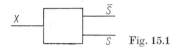

Fig. 15.1

(2) *Series complementation circuit (true complementation)*

The number being presented in series, least significant digits leading, the circuit is to supply, in series and without delay, the true complement. Let:

$$\begin{aligned} X &= x_n, \ldots, x_k, \ldots x_1\, x_0 \\ X' &= x_n', \ldots, x_k', \ldots, x_1'\, x_0' \end{aligned} \tag{15.2}$$

the number to be complemented and its complement respectively. Then:

$$x_k' = \bar{x}_k \oplus \prod_{i=0}^{i=k-1} \bar{x}_i = x_k \oplus 1 \oplus \prod_{i=0}^{i=k-1} \bar{x}_i = x_k \oplus \left(\sum_{i=0}^{i=k-1} x_i \right) \tag{15.3}$$

Putting:

$$\sum_{i=0}^{i=k-1} x_i = Z_k,$$

this gives:

$$\begin{cases} Z_{k+1} = Z_k + x_k \\ x_k' = x_k \oplus Z_k \end{cases} \tag{15.4}$$

A type SR flip-flop, whose input R is maintained at 0, provides the circuit of Fig. 15.2 (initial state $Z = 0$).

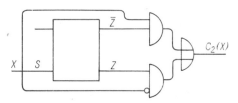

Fig. 15.2. Series complementer.

15.2 Direct Method. Examples

(3) *Series comparator in pure binary.*

The construction of a network which will allow the comparison of 2 binary numbers presented in series, least significant digits leading, is required. The output indicates whether $X - Y \geq 0$ or $X - Y < 0$.

Commencing from the formula (9.73) of § 9.4.7 gives ($S_k = 1$ if $X - Y \geq 0$, and otherwise 0).

$$S_k \oplus S_{k-1} = \left[X_k \overline{Y}_k + \left(X_k \oplus \overline{Y}_k\right) S_{k-1}\right] \oplus S_{k-1}$$
$$= \left[X_k \overline{Y}_k \oplus \left(X_k \oplus \overline{Y}_k\right) S_{k-1}\right] \oplus S_{k-1} \tag{15.5}$$
$$= X_k \overline{Y}_k \oplus (X_k \oplus Y_k) S_{k-1} = S_{k-1} \overline{X}_k Y_k + \overline{S}_{k-1} X_k \overline{Y}_k \tag{15.6}$$

It suffices to present a T type flip-flop with the signal $S_k \oplus S_{k-1}$ (Fig. 15.3a). The output at the instant k gives the result of the comparison for the first $k+1$ figures received.

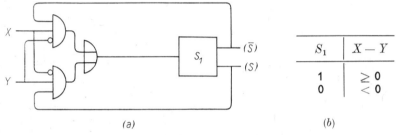

Fig. 15.3. Series comparator.

(4) *Comparator circuit showing, under the same conditions, whether $X = Y$ or $X \neq Y$.*

Here again, the formula 9.70 of § 9.4.7 gives an immediate answer. By taking a T type flip-flop for example:

$$E_k \oplus E_{k-1} = \overline{(X_k \oplus Y_k)} \, E_{k-1} \oplus E_{k-1} = (X_k \oplus Y_k) \, E_{k-1} \tag{15.7}$$

whence the circuit of Figure 15.4.

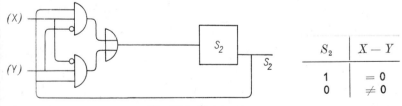

Fig. 15.4. Comparator.

(5) *Comparator circuit, showing, under the same conditions, the 3 cases* $X - Y > 0$, $X - Y = 0$ and $X - Y < 0$.

In this case, in order to code the 3 results (>0, $=$, , <0) at least 2 binary outputs are necessary. It is possible, for example, to combine the two circuits of the above examples: the vector $S_1 S_2$ directly indicating the output (Table 15.5):

Table 15.5

S_1 S_2	Output
0 0	< 0
0 1	$= 0$
1 0	> 0
1 1	—

(The fourth combination is never produced.)

(6) *Pure binary/reflected binary comparator*

Assume it is required to decode a binary number into an analog rotation θ (representing for example a shaft position). A classical method of solving this particular problem is to use an electro-mechanical feedback control system for which the feedback loop includes an encoder. The error term $\mathscr{E}$ is determined by means of a comparator from a number (assumed here to have been supplied by a computer) and from the coded output value of the shaft position. After suitable transformation and conversion (to analog for example) the error signal actuates a motor which positions the output shaft. Here it is required to design the comparator based on the following hypotheses: the number (B) to be converted in analog form is supplied in pure binary and in series with the least significant digits leading; the number (R) is supplied by a Gray encoder, also in series and again with the least significant bits leading. The error signal is to be supplied at the instant of read-out of the last figure (time n) and may take the value $+1$, 0, -1 corresponding respectively to the cases $|B| - |R| > 0$, $= 0$, and < 0. It is assumed that the computer is able to supply the timing signals at the instants -1 and n.

Method: A first method could consist of reducing the 2 numbers to a single code, pure binary for example, and making a comparison by means of a comparator operating on the numbers in pure binary. This rough method would tend to mask a certain number of simplifications and lead to a fairly complex circuit. It will be noted that, although sufficient, this preliminary conversion, which constitutes a possible intermediate step, is by no means necessary. An attempt can be made to integrate the tasks of conversion and comparison in the hope that the overall circuit corresponding to these two functions will be more simple than the sum of two separate circuits for these tasks. This is a very general situation and here, as will be seen, the method 'pays off'. There are two formulae suitable for the conversions:

$$B_k' = \oplus \sum_{i=k}^{i=n} R_i \qquad (15.8)$$

whence

$$B_k' = B'_{k+1} \oplus R_k \qquad (i < n) \qquad (15.9)$$

or

$$B'_{k+1} = B_k' \oplus R_k \qquad (i < n) \qquad (15.10)$$

The difficulty, with respect to the use of this last formula is precisely that, pending a complete exploration of the reflected word R, the value of the lowest weight B_0' is not known.

$$\left(B_0' = \oplus \sum_{i=0}^{i=n} R_i\right).$$

The following strategy will now be tried:

In order to make the comparison, the two values 0 and 1 will be used for B_0' which, so long as the 2 words B and R have not been processed completely, will play the role of a conditional variable for which the verification of the true value will be reserved until the time n, which will be sufficiently early to supply the error signal. From Fig. 15.6a, a':

$$Z_k = \oplus \sum_{i=0}^{i=k-1} R_i \qquad (15.11)$$

$$S_k = \oplus \sum_{i=0}^{i=k} R_i = Z_k \oplus R_k \qquad (15.12)$$

The flip-flop Z is reset to zero at the time T_{-1} the $n+1$ symbols of B or R being applied at the times $0, 1, 2, \ldots n$). The signal at time n serves to sample S at this instant. The outputs Z and $\bar{Z}$ supply two binary words which are the "conditional" equivalents of R in pure binary. Each of these two numbers is compared with B by means of a comparator of the type described in § 15.2(5) (Fig. 15.6b). After the comparison (time 0 to $n-1$) the circuit resolves the ambiguity at time n as a function of S_n ($=B_0'$) (Fig. 15.6c, c'). It will be noted that if T_n were not supplied (a rare case in practice) it would be necessary to integrate a digit counting function within the circuit.

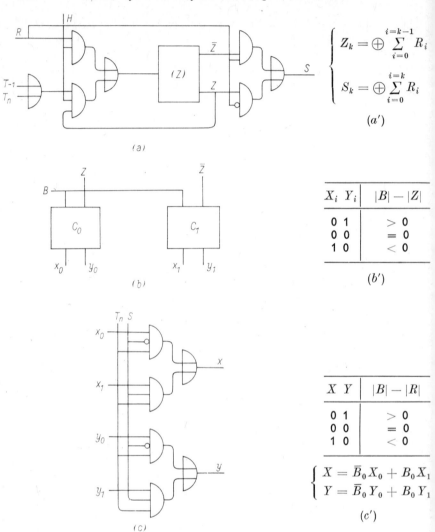

Fig. 15.6. Pure binary/reflected binary comparator.

(7) *Sequence detection*

It is required to construct a circuit having 2 inputs and 1 output, which supplies a 1 at the output every time that the sequence of input combinations formed from the preceding and from the present inputs contains the combinations (10) or (01).

An attempt can be made to construct the required circuit with the aid of gates and flip-flops. As only the preceding clock time $(n-1)$ and the present time (n) come into play, an obvious method consists of storing the 2 digit input $E_1 E_2$ in two flip-flops. At the time n a combinatorial network will make use of the flip-flop outputs B_1 and B_2, where $(B_1 B_2)_n = (E_1 E_2)_{n-1}$ and of the input $E_1 E_2$.

$$S_n = \left[E_1 \bar{E}_2 + \bar{E}_1 E_2 \right]_n + \left[B_1 \bar{B}_2 + \bar{B}_1 B_2 \right]_n \tag{15.13}$$

$$= (E_1 \oplus E_2)_n + (B_1 \oplus B_2)_n \tag{15.14}$$

This expression permits the construction of a circuit having 2 inputs, 1 output and 4 internal states (2 flip-flops). It can, however, be seen that, in fact, S_n depends only on $(B_1 \oplus B_2)_n$, for which a single flip-flop will suffice. In this case:

$$\begin{cases} S_n = (E_1 \oplus E_2)_n \oplus Q_n \\ Q_{n+1} = (E_1 \oplus E_2)_n \end{cases} \tag{15.15}$$

The task of drawing up the corresponding layout is left to the reader.

Note: It is sufficient to store the fact that 01 or 10 has appeared, but it is not necessary to retain which particular combination 01 or 10 has actually been produced.

15.3. State Diagram Method

The initiation of the first step, that of the construction of a state diagram, again depends on the language in which the specifications for the system are formulated. Apart from the regular expressions for which there exist various sythesis algorithms, this examination will be limited to a certain number of problems for which the diagram will be established by direct reasoning based on the specifications. As with the problem of decision, without doubt this method may include the formulations for rather diverse problems. A start will be made with the use of the direct reasoning method, leaving until later in this chapter the description of algorithms based on the use of regular expressions. Certain of the examples chosen have already been examined in § 15.2 by the direct method and consequently, to facilitate the comparison, the construction of the network will be examined (with coding, transition and output functions).

Construction of a state diagram by direct reasoning — Examples:

In the examples which follow use will be made of state diagrams (rather than tables) by reason of their intuitive appeal. In principle, the same results can be obtained from the use of tables.

(0) It is required to construct a synchronous system capable of detecting the illicit combinations (Chap. 3) in the excess 3 code; The figures comprising the words to be examined are presented in series on a single input.

Two slightly different problems are to be examined.

Problem 1

Starting at an "initial time" a sequence limited to 4 binary figures is applied to the circuit placed in an initial state, starting at an initial instant. After passage of the sequence the system must always supply a 1 if there has been a forbidden (illicit) sequence. It is assumed that, after this stage the system can be restored to its original state, by some external action (not described here), in order that it may accept a new sequence.

Set of sequences to be detected: (illicit sequences)

$$\begin{Bmatrix} 0\ 0\ 0\ 0 \\ 0\ 0\ 0\ 1 \\ 0\ 0\ 1\ 0 \\ 1\ 1\ 0\ 1 \\ 1\ 1\ 1\ 0 \\ 1\ 1\ 1\ 1 \end{Bmatrix}$$

The different digits which comprise the word serve to determine the character, ie illicit or licit, of the sequence; moreover they are presented at successive instants. It will be realised that there is a need for a memory store and an attempt will be made to produce the required system with the aid of a Moore sequential circuit. Basically, the circuit must be able to identify the partial sequences which serve to commence an illicit total sequence.

It will suffice (but will be seen to be superfluous) to have a memory capable of storing *all* the sequences of 1, 2, 3 and 4 binary digits; at the end of each of these it will be possible to associate a certain memory '*state*' (coded, for example, in binary with the aid of the sequence under consideration). Thus in, (Fig. 15.7a), have been represented all the sequences in question in the form of the branches of a tree. Each circle contains the sequence which has led to it from the vertex.

The tree thus constructed, and on which appear all the illicit sequences, leads quite naturally to the *state diagram* (Fig. 15.7b). On this diagram the circles represent the memory states. With each memory state is associated an output. The significance of these states can, in each case, be summarised in the following manner:

(1) Initial state, before presentation of the sequence at the input.
(2) The digit 0 has been received. On the basis of this information alone the circuit cannot yet decide, since the following symbols may or may not lead to illicit sequences, depending on their values.

15.3 State Diagram Method

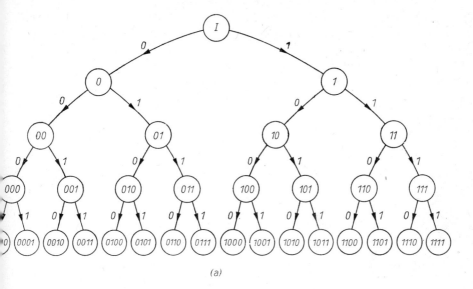

(a)

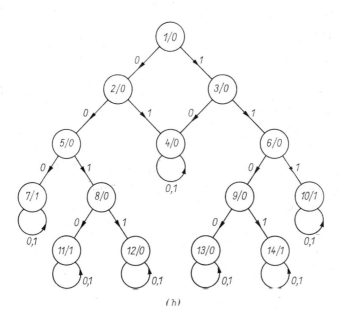

(b)

Fig. 15.7 a) and b).

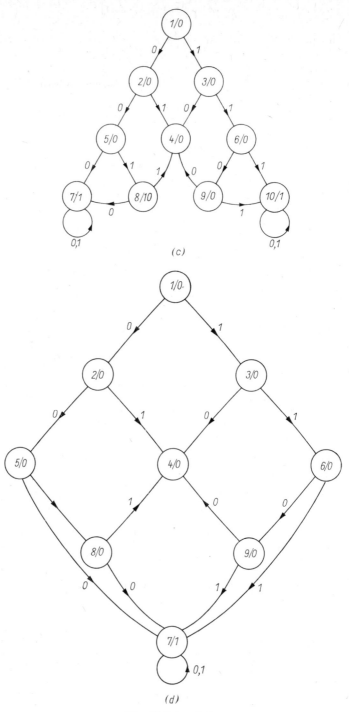

Fig. 15.7 c) and d).

15.3 State Diagram Method

(3) The symbol 1 has been received. Same remarks.
(4) The partial sequences 01 or 10 have been received: it is now certain that the sequence will be licit.
(5) Nothing further can be said and it is necessary to wait.
(6) Nothing further can be said and it is necessary to wait.
(7) Without waiting for the next digit, it is certain that the sequence will be illicit.
(10) Without waiting for the next digit, it is certain that the sequence will be illicit.
(8) 001 has been received. The last digit must be awaited.
(9) 110 has been received. The last digit must be awaited.
(11) The sequence 0010 has been received. It is illicit.
(12) The sequence 0011 has been received. It is licit.
(13) The sequence 1100 has been received. It is licit.
(14) The sequence 1101 has been received. It is illicit.

This has produced a circuit with 14 memory states. But, by hypothesis:

— there is no necessity to indicate after what time a sequence has been declared to be illicit or not;
— there is no necessity to distinguish between the different sequences of either of the two categories of sequences (licit, illicit).

Thus the diagram of states (Fig. 15.7b, c) can be simplified. In the version represented by Fig. 15.7d the system possesses 9 internal states.

Problem 2

Commencing with an initial time and an initial state, sequences of 4 consecutive numbers are applied without a separating interval. It is required to detect the illicit combinations present among the successive groups of 4 figures. The signal does not appear until the end of these groups, at the moment when the fourth figure is applied.

This time it is necessary to design a machine where the output depends on the input as well as on the internal states at the same clock time.

Proceeding in a similar manner the system of Fig. 15.8a is obtained. This, too, can be simplified (Fig. 15.8b)

(1) *Sequence detector*

Problem: To synthesise a circuit possessing the following properties: it possesses 2 inputs $E_1 E_2$, 1 output. The input E_1 receives the information and when $E_2 = 0$, the system provides an output of value 1 when

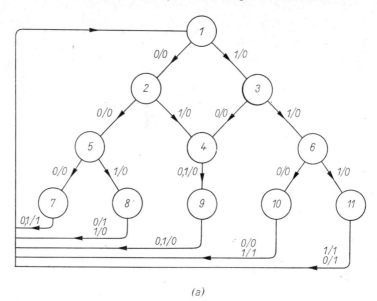

Fig. 15.8

15.3 State Diagram Method

the sequence formed by the two preceding symbols and by the present digit is one of the sequences 000, 101, 111 (at times $n-2$, $n-1$, n). The input E_2 serves to return the system to its initial state: this occurs when $E_2 = 1$.

It is necessary to introduce an initial state I. It is also necessary to consider the states of the memory store with which are associated the sequences formed by the last digit received, then by the last two, which are coded temporarily by the binary word representing the sequence. Thus the state 00 signifies that the last two symbols received were 00, etc. The states diagram can, therefore, be constructed as follows: before the start of the sequence the system is in the state I. If the first symbol is 0, it passes into the state 0, if it is 1, it passes into the state 1. After receipt of the second symbol, the system is found to be in one of the states 00, 01, 10, and 11, obtained in a similar manner (Fig. 15.9a). Once there, the system continues its evolution among the 4 states until $E_2 = 0$. The figures represent the diagrams corresponding to $E_2 = 0$ and $E_2 = 1$. From Fig. 15.9a it will be seen that the diagram can be considered as being formed of 2 parts separated on the figure by a dotted line and corresponding respectively to the transient and permanent regimes. This first diagram $(E_2 = 0)$ is an example of an irreversible system (vis-à-vis E_1): no sequence applied to E_1 will cause a return to

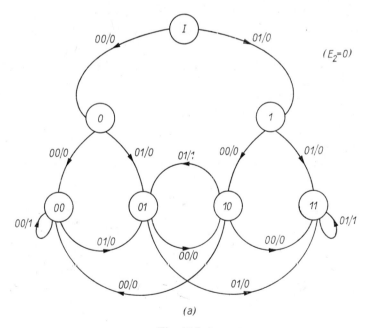

(a)

Fig. 15.9 a).

448 15. The Synthesis of Synchronous Sequential Networks

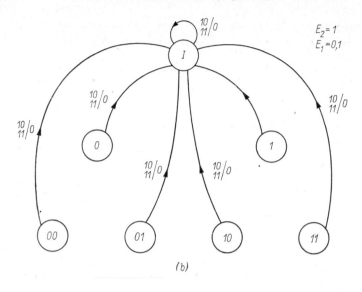

(b)

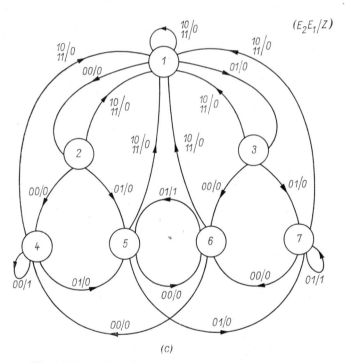

(c)

Fig. 15.9 b) and c). A sequence detector with zero re-set

15.3 State Diagram Method

the initial state I. By numbering the states and uniting the two partial diagrams is obtained the Fig. 15.9c, with which corresponds a reduced table (Table 15.11):

Table 15.10

Q \ E	0 0	0 1	1 1	1 0
1	2, 0	3, 0	1, 0	1, 0
2	4, 0	5, 0	1, 0	1, 0
3	6, 0	7, 0	1, 0	1, 0
4	4, 1	5, 0	1, 0	1, 0
5	6, 0	7, 0	1, 0	1, 0
6	4, 0	5, 1	1, 0	1, 0
7	6, 0	7, 1	1, 0	1, 0

Table 15.11

Q \ E	0 0	0 1	1 1	1 0
1	2, 0	3, 0	1, 0	1, 0
2	4, 0	3, 0	1, 0	1, 0
3	6, 0	7, 0	1, 0	1, 0
4	4, 1	3, 0	1, 0	1, 0
6	4, 0	3, 1	1, 0	1, 0
7	6, 0	7, 1	1, 0	1, 0

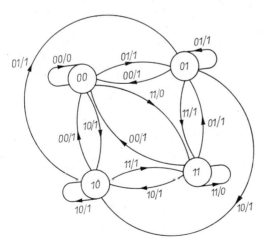

Fig. 15.12

(2) Return to the example of § 15.2.7 and, by this method, consider a diagram with 4 internal states, associated with 4 inputs whose values are liable to be placed in the memory store: an arrow leading to a state as well as the state itself carry the same symbol (i.e. the input is memorized) (Fig. 15.12). Whence Table 15.13:

Table 15.13

Q \ E	0 0	0 1	1 0	1 1
0 0	0 0/0	0.1/1	1 0/1	1 1/0
0 1	0 0/1	0 1/1	1 0/1	1 1/1
1 0	0 0/1	0 1/1	1 0/1	1 1/1
1 1	0 0/1	0 1/1	1 0/1	1 1/0

The equivalence of the states 01 and 10 is verified, as is that for the states 00 and 01. Hence the reduced table (Table 15.14):

Table 15.14

Q \ E	0 0	0 1	1 0	1 1
0 0	0 0/0	0 1/1	1 0/1	1 1/0
0 1	0 0/1	0 1/1	1 0/1	1 1/1

A single binary variable suffices to code the states 00 and 01: according to the code employed one of the 2 tables (a) and (b) of Table 15.15 is obtained:

Table 15.15

Q \ E	0 0	0 1	1 0	1 1	Q \ E	0 0	0 1	1 0	1 1
0	0/0	1/1	1/1	0/0	1	1/0	0/1	0/1	1/0
1	0/1	1/1	1/1	0/1	0	1/1	0/1	0/1	1/1

From Table 15.15a are deduced:

$$\begin{cases} Q_{n+1} = \bar{Q}_n(E_1\bar{E}_2 + \bar{E}_1 E_2)_n + Q_n(E_1\bar{E}_2 + \bar{E}_1 E_2)_n \\ \qquad = (\bar{Q}_n + Q_n)(E_1 \oplus E_2)_n = (E_1 \oplus E_2)_n \qquad (15.16) \\ S_n \;\; = \bar{Q}_n(E_1\bar{E}_2 + \bar{E}_1 E_2)_n + Q_n = (E_1\bar{E}_2 + \bar{E}_1 E_2)_n + Q_n \\ \qquad = (E_1 \oplus E_2)_n + Q_n \qquad (15.17) \end{cases}$$

which are the expressions obtained above by the direct method.

15.3 State Diagram Method

(3) *Binary comparator*

This is the example of § 15.2.4.

Evidently 2 memory states will suffice ("equal"), ("unequal"). Also available are Figure 15.16 and Table 15.17.

State 1 signifies: that up to the rank k there has been no equality.

State 2 signifies: that up to rank k inequality has been detected at least once and that the numbers are certainly unequal.

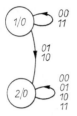

Fig. 15.16

Table 15.17

E Q	0 0	0 1	1 0	1 1
1	1	2	2	1
2	2	2	2	2

Coding of internal states:

Table 15.18 is obtained by use of the coding $1 \to 1$ and $2 \to 0$ and the relations:

$$\begin{cases} Q_{n+1} = Q_n \overline{(E_1 \oplus E_2)_n} \\ S_n = Q_n \end{cases} \quad (15.18)$$

Table 15.18

E Q	0 0	0 1	1 0	1 1
1	1	0	0	1
0	0	0	0	0

(a)

Q	S
0	0
1	1

(b)

(4) *Comparator*

This is the example of § 15.2.5. The reasoning is as follows: given 2 numbers of the form:

$$\begin{cases} X(k) = 1\,X(k-1) \\ Y(k) = 0\,Y(k-1) \end{cases}$$

where $X(k)$, $Y(k)$ denote the words truncated to the $k+1$ first figures, then:

$$X(k) - Y(k) > 0. \quad (15.19)$$

Indeed:

$$\begin{cases} X(k) = 2^k + x_{k-1} & 0 \leq x_{k-1} \leq 2^k - 1 \\ Y(k) = 0 + y_{k-1} & 0 \leq y_{k-1} \leq 2^k - 1 \end{cases} \qquad (15.20)$$

whence:

$$X(k) - Y(k) = 2^k + (x_{k-1} - y_{k-1}) \qquad (15.21)$$

but:

$$|x_{k-1} - y_{k-1}| \leq 2^k - 1 \qquad (15.22)$$

Consequently:

$$X(k) - Y(k) \geq 2^k - (2^k - 1) = 1 > 0 \qquad (15.23)$$

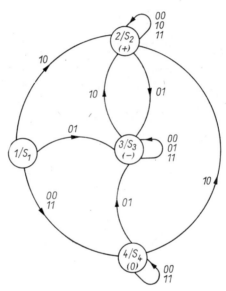

Fig. 15.19

Consideration can be given to a diagram (Fig. 15.19) having an initial state 1 (the state prior to the start of the words X and Y), from which it is possible to attain 3 other states (2, 3, 4) associated with the following 3 circumstances:

$$2(+): X(k) - Y(k) > 0$$
$$3(-): X(k) - Y(k) < 0$$
$$4(0): X(k) - Y(k) = 0$$

15.3 State Diagram Method

It will be noted that the state 1 can only be attained through an external action not represented here. Table 15.20 can be obtained by associating with each state an output which serves to identify it (Moore network):

Table 15.20

Q \ E	0 0	0 1	1 0	1 1		Q	S
1	4	3	2	4		1	S_1
2	2	3	2	2		2	S_2
3	3	3	2	3		3	S_3
4	4	3	2	4		4	S_4

Simplification: If it be supposed that $S_1 = S_4$, the states 1 and 4 are equivalent and can be merged into a unique state. This can be interpreted as follows: the state 1 plays the rôle of a state attained when the preceding symbols were zero for the two words to be compared. Hence the (reduced) table which follows (Table 15.21):

Table 15.21

Q \ E	0 0	0 1	1 0	1 1
1	1	3	2	1
2	2	3	2	2
3	3	3	2	3

Binary coding of states.

Two flip-flops suffice; Table 15.22 constitutes an example of the tables that can be obtained.

Table 15.22

Q \ E	0 0	0 1	1 0	1 1
0 0	0 0	1 0	0 1	0 0
0 1	0 1	1 0	0 1	0 1
1 0	1 0	1 0	0 1	1 0

Q_1 and Q_2 being states of the flip-flops, this gives for example:

$$\begin{cases} (Q_1)_{n+1} = \left[\bar{E}_1 E_2 + Q_1(\bar{E}_1 + E_2)\right]_n \\ (Q_2)_{n+1} = \left[E_1 \bar{E}_2 + Q_2(E_1 + \bar{E}_2)\right]_n \end{cases} \quad (15.24)$$

and it can be said, for example, that:

$$S_n = (S_1 S_2)_n = (Q_1 Q_2)_n \quad (15.25)$$

Remark: The state diagram of Figure 15.23 can be obtained by assuming that the numbers proposed are presented with the least significant digits leading. In this case comparison stops as soon as the first inequality between two figures of the same rank is revealed. By contrast, in the other case it is not known which is the longest word, hence the procedure indicated.

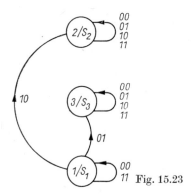

Fig. 15.23

(5) *Comparison of two numbers to within a tolerance of one digit of least significant weight.*

The circuit receives the first two numbers X and Y in series, least significant digits leading, and must be capable of interrogation at the end of a word to say whether $|X - Y| \leq 1$, or $|X - Y| > 1$. The following reasoning can be adopted: there are several cases:

1. X terminates in 1. X is, therefore, of the form:

$$X = \overbrace{X'011 \cdots 11}^{k} = X''1.$$

so that:

$$\begin{cases} X + 1 = \overbrace{X'100 \cdots 00}^{k} \\ X - 1 = X'011 \cdots 10 = X''0 \end{cases} \quad (15.26)$$

15.3 State Diagram Method

2. X terminates in 0, giving:

$$\begin{cases} X = X'\overbrace{100 \cdots 00}^{m} = X''0 \\ X + 1 = X'100 \cdots 01 = X''1 \end{cases} \quad (15.27)$$

Finally $X - Y = 1$ if an input sequence (of 2 symbols XY) of one of the following types has been received:

$$\begin{cases} X'011 \cdots 1 \\ X'100 \cdots 0 \end{cases}$$

or

$$\begin{cases} X'1 \\ X'0 \end{cases}$$

or

$$\begin{cases} X'100 \cdots 00 \\ X'011 \cdots 11 \end{cases}$$

or

$$\begin{cases} X'0 \\ X'1 \end{cases}$$

whence the Table 15.24:

Table 15.24

Q \ E	00	01	10	11	Z
1	5	7	2	5	–
2	5	4	3	5	–
3	6	4	3	6	–
4	5	6	6	5	–
5	5	6	6	5	0
6	6	6	6	6	1
7	5	8	9	5	–
8	6	8	9	6	–
9	5	6	6	5	–

The signification of states:

1: initial state;
2: the input 10 has been received at time 1;
3: the input 01 has been received two or more times in succession, from the start;

4: the input sequence corresponds to $|X - Y| = 1$ up to the present, but in order that this be true, the sequence to follow must be formed of 00 or 11;
5: $|X - Y| \leq 1$ will be obtained if the remainder of the sequence is formed of 00 or 11;
6: $|X - Y| > 1$;
7 & 8: have significations analogous to those of 2 and 3 by replacement of 10 by 01 and vice-versa;
9: analogous to 4.

Simplification: 4 and 9 are equivalent, as are 4 and 5. Whence a reduction of Table 15.24 which is left to the reader by way of an exercise.

15.4 Diagrams Associated with a Regular Expression

15.4.1 Development of a Regular Expression from a Base

It has been shown (Chapter 13) that with every state diagram for an automaton having one binary output, can be associated a system of linear equations in the algebra of regular expressions. It has been shown how the system can be solved and how the expression E_i, associated with an initial state q_i can be determined. For example, by the use of the Arden method this resolution can be achieved in a finite number of steps involving a finite number of operations in the algebra of regular expressions. Thus with every state q_i of a single-output automaton, can be associated an expression E_i, which is therefore regular and which denotes the set of input sequences which give rise to the production of a 1 at the output when the system is initially placed in the state q_i. In the more general case of an automaton having n outputs, for each state q_i and for each output Z_r, the above operation can be repeated and each state be associated with an orderly sequence of regular expressions E_i^r, the expression E_i^r being the set of input sequences which cause the value 1 to appear as the output Z_r.

This ordered set of regular expressions $\{E_i^r\}$ $(r = 1, \ldots n)$ is known as the *regular vector* associated with q_1. A given input sequence S may or may not belong to each E_i^r. For every E_i^r to which S does belong, $Z_r = 1$ and, conversely, for every E_i^r to which S does not belong, $Z_r = 0$.

Summarizing, every event which can be represented in a 1 binary output automaton is regular and the Arden procedure supplies a method by which its algebraic expression can be obtained, and a proof of its regular character at one and the same time.

More generally, it supplies a description of the performance of an (≥ 1) output circuit with the aid of n regular expressions.

15.4 Diagrams Associated with a Regular Expression

Consider now the reverse problem: for a given set n of regular expressions E_i^r ($n \geq 1$), does there exist an n-output automaton such that the output Z_r takes the value 1 for the sequences of E_i^r (and for these alone) for every r ($r = 1, ..., n$) and, if so, construct the states diagram. To simplify the presentation a study will first be made of a single output circuit.

Consider then a regular expression E, with an alphabet A_0 having $k = 2^n$ input symbols. Does there exist a finite automaton to represent this event? It can be seen that one will exist if it is possible to find a finite set of m regular expressions E_i ($i = 1, 2, ..., m$) such that:

$$E_i = a_1 E_{i_1} \cup a_2 E_{i_2} \cup \cdots \cup a_j E_{i_j} \cup \cdots \cup a_k E_{i_k} \cup \delta(E_i) \quad (15.28)$$
$$(i = 1, ..., m), \quad (j = 1, ..., k)$$

E being equal to one of the E_i, it will be assumed to be E_1 in the text which follows. Indeed, it is then possible to obtain a Moore or a Mealy automaton, representing $E (E = E_1)$ in the following manner [141]:

Moore diagram

The expressions E_i are taken to be the "states".

The successor to a state E_1 is the state E_{ij}, the output associated with the state E_i is 1 if $\delta(E_i) = \lambda$ and 0 if $\delta(E_i) = \varnothing$.

Thus this automaton possesses k inputs, m internal states and one output. If it is assumed to be placed initially in the *state* E_i, it adequately represents the event E_i. Indeed let us apply to this initial state E_i a sequence S of length l: $S = a_{i_1}, a_{i_2}, ... a_{i_l}$. If a sequence $a_{i_1}, ... a_{i_k}$ brings the circuit to the state E_u, then the sequence $a_{i_1}, ... a_{i_k}, a_{i_{k+1}}$ will put it into the state E_v and will give:

$$D_{a_{i_{k+1}}}(E_u) = E_v \quad (15.29)$$

Whence, step-by-step:

$$D_S(E_i) = E_i s \quad (15.30)$$

($E_i S$ denotes the successor of E_i by S). Thus, from the initial state E_i, the sequence S will take the state to $D_S E_i$. But, if a sequence S belongs to E_i this is equivalent to $\lambda \in D_S S_i$, i.e.

$$\delta(D_S E_i) = \lambda, \quad Z(D_S E_i) = 1.$$

Conversely, if $S \notin E_i$, $\lambda \notin D_S S_i$

$$\delta(D_S S_i) = \varnothing, \quad Z = 0.$$

Thus the automaton placed in the state E_i will accept the sequences of E_i and these only. Thus it adequately represents the event E_i.

Mealy diagram

Here again E_i is taken to represent the states. The successor of E_i through a_j is E_{i_j}, the output associated with the pair (E_i, a_j) is 1 if E_{i_j} contains $\lambda \bigl(\delta(E_{i_j}) = \lambda\bigr)$ and 0 in the contrary case $\bigl(\delta(E_{i_j}) = \varnothing\bigr)$.

With this modification, it will be seen that by use of the above reasoning, the automaton placed initially in the state E_i, also represents the event E_i.

Reverting to the equations (15.28) Spivak [141] states that the *finite* set of expressions E_i constitutes a *base* $\{E_i\}$ and that the equations (15.28) represent the development of E on this base $\{E_i\}$. It has just been shown that if such a base can be found, it is possible to derive from it a Mealy or a Moore diagram to represent E.

These diagrams are, however, not necessarily minimal. In order that they should be so it is necessary that the E_i of which the base is composed satisfy certain conditions to be described later. First it must be shown that every regular expression E can be decomposed in accordance with the base. For this, it will be recalled, a regular expression E is constructed by a *finite* number of applications of the operations $\cup$, $\cdot$, *, from alphabetical symbols, from λ and from $\varnothing$. It will now be shown that by recurrence the regular operators maintain the *finite character* of the base.[1]

(1) *Events* $\varnothing$, λ *and* $a_i (i = 1, 2, \ldots, k)$.

Having:

a) $\qquad \varnothing = \delta(\varnothing) \cup a_1 \varnothing \cup a_2 \varnothing \cdots \cup a_k \varnothing$ (15.31)

Thus $\varnothing$ is a base for $\varnothing$.

b) $\begin{cases} \lambda = \delta(\lambda) \cup a_1 \varnothing \cup a_2 \varnothing \cup \cdots \cup a_k \varnothing & (15.32) \\ \varnothing = \delta(\varnothing) \cup a_1 \varnothing \cup a_2 \varnothing \cup \cdots \cup a_k \varnothing & (15.33) \end{cases}$

The set $\{\lambda, \varnothing\}$ therefore constitutes a base for the regular expression λ.

c) $\qquad\qquad\qquad E = a_i.$

$a_i = \delta(a_i) \cup a_1 \varnothing \cup a_2 \varnothing \cdots \cup a_{i-1} \varnothing \cup a_i \lambda \cup a_{i+1} \varnothing \cdots a_k \varnothing$ (15.34)

λ and $\varnothing$ are still expressed by (15.32) and (15.33)

Thus the set $(a_i, \lambda, \varnothing)$ constitutes a base for a_i.

---

[1] Cf. *also* A. *Salomaa:* Two Complete Axiom systems for the Algebra of Regular Events, Journal of the Association for Computing Machinery, Vol 13; N° 1. pp 158–169 (January 1966).

15.4 Diagrams Associated with a Regular Expression

(2) Union $E \cup F$.

For two given regular expressions E and F which are developed on the bases $\{E_i\}$ and $\{F_j\}$ ($i = 1, \ldots, m_E$, $j = 1, \ldots, m_F$) with $E = E_1$, $F = F_1$ the expression $E \cup F$ can be developed on the base $E_i \cup F_j$. From:

$$E_i = a_1 E_{i_1} \cup a_2 E_{i_2} \cup \cdots \cup a_k E_{i_k} \cup \delta(E_i) \qquad (15.35)$$

and

$$F_j = a_1 F_{j_1} \cup a_2 F_{j_2} \cup \cdots \cup a_k F_{j_k} \cup \delta(F_j) \qquad (15.36)$$

it can be deduced that:

$$E_i \cup F_j = a_1(E_{i_1} \cup F_{j_1}) \cup a_2(E_{i_2} \cup F_{j_2}) \cup \cdots \cup a_k(E_{i_k} \cup F_{j_k}) \cup \delta(E_i \cup F_j) \qquad (15.37)$$

which properly constitutes the system of equations for a base, with $E_1 \cup F_1 = E \cup F$. It will be noted that the new base has, at most, $m_E \times m_F$ distinct elements.

(3) Event $\bar{E}$

For a given expression E decomposable with respect to a base E_i ($i = 1, \ldots, m$), the expression $\bar{E}$ is decomposable with respect to a base $\bar{E}_i$. In effect:

$$\begin{aligned} E_i &= a_1 E_{i_1} \cup a_2 E_{i_2} \cup \cdots \cup a_k E_{i_k} \cup \delta(E_i) \\ E_1 &= E; \quad (i = 1, \ldots, m). \end{aligned} \qquad (15.38)$$

Consider next the following expression K_i:

$$K_i = a_1 \bar{E}_{i_1} \cup a_2 \bar{E}_{i_2} \cup \cdots a_k \bar{E}_{i_k} \cup \left(\delta \bar{E}_i\right) \qquad (15.39)$$

then:

$$\begin{aligned} E_i \cup K_i &= a_1\left(E_{i_1} \cup \bar{E}_{i_1}\right) \cup \cdots \cup a_k\left(E_{i_k} \cup \bar{E}_{i_k}\right) \cup \delta\left(E_i \cup \bar{E}_i\right) \\ &= a_1 i^* \cup \cdots \cup a_k i^* \cup \delta(i^*) \\ &= (a_1 \cup a_2 \cup \cdots \cup a_k) i^* \cup \lambda \\ &= i i^* \cup \lambda = i^* \end{aligned} \qquad (15.40)$$

and, on the other hand:

$$\begin{aligned} E_i \cap K_i &= a_1\left(E_{i_1} \cap \bar{E}_{i_1}\right) \cup \cdots \cup a_k\left(E_{i_k} \cup \bar{E}_{i_k}\right) \cup \delta\left(E_i \cap \bar{E}_i\right) \\ &= a_1 \varnothing \cup a_2 \varnothing \cup \cdots \cup a_k \varnothing \cup \varnothing = \varnothing \end{aligned}$$

Therefore:
$$K_i = \bar{E}_i \quad (i = 1, \ldots, m) \tag{15.41}$$

The new base has, therefore, at most a number $m_{\bar{E}}$ of elements equal to m_E.

(4) *Arbitrary Boolean expression* $B(E, F)$

From the previous two operations, any Boolean expression $B(E, F)$ can be constructed in a finite number of steps. This expression will, itself, be decomposable with respect to a finite base.

(5) *Product* $E \cdot F$

Suppose E and F are decomposable with respect to the bases $\{E_j\}$ and $\{F_j\}$ characterised by the equations (15.35, 15.36). Consider then the set of terms of the form:

$$K_{ij} = E_i F \cup \left(\bigcup_{l=1}^{l=m_F} \alpha_l F_l \right) \tag{15.42}$$

where α_l takes the value λ or $\varnothing$.

If $\delta(E_i) = \varnothing$, then:

$$K_{i_\alpha} = a_1 \left(E_{i_1} F \cup \bigcup_{l=1}^{l=m_F} \alpha_l F_{l_1} \right) \cup \cdots \cup a_k \left(E_{i_k} F \cup \bigcup_{l=1}^{l=m_F} \alpha_l F_{l_k} \right) \cup \delta\left(K_{i_\alpha}\right) \tag{15.43}$$

If $\delta(E_i) = \lambda$, then:

$$K_{i_\alpha} = a_1 \left(E_{i_1} F \cup F_{1_1} \cup \bigcup_{l=1}^{l=m_F} \alpha_l F_{l_1} \right) \cup \cdots \cup a_k \left(E_{i_k} F \cup F_{1_k} \cup \bigcup_{l=1}^{l=m_F} \alpha_l F_{l_k} \right)$$
$$\cup \delta\left(K_{i_\alpha}\right) \tag{15.44}$$

The expressions thus form a base for which the number of distinct terms is at most:
$$m_E 2^{m_F}$$

(6) *Iteration* E^*

We have:
$$E = E_1 = a_1 E_{1_1} \cup a_2 E_{1_2} \cup \cdots \cup a_k E_{1_k} \cup \delta(E)$$
$$= a_1 D_{a1} E \cup a_2 D_{a2} E \cup \cdots \cup a_k D_{a_k} E \cup \delta(E) \tag{15.45}$$
$$E^* = a_1 D_{a1} E^* \cup a_2 D_{a2} E^* \cup \cdots \cup a_k D_{a_k} E^* \cup \lambda$$
$$= \left(a_1 D_{a1} E \cup a_2 D_{a2} E \cup \cdots \cup a_k D_{a_k} E \right) E^* \cup \lambda \tag{15.46}$$

15.4 Diagrams Associated with a Regular Expression

Next consider the set of expressions of form $K_\alpha E^*$, with

$$K_\alpha = \bigcup_{l=1}^{l=m_F} \alpha_l E_l, \quad (\alpha_l = \lambda \text{ ou } \varnothing) \tag{15.47}$$

If $\delta(K_\alpha) = \varnothing$, then

$$K_\alpha E^* = a_1 \left(\bigcup_{l=1}^{l=m_E} \alpha_l E_{l_1} \right) E^* \cup \cdots \cup a_k \left(\bigcup_{l=1}^{l=m_E} \alpha_l E_{l_k} \right) E^* \tag{15.48}$$

If $\delta(K_\alpha) = \lambda$, then

$$K_\alpha E^* = a_1 \left(\bigcup_{l=1}^{l=m_E} \alpha_l E_{l_1} \cup E_{1_1} \right) E^* \cup \cdots \cup a_k \left(\bigcup_{l=1}^{l=m_E} \alpha_l E_{l_k} \cup E_{1_k} \right) E^* \cup \lambda \tag{15.49}$$

The expressions $K_\alpha E^*$ thus constitute a base having at most $2^{m_E} + 1$ distinct elements ($K_0 E^* = E^*$).

Consider then the regular expressions formed with the aid of n applications of the regular operations and assumed to be capable of development with respect to a base. Every regular operation on one or two expressions of this type supplies an expression with $n + 1$ operations which can still be developed with respect to a base. This is true for the events $\varnothing$, λ and a_i ($i = 1, 2, \ldots k$). It is therefore true, step-by-step, for every regular expression. Consequently, with every regular expression can be associated a Moore or a Mealy diagram of states.

Finally, every regular expression is *decomposable* for a finite base and can, therefore, be represented by a finite automaton.

It can be said that the expressions E_i on a base do not a priori satisfy any particular assumption, except for being regular, of finite number, and for satisfying equations (15.28). In practice two more restrictive types of assumptions can often be made:

a) $E_i \neq E_j$ when $i \neq j$ \hfill (15.50)

then, the corresponding Moore diagram is minimal.

Indeed, the events represented by the automaton initially in the *state* E_i or E_j are the *events* E_i or E_j respectively which are assumed to be different. The states E_i and E_j are therefore not equivalent and cannot be confused. This being true for every pair of states, the diagram is minimal.

b) $E \neq E_j$ and $E_i \neq E_j \cup \lambda$ and $E_j \neq E_i \cup \lambda$ when $i \neq j$. (15.51)

If one of these conditions has not been satisfied, then:

$$D_a(E_i) = D_a(E_j)$$

i.e. all the future states obtained from E_i and E_j are the same, and furthermore, the transitions from E_i to $D_a(E_i)$ and E_j to $D_a(E_j)$ occur for every input a and, during the application of a, produce the same output, which signifies that the two states E_i and E_j are equivalent.

Conversely, if the three conditions of (15.51) are satisfied, this means that there is at least one sequence of length ≥ 1 which belongs to one but not to the other. If $s \in E_i$ and $\notin E_j$, for example,

$$\begin{cases} \lambda \in D_s E_i \\ \lambda \in D_s \bar{E}_j = \overline{D_s E_j} \end{cases} \qquad (15.52)$$

In other words, the application of S to E_i or E_j produces the output 1 in one case and 0 in the other, and consequently E_i and E_j are not equivalent.

Summarizing: 1. for a Moore diagram to be minimal it is necessary and sufficient to associate a single states with *equal* expressions.

2. for a Mealy diagram to be minimal it is necessary and sufficient to attribute a single state to equal expressions, or to those which differ only by λ — ("within λ").

c) The expressions E_i are not similar.

(Two expressions are said to be similar [23] if they can be reduced the one to the other by the application of the rules 13.22, 23, 26, 27, 28, 29 only. Two expressions which are similar are necessarily equal).

There is no pair E_i, E_j such that E_i and E_j are similar. Although this restriction is weaker than the preceding restrictions, it can be seen that the bases thus obtained remain finite.

Consider n operations on the regular expressions and assume that they can be represented by such and such a base. If an expression be formed on $n + 1$ operations, it is known that a new base for the expression can be obtained by application of the rules given above. This base is finite and if the relations of idempotence, associativity and commutativity of the operation $\cup$ are applied, the number of elements of the base which turn out to be dissimilar can only be less than the number of elements indicated above, and therefore finite.

Moreover, for the expressions λ, $\varnothing$, a_i, an examination of the formulae reveals that a finite base can be found for which the elements are not similar. Consequently, step-by-step, this will be true for every expression: a base can always be found such that the elements are dissimilar. Thus this condition maintains the finite character of the base obtained.

15.4 Diagrams Associated with a Regular Expression

This property is of great importance: to be able to recognise that two expressions are similar is very easy, and if recognition of equality of expressions be limited to that case only, the diagram obtained will remain finite (although, in general, not minimal).

15.4.2 Interpretation in Terms of Derivatives

Assume that $E = E_1$. It has already been shown that the application of a sequence S of length ≥ 0 from the *state* E leads to a state represented by $D_S E$. This state is equal to an expression E_i of base E.

From this, then, the following interpretation can be deduced:

1. Every E_{ij} is equal to an E_i of the base.
2. Since the E_i are of *finite number*, the distinct derivatives of a regular expression are of *finite number*.

Thus, although there is an *infinity* of sequences S with respect to which the derivatives $D_S E$ can be calculated, there is only a *finite* number of distinct derivatives, since this number is necessarily less than that of the distinct expressions in the base which, moreover, always exists for a given regular expression E. The finite set of expressions denoted by $D_S E$ can, therefore, only be separated into a finite number of classes.

It is seen that with each class is associated a certain *type* of derivative in the sense that, knowing one derivative of a class, it can automatically be deduced that the others, where they exist, possess very precise properties (they are equal to this first derivative, or similar, according to the case under consideration). Consequently, it is convenient to write each class with a derivative (and one only) taken from within this class. This is known as the *characteristic derivative* of the class in question. For every derivative $D_S E$ the class to which it belongs or the characteristic derivative with which it can be associated can be determined through the relation (equality, similarity) under consideration.

Two given derivatives $D_S E$ and $D_t E$ are said to be *equivalent* (i.e. equal, similar, as the case may be) when they are associated with the same *characteristic derivative*, and to be *distinct* when they are associated with two different characteristic derivatives.

Expressed in terms of derivatives, the results obtained above can be summarized thus:

1. The determination of the derivatives $D_s E$ will only yield a finite number of distinct expressions, if one or other (at choice) of the following two rules be applied:

Rule 1. Two derivatives are not considered to be distinct when they are similar.

Rule 2. Two derivatives are not considered to be distinct when they are equal (i.e. represent the same set).

2. The derivatives thus determined satisfy a system of equations of the type (15.28) from which a Mealy or a Moore (at choice) diagram of states can be deduced.

3. The states diagrams thus obtained are not, in general, minimal. They are, however, certainly minimal in the following two cases:

   3.1 Where application of rule 2 has established a Moore diagram by associating an expression with a state.

   3.2 After the application of rule 2, each time that two of the expressions obtained being equal (or differing at most by λ) have been associated with a single state.

This leads to the algorithms of synthesis which will be stated for the 2 cases (Moore machine, Mealy machine).

Firstly, in order to simplify the language, the following convention will be adopted: each time that is obtained a derivative to be compared with a number of derivatives already determined (and thus, in the last resort, to be to be compared with a certain number of *characteristic* derivatives) it will be said to be *'significant'* if it can be associated with a new characteristic derivative, and only in this case.

15.4.3 Algorithm for the Synthesis of a Moore Machine [23]

1. Designate as q_λ the initial state associated with the derivative $D_\lambda(R) = R$.

2. Compute the derivatives $D_S(R)$ relative to the sequences of length $1, 2, \ldots, k$ etc., and, for each of these derivatives, determine whether or not they are significant. Each time a new characteristic derivative is thus obtained introduce a new internal state. If S is the sequence associated with this new derivative with $S = S'x$, where x is the symbol such that $D_{S'}R$ was the characteristic derivative and $D_{S'x}$ is the new one, introduce a transition from the state associated with $D_{S'}R$ to the state associated with $D_S R$, for the input x.

Furthermore, for every state thus introduced, the associated output Z equals 1 when $D_S(R)$ contains λ and only in this case. For every length l of sequences S with respect to which derivation is performed is obtained a certain set of characteristic derivatives which, may possibly be empty. If empty, the construction of the diagram is terminated. If not, the derivatives for sequences of length $l + 1$ are then determined. To this end, only the characteristic derivatives obtained for the length l will be used, discarding the others until such time as a certain length no longer provides new characteristic derivatives; this will certainly occur as was shown above.

15.4.4 Algorithm for the Construction of a Mealy Machine

On the whole the method is the same.

1. An initial state q_λ associated with the derivative $D_\lambda(R)$ is introduced.

2. The successive derivatives are determined for the lengths $1, 2, \ldots l$, etc. as for the preceding algorithm. Here, however, a *single state* is associated with two identical derivatives or with two derivatives which differ only in the term λ. The presence of λ will suffice to indicate whether the value of the output is 1 when passing from a state q_u (associated with $D_u(R)$), to a state q_{ux} (associated with $D_{ux}(R)$) when $D_{ux}(R)$ is found to be characteristic and thus able to distinguish between two sequences which differ only during this transition, without regard to the next states q_{ux} and $q_{ux'}$.

Remark: The above two algorithms give states diagrams which represent *completely specified* machines. It has already been seen that if it is possible to recognize the equality of two derivatives, or equality except for a term λ, the Moore and Mealy diagrams obtained are *minimal*. No attempt is made to check for the equality of each new derivative with that of the derivatives previously obtained, and if the study is limited to a check on the similarity of the derivatives, diagrams will be obtained which will, in general, not be minimal. They can conveniently be minimised by use of the method of Chapter 14 which produces in this case (completely specified machines) a minimum machine (unique up to choice of symbols).

In practice, checking for equality will yield this very situation with the exception of two cases, those of *finite* events and of *definite* events. It should be noted, however, that if for certain derivatives it can be established that they are equal to derivatives already determined, without being similar, advantage can always be taken of this circumstance which already reduces the number of states for the diagram which has, ultimately, to be minimised.

In short, the derivatives method presents the following advantages:

— from the *theoretical* viewpoint it provides an interpretation of the states.

— from the *practical* viewpoint, it supplies a diagram which represents the expression under consideration. In addition it is obtained directly in a minimal form provided that the equality or inequality of each derivative is checked against the derivatives previously determined. This operation which, in general, can be attempted from the algebraic expressions for the derivatives, reduces to a comparison of finite sets in the particular cases of finite or definite events.

For these it will be of interest to determine the minimal diagrams directly. The following examples will serve to illustrate the above two algorithms. Certain among them have already been treated, by other methods, at the beginning of this chapter.

15.4.5 Examples of Synthesis

(1) **Example**

It is required to construct a states diagram for the following regular expression:
$$R = (0^* 10^* 1)^+ 0^*. \tag{15.53}$$

For the construction of a Moore diagram, it has been shown (13.2.7) that R has only three distinct derivatives: $D_\lambda(R)$, $D_1(R)$ and $D_{11}(R)$.
First
$$D_\lambda(R) = R$$
and $D_\lambda(R)$ is associated with the initial state q_λ.
Then
$$D_0(R) = R = D_\lambda(R)$$

i.e. the output 0 applied to the state q_λ maintains the system in this state.
Consequently:
$$D_1(R) = 0^* 1 (0^* 10^* 1)^* 0^*. \tag{15.54}$$

It has been shown that $D_1(R)$ differs from R and the state q_1 will, therefore, be attributed to it. It will be recalled that it is always possible to assume two derivatives are distinct, and test only later any equivalence that might exist. Moreover $D_1(R)$ does not contain λ since its

15.4 Diagrams Associated with a Regular Expression

sequences are of length ≥ 1 (due to the initial term $0^* 1$). Thus $Z(q_1) = 0$.

Next: $D_{10}(R) = D_1(R)$. The input 0 maintains the system in the state q_1 when it is already there.

Finally, $D_{11}(R)$ supplies the third and last state q_{11} of the diagram. From q_1 this is attained with the input 1, maintained in this state by the input 0, (since $D_{110}(R) = D_{11}(R)$) and abandoned with the output 1 to revert to q_1, (since $D_{111}(R) = D_1(R)$). Thus is obtained the diagram of Fig. 15.25.

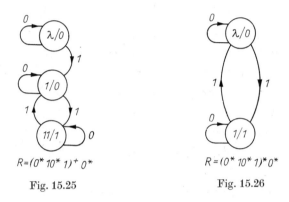

$R = (0^* 10^* 1)^+ 0^*$

Fig. 15.25

$R = (0^* 10^* 1)^* 0^*$

Fig. 15.26

Remark: If consideration had been given to the event:

$$R = (0^* 10^* 1)^* 0 \qquad (15.55)$$

(which is produced when the sequence contains an even or null number of 1), only two distinct derivatives would have been obtained: $D_\lambda R$ and $D_1 R$, whence the diagram of Fig. 15.26.

(2) Example

Reverting here, with a slight variation, to the example of § 15.3 (0). The event to be considered can be expressed:

$$E = 00(00 \cup 01 \cup 10) \cup 11(01 \cup 10 \cup 11) \qquad (15.56)$$

(definite initial event)

The determination of the derivatives follows immediately from the definition. They are given below, alongside the symbol for the associated state and the value of the output Z.

Thus:
$$(15.57)$$

$D_\lambda E = E$ \hfill $(q_\lambda, Z = 0)$

$D_0 E = \left\{\begin{array}{c} 0\ 0\ 0 \\ \cup\ 0\ 0\ 1 \\ \cup\ 0\ 1\ 0 \end{array}\right\}$ \hfill $(q_0, Z = 0)$

$D_1 E = \left\{\begin{array}{c} 1\ 0\ 1 \\ \cup\ 1\ 1\ 0 \\ \cup\ 1\ 1\ 1 \end{array}\right\}$ \hfill $q_1, Z = 0)$

$D_{00} E = \left\{\begin{array}{c} 0\ 0 \\ \cup\ 0\ 1 \\ \cup\ 1\ 0 \end{array}\right\}$ \hfill $(q_{00}, Z = 0)$

$D_{01} E = \emptyset$ \hfill $(q_{01}, Z = 0)$

$D_{10} E = \emptyset$ \hfill $(q_{01}, Z = 0)$

$D_{11} E = \left\{\begin{array}{c} 0\ 1 \\ \cup\ 1\ 0 \\ \cup\ 1\ 1 \end{array}\right\}$ \hfill $(q_{11}, Z = 0)$

$D_{000} E = \{0 \cup 1\} = i$ \hfill $(q_{000}, Z = 0)$

$D_{001} E = \{0\}$ \hfill $(q_{001}, Z = 0)$

$D_{010} E = \emptyset$
$D_{011} E = \emptyset$
$D_{100} E = \emptyset$ \hfill $(q_{01}, Z = 0)$
$D_{101} E = \emptyset$

$D_{110} E = \{1\}$ \hfill $(q_{110}, Z = 0)$

$D_{111} E = \left\{\begin{array}{c} 0 \\ 1 \end{array}\right\} = \{i\}$ \hfill $(q_{000}, Z = 0)$

$D_{0000} E = \lambda$
$D_{0001} E = \lambda$ \hfill $(q_{0000}, Z = 1)$
$D_{0010} E = \lambda$

$D_{0011} E = \emptyset$ \hfill $(q_{01}, Z = 0)$
$D_{1100} E = \emptyset$ \hfill $(q_{01}, Z = 0)$

$D_{1101} E = \lambda$
$D_{1110} E = \lambda$ \hfill $(q_{0000}, Z = 1)$
$D_{1111} E = \lambda$

and there are no further new characteristic derivatives of order 5, the derivation of expressions of order 4 can only provide ∅, which has already been obtained and associated with q_{01}.

(3) Example

Reverting this time, with the derivatives method, to the example of § 15.3(1):

a) *System without zero-reset.*

Under these conditions, if the input E_2 is not allowed to intervene it will be seen that the event in question is a non-initial definite event which is represented by the following regular expression:

$$R = i^* (000 \cup 101 \cup 111). \qquad (15.58)$$

The successive derivatives are now to be determined and the internal states progressively introduced in accordance with the rules given above in a manner so as to obtain the desired system, this time, in the form of a Mealy machine.
Thus:

$$(15.59)$$

$$
\begin{aligned}
D_\lambda(R) &= R & & q_\lambda \\
D_0(R) &= 00 \cup i^* F & & (q_0, Z(q_\lambda, 0) = 0) \\
D_1(R) &= 01 \cup 11 \cup i^* F & & (q_1, Z(q_\lambda, 1) = 0) \\
D_{00}(R) &= 0 \cup 00 \cup i^* F & & (q_{00}, Z(q_0, 0) = 0) \\
D_{01}(R) &= 01 \cup 11 \cup i^* F & & \\
& \quad (D_{01}(R) = D_1(R)) & & (q_1, Z(q_0, 1) = 0) \\
D_{10}(R) &= 1 \cup 00 \cup i^* F & & (q_{10}, Z(q_1, 0) = 0) \\
D_{11}(R) &= 1 \cup 01 \cup 11 \cup i^* F & & (q_{11}, Z(q_1, 1) = 0) \\
D_{000}(R) &= \lambda \cup 0 \cup 00 \cup i^* F & & \\
& \quad (D_{000}(R) \cup \lambda = D_{00}(R)) & & (q_{00}, Z(q_{00}, 0) = 1) \\
D_{001}(R) &= 01 \cup 11 \cup i^* F & & (q_1, Z(q_{00}, 1) = 0) \\
D_{100}(R) &= 0 \cup 00 \cup i^* F & & (q_{00}, Z(q_{10}, 0) = 0) \\
D_{101}(R) &= \lambda \cup 01 \cup 11 \cup i^* F & & (q_1, Z(q_{10}, 1) = 1) \\
D_{110}(R) &= 1 \cup 00 \cup i^* F & & (q_{10}, Z(q_{11}, 0) = 0) \\
D_{111}(R) &= \lambda \cup 1 \cup 01 \cup 11 & & (q_{11}, Z(q_{11}, 1) = 1)
\end{aligned}
$$

There are no new derivatives with respect to the sequences of length 3 and the process is halted. Thus is obtained the states diagram of Figure 15.27 which comprises 6 states. It is easy to see that it is minimal.

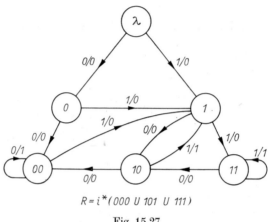

$R = i^*(000 \cup 101 \cup 111)$

Fig. 15.27

b) *System with reset to zero*

Consider now the system complete with zero reset, such as was studied, in fact, in § 15.3. It will be convenient to represent an input $E_2 E_1$ by its equivalent decimal representation.

Under these conditions, *one* way of specifying the system consists in saying that it will produce an output equal to 1 when, *from a passage through the initial state*, there has been a sequence terminating in 000, 101 or 111; the initial state being attained at the instant 0 or at other later times following the application of the input E_2. The value $E_2 = 1$ may be repeated several times. The initial state is attained after a sequence of length null (λ) or again, after any sequence of 2's or 3's ($E_2 = 1$), this latter being of length ≥ 1 or, again, by two successive sequences of the preceding type, or by three etc.

It will be seen that this can be expressed:

$$R' = \{\lambda \cup (1 \cup 0)^* (2 \cup 3)^+ \cup [(1 \cup 0)^* (2 \cup 3)^+]^2$$
$$\cup [(1 \cup 0)^* (2 \cup 3)^+]^3 \cup \cdots\} (1 \cup 0)^* S, \qquad (15.60)$$
$$(S = 000 \cup 101 \cup 111).$$

whence:

$$R' = [(1 \cup 0)^* (2 \cup 3)^+]^* (1 \cup 0)^* S \qquad (15.61)$$

15.4 Diagrams Associated with a Regular Expression

i.e., after a number ≥ 0 of sojourns in the initial state, the last of which must be followed by an input sequence terminating in 000, 101 or 111.

It can also be said that:

$$R' = [(1 \cup 0)^* (2 \cup 3)^*]^* (1 \cup 0)^* S. \tag{15.62}$$

Indeed

$$(1 \cup 0)^* (2 \cup 3)^* = (1 \cup 0)^* [(2 \cup 3)^+ \cup \lambda]$$

$$= (1 \cup 0)^* (2 \cup 3)^+ \cup (1 \cup 0)^* \tag{15.63}$$

then by application of the identity $(A^* \cdot B^*)^* = (A \cup B)^*$

$$[(1 \cup 0)^* (2 \cup 3)^+ \cup (1 \cup 0)^*]^* = [(1 \cup 0)^* (2 \cup 3)^+]^* [(1 \cup 0)^*]^*$$

$$= [(1 \cup 0)^* (2 \cup 3)^+]^* (1 \cup 0)^* \tag{15.64}$$

whence:

$$[(1 \cup 0)^* (2 \cup 3)^*]^* (1 \cup 0)^* S = [(1 \cup 0)^* (2 \cup 3)^+]^* (1 \cup 0)^* (1 \cup 0)^* S$$

$$= [(1 \cup 0)^* (2 \cup 3)^+]^* (1 \cup 0)^* S = R'. \tag{15.65}$$

On the basis of this new expression, it can be written that:

$$R' = [(1 \cup 0) \cup (2 \cup 3)]^* (1 \cup 0)^* S$$

$$= [(2 \cup 3) \cup (1 \cup 0)]^* (1 \cup 0)^* S \tag{15.66}$$

$$= [(2 \cup 3)^* (1 \cup 0)^*]^* (1 \cup 0)^* S$$

$$= \{\lambda \cup (2 \cup 3)^* (1 \cup 0)^* \cup [(2 \cup 3)^* (1 \cup 0)^*]^2 \cup \cdots\} (1 \cup 0)^* S$$

$$= \{\lambda (1 \cup 0)^* \cup (2 \cup 3)^* (1 \cup 0)^* \cup [(2 \cup 3)^* (1 \cup 0)^*]^2 (1 \cup 0)^* \cup \cdots\} S \tag{15.67}$$

Obviously:

$$[(2 \cup 3)^* (1 \cup 0)^*]^n (1 \cup 0)^* = [(2 \cup 3)^* (1 \cup 0)^*]^n, \tag{15.68}$$

whence:

$$R' = \{\lambda (1 \cup 0)^* \cup (2 \cup 3)^* (1 \cup 0)^* [(2 \cup 3)^* (1 \cup 0)^*]^2 \cup \cdots\} S \tag{15.69}$$

Also:

$$\lambda (1 \cup 0)^* \cup (2 \cup 3)^* (1 \cup 0)^* = [\lambda \cup (2 \cup 3)^*] (1 \cup 0)^*$$
$$= (2 \cup 3)^* (1 \cup 0)^*$$
$$= \lambda \cup (2 \cup 3)^* (1 \cup 0)^* \quad (15.70)$$

$$R' = \{\lambda \cup (2 \cup 3)^* (1 \cup 0)^* \cup [(2 \cup 3)^* (1 \cup 0)^*] \cup \cdots\} S$$
$$R' = [(2 \cup 3)^* (1 \cup 0)^*]^* S = [(2 \cup 3) \cup (1 \cup 0)]^* S = i^* S$$
$$R' = i^* S \, (i = 0 \cup 1 \cup 2 \cup 3). \quad (15.71)$$

Thus is obtained an expression analogous to that of R with, however, a different significance for the letter i.

Remark: The relationships (15.65) and (15.71) are, moreover, easily obtained by direct reasoning.

Interpretation of this is easy if it be noted, for example, for (15.65) that:

$$[(1 \cup 0)^* (2 \cup 3)^*]^n (1 \cup 0)^* S = \left\{ \lambda \cup \bigcup_{k=1}^{k=n} [i^* (2 \cup 3)^+]^k \right\} (1 \cup 0)^* S$$
$$(15.72)$$

As for (15.71) this can also be established in the following manner (cf. 15.62):

a) $i^* (0 \cup 1)^* \cup i^* = i^* [(0 \cup 1)^* \cup \lambda] = i^* (1 \cup 0)^* \quad (15.73)$

consequently:

$$i^* (0 \cup 1)^* \supset i^* \quad (15.74)$$

b) $\quad\quad\quad\quad i^* = i^* i^* \supset i^* (1 \cup 0)^* \quad (15.75)$

$$i^* \supset i^* (1 \cup 0)^* \quad (15.76)$$

whence:

$$i^* \subset i^* (1 \cup 0)^* \subset i^* \quad (15.77)$$

$$i^* (1 \cup 0)^* = i^*. \quad (15.78)$$

But, it has been shown that:

$$[(1 \cup 0)^* (2 \cup 3)^*]^* = [0 \cup 1 \cup 2 \cup 3]^* = i^*.$$

From the expression (15.71) the derivatives can again be calculated, thus giving:

15.4 Diagrams Associated with a Regular Expression

$$(15.79)$$

$$
\begin{aligned}
D_\lambda(R') &= R' & &(q_\lambda) \\
D_0(R') &= 00 \cup i^*S & &(q_0,\ Z(q_\lambda,\ 0) = 0) \\
D_1(R') &= 01 \cup 11 \cup i^*S & &(q_1,\ Z(q_\lambda,\ 1) = 0) \\
D_2(R') &= i^*S = R' & &(q_\lambda,\ Z(q_\lambda,\ 2) = 0) \\
D_3(R') &= i^*S = R' & &(q_\lambda,\ Z(q_\lambda,\ 3) = 0) \\
D_{00}(R') &= 0 \cup 00 \cup i^*S & &(q_{00},\ Z(q_0,\ 0) = 0) \\
D_{01}(R') &= 01 \cup 11 \cup i^*S & &(q_1,\ Z(q_0,\ 1) = 0) \\
&\quad \bigl(D_{01}(R') = D_1(R')\bigr) \\
D_{02}(R) &= i^*S = R' & &(q_\lambda,\ Z(q_0,\ 2) = 0) \\
D_{03}(R') &= i^*S = R' & &(q_\lambda,\ Z(q_0,\ 3) = 0) \\
D_{10}(R') &= 1 \cup 00 \cup i^*S & &(q_{10},\ Z(q_1,\ 0) = 0) \\
D_{11}(R') &= 1 \cup 01 \cup 11 \cup i^*S & &(q_{11},\ Z(q_1,\ 1) = 0) \\
D_{12}(R') &= R' & &(q_\lambda,\ Z(q_1,\ 2) = 0) \\
D_{13}(R') &= R' & &(q_\lambda,\ Z(q_1,\ 3) = 0) \\
D_{000}(R') &= \lambda \cup 0 \cup 00 \cup i^*S & &(q_{00},\ Z(q_{00},\ 0) = 1) \\
D_{001}(R') &= 01 \cup 11 \cup i^*S & &(q_1,\ Z(q_{00},\ 1) = 0) \\
D_{002}(R') &= R' & &(q_\lambda,\ Z(q_{00},\ 2) = 0) \\
D_{003}(R') &= R' & &(q_\lambda,\ Z(q_{00},\ 3) = 0) \\
D_{100}(R') &= 0 \cup 00 \cup i^*S & &(q_{00},\ Z(q_{10},\ 0) = 0) \\
D_{101}(R') &= 0 \cup 00 \cup i^*S & &(q_1,\ Z(q_{10},\ 1) = 1) \\
D_{102}(R') &= R' & &(q_\lambda,\ Z(q_{10},\ 2) = 0) \\
D_{103}(R') &= R' & &(q_\lambda,\ Z(q_{10},\ 3) = 0) \\
D_{110}(R') &= 1 \cup 00 \cup i^*S & &(q_{10},\ Z(q_{11},\ 0) = 0) \\
D_{111}(R') &= \lambda \cup 1 \cup 01 \cup 11 & &(q_{11},\ Z(q_{11},\ 1) = 1) \\
D_{112}(R') &= R' & &(q_\lambda,\ Z(q_{11},\ 2) = 0) \\
D_{113}(R') &= R' & &(q_\lambda,\ Z(q_{11},\ 3) = 0)
\end{aligned}
$$

Taking due account of the notation, the reader will be able to verify that Figure 15.9c does indeed hold.

15.4.5.4 Example: Synthesis of a Network Reporting a Composite Definite Event

Given: one input, one output

$$R = E \cup i^*F = (01 \cup 10 \cup 110) \cup i^*(011 \cup 100 \cup 1010) \qquad (15.80)$$

1. No term in E nor in F terminates with a term in F.
2. None of the sets i^*01, i^*10 nor i^*110 is contained in R.

Indeed
$$101 \in i^* \; 01 \text{ and } \; 101 \notin R$$
$$1110 \in i^* \; 10 \text{ and } \; 1110 \notin R$$
$$1110 \in i^* \; 110 \text{ and } \; 1110 \notin R$$

The regular expression in question satisfies the conditions of § 13.2.6. Thus the minimal diagrams can be sought directly.

Moore machine

We have the following derivatives and states:

(15.81)

$$
\begin{aligned}
D_\lambda(R) &= R & (q_\lambda, \; 0) \\
D_0(R) &= (1 \cup \varnothing \cup \varnothing) \cup (11) \cup i^*F \\
&= (1 \cup 11) \cup i^*F & (q_0, \; 0) \\
D_1(R) &= (0 \cup 10) \cup (00 \cup 010) \cup i^*F \\
&= (0 \cup 00 \cup 10 \cup 010) \cup i^*F & (q_1, \; 0) \\
D_{00}(R) &= (\varnothing \cup \varnothing) \cup 11 \cup i^*F \\
&= 11 \cup i^*F & (q_{00}, \; 0) \\
D_{01}(R) &= (\lambda \cup 1) \cup (00 \cup 010) \cup i^*F \\
&= (\lambda \cup 1 \cup 00 \cup 010) \cup i^*F & (q_{01}, \; 1) \\
D_{10}(R) &= (\lambda \cup 0 \cup \varnothing \cup 10) \cup (11) \cup i^*F \\
&= (\lambda \cup 0 \cup 10 \cup 11) \cup i^*F & (q_{10}, \; 1) \\
D_{11}(R) &= (\varnothing \cup \varnothing \cup 0 \cup \varnothing) \cup (00 \cup 010) \cup i^*F \\
&= (0 \cup 00 \cup 010) \cup i^*F & (q_{11}, \; 0) \\
D_{000}(R) &= \varnothing \cup 11 \cup i^*F = 11 \cup i^*F & (q_{00}, \; 0) \\
D_{001}(R) &= (1 \cup 00 \cup 010) \cup i^*F & (q_{001}, \; 0) \\
D_{010}(R) &= (\varnothing \cup \varnothing \cup 0 \cup 10) \cup 11 \cup i^*F \\
&= (0 \cup 10 \cup 11) \cup i^*F & (q_{010}, \; 0) \\
D_{011}(R) &= (\varnothing \cup \lambda \cup \varnothing \cup \varnothing) \cup (00 \cup 010) \cup i^*F \\
&= (\lambda \cup 00 \cup 010) \cup i^*F & (q_{011}, \; 1) \\
D_{100}(R) &= (\varnothing \cup \lambda \cup \varnothing \cup \varnothing) \cup (11) \cup i^*F \\
&= (\lambda \cup 11) \cup i^*F & (q_{100}, \; 1) \\
D_{101}(R) &= (\varnothing \cup \varnothing \cup 0 \cup 1) \cup (00 \cup 010) \; i^*F \\
&= (0 \cup 1 \cup 00 \cup 010) \cup i^*F & (q_{101}, \; 0) \\
D_{110}(R) &= (\lambda \cup 0 \cup 10) \cup (11) \cup i^*F \\
&= \lambda \cup 0 \cup 10 \cup 11 & (q_{10}, \; 1) \\
D_{111}(R) &= (\varnothing \cup \varnothing \cup \varnothing) \cup (00 \cup 010) \cup i^*F \\
&= (00 \cup 010) \cup i^*F & (q_{111}, \; 0) \\
D_{0010}(R) &= (\varnothing \cup 0 \cup 10) \cup (11) \cup i^*F \\
&= (0 \cup 10 \cup 11) \cup i^*F & (q_{010}, \; 0)
\end{aligned}
$$

15.4 Diagrams Associated with a Regular Expression

$$D_{0011}(R) = (\lambda \cup \emptyset \cup \emptyset) \cup (00 \cup 010) \cup i^*F$$
$$= (\lambda \cup 00 \cup 010) \cup i^*F \qquad (q_{011}, 1)$$

$$D_{0100}(R) = (\lambda \cup \emptyset \cup \emptyset) \cup (11) \cup i^*F$$
$$= (\lambda \cup 11) \cup i^*F \qquad (q_{100}, 1)$$

$$D_{0101}(R) = (\emptyset \cup 0 \cup 1) \cup (00 \cup 010) \cup i^*F$$
$$= (0 \cup 1 \cup 00 \cup 010) \cup i^*F \qquad (q_{101}, 0)$$

$$D_{0110}(R) = (\emptyset \cup 0 \cup 10) \cup (11) \cup i^*F$$
$$= (0 \cup 10 \cup 11) \cup i^*F \qquad (q_{010}, 0)$$

$$D_{0111}(R) = (\emptyset \cup \emptyset \cup \emptyset) \cup (00 \cup 010) \cup i^*F$$
$$= (00 \cup 010) \cup i^*F \qquad (q_{111}, 0)$$

$$D_{1000}(R) = (\emptyset \cup \emptyset) \cup (11) \cup i^*F$$
$$= 11 \cup i^*F \qquad (q_{00}, 0)$$

$$D_{1001}(R) = (\emptyset \cup 1) \cup (00 \cup 010) \cup i^*F$$
$$= (1 \cup 00 \cup 010) \cup i^*F \qquad (q_{001}, 0)$$

$$D_{1010}(R) = (\lambda \cup \emptyset \cup 0 \cup 10) \cup (11) \cup i^*F$$
$$= (\lambda \cup 0 \cup 10 \cup 11) \cup i^*F \qquad (q_{10}, 1)$$

$$D_{1011}(R) = (\emptyset \cup \lambda \cup \emptyset \cup \emptyset) \cup (00 \cup 010) \cup i^*F$$
$$= (\lambda \cup 00 \cup 010) \cup i^*F \qquad (q_{011}, 1)$$

$$D_{1110}(R) = (0 \cup 10) \cup (11) \cup i^*F$$
$$= (0 \cup 10 \cup 11) \cup i^*F \qquad (q_{010}, 0)$$

$$D_{1111}(R) = (\emptyset \cup \emptyset) \cup (00 \cup 010) \cup i^*F$$
$$= (00 \cup 010) \cup i^*F \qquad (q_{111}, 0)$$

Whence Table 15.28 (the states are represented by the corresponding indices).

Table 15.28

Q \ E	0	1	Z
λ	0	1	0
0	0 0	0 1	0
1	1 0	1 1	0
0 0	0 0	0 0 1	0
0 1	0 1 0	0 1 1	1
1 0	1 0 0	1 0 1	1
1 1	1 0	1 1 1	0
0 0 1	0 1 0	0 1 1	0
0 1 0	1 0 0	1 0 1	0
0 1 1	0 1 0	1 1 1	1
1 0 0	0 0	0 0 1	1
1 0 1	1 0	0 1 1	0
1 1 1	0 1 0	1 1 1	0

Mealy machine

The determination of the derivatives proceeds in a similar manner with, however, the modification already indicated: two derivatives which differ only by the presence of λ are associated with a single state; the transition $F(q_x \cdot e) = q$ gives the output $Z = 1$ if $\lambda \in D_{xe}$ where x is the sequence identifying the significant derivative associated with q_x and e the input symbol provoking the transition.

(15.82)

$D_\lambda(R) = R$ \qquad q_λ

$D_0(R) = (1 \cup 11) \cup i^*F$ \qquad $\big(q_0,\ Z(q_\lambda,\ 0) = 0\big)$

$D_1(R) = (0 \cup 00 \cup 10 \cup 010) \cup i^*F$ \qquad $\big(q_1,\ Z(q_\lambda,\ 1) = 0\big)$

$D_{00}(R) = 11 \cup i^*F$ \qquad $\big(q_{00},\ Z(q_0,\ 0) = 0\big)$

$D_{01}(R) = (\lambda \cup 1 \cup 00 \cup 010) \cup i^*F$ \qquad $\big(q_{01},\ Z(q_0,\ 1) = 1\big)$

$D_{10}(R) = (\lambda \cup 0 \cup 10 \cup 11) \cup i^*F$ \qquad $\big(q_{10},\ Z(q_1,\ 0) = 1\big)$

$D_{11}(R) = (0 \cup 00 \cup 010) \cup i^*F$ \qquad $\big(q_{11},\ Z(q_1,\ 1) = 0\big)$

$D_{000}(R) = 11 \cup i^*F$ \qquad $\big(q_{00},\ Z(q_{00},\ 0) = 0\big)$
$\big(D_{000}(R) = D_{00}(R)\big)$

$D_{001}(R) = (1 \cup 00 \cup 010) \cup i^*F$ \qquad $\big(q_{01},\ Z(q_{00},\ 0) = 0\big)$
$\big(D_{001}(R) \cup \lambda = D_{01}(R)\big)$

$D_{010}(R) = (0 \cup 10 \cup 11) \cup i^*F$ \qquad $\big(q_{10},\ Z(q_{01},\ 0) = 0\big)$
$\big(D_{010}(R) \cup \lambda = D_{10}(R)\big)$

$D_{011}(R) = (\lambda \cup 00 \cup 010)$ \qquad $\big(q_{011},\ Z(q_{01},\ 1) = 1\big)$

$D_{100}(R) = (\lambda \cup 11) \cup i^*F$ \qquad $\big(q_{00},\ Z(q_{01},\ 0) = 1\big)$
$\big(D_{100}(R) = \lambda \cup D_{00}(R)\big)$

$D_{101}(R) = (0 \cup 1 \cup 00 \cup 010) \cup i^*F$ \qquad $\big(q_{101},\ Z(q_{10},\ 1) = 0\big)$

$D_{110}(R) = (\lambda \cup 0 \cup 10 \cup 11)$ \qquad $\big(q_{10},\ Z(q_{11},\ 0) = 1\big)$
$\big(D_{110}(R) = D_{10}(R)\big)$

$D_{111}(R) = (00 \cup 010) \cup i^*F$ \qquad $\big(q_{011},\ Z(q_{11},\ 1) = 0\big)$
$\big(D_{111}(R) \cup \lambda = D_{011}(R)\big)$

$D_{0110}(R) = (0 \cup 10 \cup 11) \cup i^*F$ \qquad $\big(q_{10},\ Z(q_{011},\ 0) = 0\big)$
$\big(D_{0110}(R) = D_{01}(R)\big)$

$D_{0111}(R) = (00 \cup 010) \cup i^*F$ \qquad $\big(q_{011},\ Z(q_{011},\ 1) = 0\big)$

$D_{1010}(R) = (\lambda \cup 0 \cup 10 \cup 11) \cup i^*F$ \qquad $\big(q_{10},\ Z(q_{101},\ 0) = 1\big)$
$\big(D_{1010}(R) = D_{10}(R)\big)$

$D_{1011}(R) = (\lambda \cup 00 \cup 010)$ \qquad $\big(q_{011},\ Z(q_{101},\ 1) = 1\big)$
$\big(D_{1011}(R) = D_{011}(R)\big)$

15.4 Diagrams Associated with a Regular Expression

$$D_{1110}(R) = (0 \cup 10 \cup 11) \cup i*F$$
$$D_{1110}(R) \cup \lambda = D_{10}(R)$$
$$D_{1111}(R) = (00 \cup 010) \cup i*F$$
$$D_{1111}(R) \cup \lambda = D_{011}(R)$$

$$(q_{10}, Z(q_{011}, 0) = 0)$$

$$(q_{011}, Z(q_{011}, 1) = 0)$$

From which the following table can be extracted:

Table 15.29

$Q \backslash E$	0	1
λ	0, 0	1, 0
0	0 0, 0	0 1, 1
1	1 0, 1	1 1, 0
0 0	0 0, 0	0 1, 0
0 1	1 0, 0	0 1 1, 1
1 0	0 0, 1	1 0 1, 0
1 1	1 0, 1	0 1 1, 0
0 1 1	1 0, 0	0 1 1, 0
1 0 1	1 0, 1	0 1 1, 1

15.4.6 Networks Having Several Binary Outputs

The generalization of the algorithms of §§ 15.4.3, 15.4.4 for the case of multiple circuits is immediate. It is based on the following considerations:

A circuit having n outputs $(n \geq 1)$ is described by n distinct expressions $R_1, R_2, \ldots, R_n$, the expression R_r being associated with the output Z_r. The value of Z_r is 1 when the output sequence belongs to the event denoted by R_r, and only in this case. The automaton in question can thus be defined as an ordered set R of regular expressions R $(r = 1, \ldots, n)$.

$$R = (R_1, R_2, \ldots, R_r, \ldots, R_m)$$

In the same way the vector of the derivatives of the R_2 is known as the "derivative vector" and denoted by $D_S(R)$ such that:

$$D_S(R) = (D_S R_1, D_S R_2, \ldots, D_S R_r, \ldots, D_S R_m) \qquad (15.83)$$

Finally, it can be said that two regular vectors are equal if all their components of the same rank are equal, i.e. denote the same set.

33*

The method of construction of a Moore or of a Mealy diagram is then entirely similar to that described above for the case of one output with, however, the following differences:

— for the Moore machine a state is associated with every characteristic ("significant") vector derived.
— The procedure is the same for the Mealy machine, with the restriction that when two vectors differ only by a vector of the components $\emptyset$ or λ, they are associated with a single state.

Under these conditions the following properties are obtained (Brzozowski).

Theorem: *A diagram constructed by the above method properly represents every input sequence by a sequence of output combinations in conformance with the specifications and, additionally, is minimal.*

It is clear that the circuit in question offers the required performance, i.e. it provides an output $Z = (Z_1, Z_2, ..., Z_n)$ for which the value of the components is 1 when the input sequence S belongs to R_r. With S corresponds a derivative $D_s R$ with which is associated a certain state q, and the output Z_r is equal to 1 if $\lambda \in D_s R_r$, i.e. if $S \in R$.

The diagram is minimal. Indeed, if two derivatives are not equal they differ in at least one component. Their associated states cannot but be different since, for a sequence belonging to one of the events — but not to the other, the output produced will be 1 in one case and zero in the other. As the states have been associated precisely with the distinct derivatives, the diagram can only be minimal since every equivalence of states would contradict the fact that these derivatives are different.

15.5 Coding of Initial States. Memory Control Circuits

Having obtained a states diagram for the required system it will be necessary to finish up with a binary circuit which necessitates, in particular, binary coding of the internal states. If the diagram comprises n internal states, k binary variables will be needed with $k \geq k_0$, where k_0 is the smallest k such that $2^k \geq n$.

Minimal coding

The problem of the determination of a coding which minimises a cost function associated with the circuit (the number of components, for example) has not found any general solution at this time. In practice, it is necessary to reason directly on the particular problem under review, and to try a certain number of solutions among which the one which appears to be the more promising is chosen. In general, the number of possible codings (a function of n and k) is too high to permit an examination of all possible solutions, even when considering as equivalent the codings which differ only in permutation or even in the complementation of one or more variables.

On a theoretical level, various methods are the subject of current research. Particular mention must be made of the methods of decomposition for which the point of view, closely akin to that adopted for functional decompositions (Chapter 10), consists of decomposing an automaton into more easily coded automata, and the statistical methods aimed at approximate optimisation of tables comprising a sufficiently large number of states. These methods, promising though they may be, have so far furnished only a beginning of a solution and will not be discussed here.

This work will be limited to a description of some solutions obtained for the particular case of counters: they will be the subject of Chapter 16[1].

Memory control circuits

It has been shown that these could take the form of calibrated delays or of flip-flops. In the first case each memory unit possesses 1 input, 1 output, and the values taken by these two leads are equal to one of the variables q_i of the internal states. In the case of the flip-flops the situation is a little more complicated: for writing in a binary value, there are in effect a certain number of control leads for which the number and nature depend on the types of flip-flop employed. Thus the problem, assuming that the coding of the internal states has been fixed (with k binary variables), is the determination of logical functions, thanks to which suitable values are applied at the input most for the flip-flops. Figure 15.30 lists the leads common types of flip-flop.

Procedure

This will be illustrated in each case by the function f as defined in Figure 15.31 a.

1. Delays (cf. Figure 15.31 a)
2. T type flip-flops. Form $Q_n \oplus Q_{n+1} = Q_n \oplus f$ (Fig. 15.31 b)
3. $S - R$ type flip-flops.

Determine the 4 functions f_{ij} associated with the transitions $Q = i \to Q = j$ ($i = 0, 1$, $j = 0, 1$) (Fig. 15.31 c). Then (m: minterms;

[1] Cf. Hartmanis, J., Stearns, R. E.: Algebraic Structure Theory of Sequential Machines. Englewood Cliffs: Prentice Hall, 1966; Haring, D. R.: Sequential Circuit Synthesis: State Assignment Aspects, Research Monograph N° 31, The MIT Press. 1966.

f: coefficients):

$$\begin{cases} S = \sum_{i} f^i_{01} m_i + \sum_{j} a_j f^j_{11} m_j \\ R = \sum_{k} f^k_{10} m_k + \sum_{l} b_l f^l_{00} m_l \end{cases} \quad (a_j, b_l = 0 \text{ or } 1, \text{ at choice}) \quad (15.84)$$

The values of the variables for which $f_{11} = 1$ play the part of unspecified combinations for S, while those for which $f_{00} = 1$ play the same rôle for R.

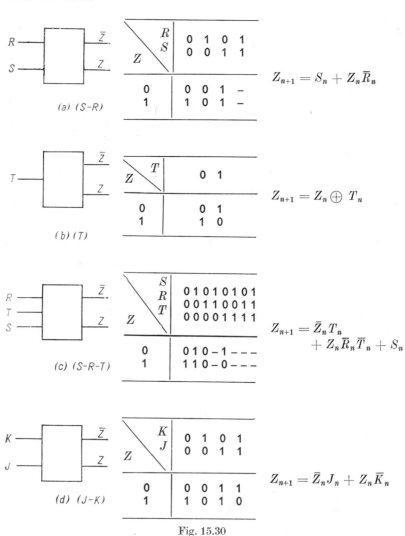

Fig. 15.30

15.5 Coding of Initial States. Memory Control Circuits

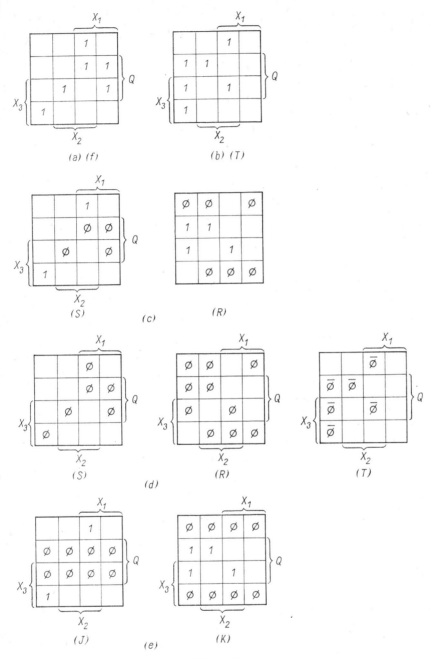

Fig. 15.31

4. Type *S-R-T* flip-flops (Fig. 15.31 d)
With the same functions f_{ij} and in the same manner:

$$\begin{cases} S = \sum_i c_i f_{01}^i m_i + \sum_j d_j f_{11}^j m_j \\ R = \sum_k e_k f_{10}^k m_k + \sum_l g_l f_{00}^l m_l \\ T = \sum_i \bar{c}_i f_{01}^i m_i + \sum_k \bar{e}_k f_{10}^k m_k \end{cases} \qquad (15.85)$$

5. Type *J-K* Flip-flops (Fig. 15.32 e)
This time:

$$\begin{cases} J = \sum_i f_{01}^i m_i + \sum_j r_j f_{11}^j m_j + \sum_k s_k f_{10}^k m_k \\ K = \sum_k f_{10}^k m_k + \sum_l u_l f_{00}^l m_l + \sum_i v_i f_{01}^i m_i \end{cases} \qquad (15.86)$$

Table 15.32 lists the functions corresponding with the example chosen:

Table 15.32

ij	f	T	S-R	S-R-T	J-K
0 0	0	0	0 –	0 – 0	0 –
0 1	1	1	1 0	x 0 $\bar{x}$	1 –
1 0	0	1	0 1	0 y $\bar{y}$	– 1
1 1	1	0	– 0	– 0 0	– 0

Chapter 16

Counters

16.1 Introduction

From a practical point of view, counters constitute an important class of sequential networks. They are mostly used just for counting, but this does not preclude their use as basic constituent elements for the realisation of more complex functions. Such is the case in a certain number of special purpose digital systems as, for example, certain "binary-decimal" converters, adders, integrators, square root extractors, etc. in which relatively important functions are obtained by the use of counting techniques. [42]

Their general operation can be described as follows: the counter possesses an input to which can be applied a binary signal E. The output and transition functions are of the type (example: Fig. 16.1):

$$F(Q_i, E): \begin{cases} F(Q_i, 0) = Q_i \\ F(Q_i, 1) = Q_{|i+1|N} \end{cases} \qquad (16.1)$$

$$S = G(Q_i) \qquad (16.2)$$

(For a counter with N internal states, the system returns to the state Q_0 if an input 1 be applied with an internal state Q_{N-1}).

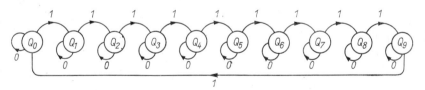

Fig. 16.1. State diagram for a counter

The synchronous counters described below will make use of the following three codes for internal states: pure binary, binary coded decimal (BCD), and reflected binary. The output is identical with this internal state.

16.2 Pure Binary Counters

16.2.1 Up-Counting and Down-Counting in Pure Binary Code

16.2.1.1 Up-Counting

The passage from a number X to a number $X + 1$ corresponds to certain Boolean relations linking the representations of the numbers X and $X + 1$.

For this it suffices to consider the addition:

$$\begin{array}{rl} X = & x_n\ x_{n-1}\ \cdots\ x_k\ \cdots\ x_1\ x_0 \\ + 1 = & 0\ \ 0\ \ \cdots\ 0\ \cdots\ 0\ \ 1 \\ \hline X' = & x_n'\ x_{n-1}'\ \cdots\ x_k'\ \cdots\ x_1'\ x_0' \end{array}$$

we have:

$$x_0' = x_0 \oplus 1 = \bar{x}_0 \qquad\qquad r_1 = x_0 1 = x_0$$
$$x_1' = x_1 \oplus r_1 = x_1 \oplus x_0 \qquad r_2 = x_1 x_0$$
$$x_2' = x_2 \oplus r_2 = x_2 \oplus x_1 x_0 \qquad r_3 = x_2 x_1 x_0$$
$$\cdots\cdots\cdots\cdots\cdots\cdots\cdots\cdots\cdots\cdots$$
$$x_k' = x_k \oplus r_k = x_k \oplus \prod_{i=0}^{i=k-1} x_i \qquad r_{k+1} = \prod_{i=0}^{i=k} x_i \qquad (16.3)$$

$$\begin{cases} x_k' = x_k \oplus \prod_{i=0}^{i=k-1} x_i & (16.4) \\[2ex] r_{k+1} = \prod_{i=0}^{i=k} x_i & (16.5) \end{cases}$$

These relations give the representation of the number $X + 1$ as a function of that of X. Their intuitive meaning is as follows:

The 'carry' propagates to the end of the block of 1's commencing from the left of the word:

$$\begin{array}{rl} & = X'\ 0\ 1\ 1\ 1\ \cdots\ 1\ 1 \\ + & \underline{\qquad\qquad\qquad\qquad 1} \\ & = X'\ 1\ 0\ 0\ 0\ \cdots\ 0\ 0 \\ R & = (0)\ 0\ 1\ 1\ 1\ \cdots\ 1\ 1\ (\text{'carries'}) \end{array}$$

16.2.1.2 Down-Counting

$$X = x_n\ x_{n-1}\ \cdots\ x_k\ \cdots\ x_1\ x_0$$
$$-1 = 0\ \ 0\ \ \cdots\ 0\ \cdots\ 0\ \ 1$$
$$X'' = x_n''\ x_{n-1}''\ \ \ x_k''\ \ \ x_1''\ x_0'' \tag{16.6}$$

$$x_0'' = x_0 \oplus 1 = \bar{x}_0 \qquad\qquad r_1 = \bar{x}_0$$
$$x_1'' = x_1 \oplus r_1 = x_1 \oplus \bar{x}_0 \qquad r_2 = \bar{x}_1 r_1 = \bar{x}_1 \bar{x}_0$$
$$x_2'' = x_2 \oplus r_2 = x_2 \oplus \bar{x}_1 \bar{x}_0 \qquad r_3 = \bar{x}_2 r_2 = \bar{x}_2 \bar{x}_1 \bar{x}_0$$
$$\ldots\ldots\ldots\ldots\ldots\ldots\ldots\ldots\ldots\ldots\ldots\ldots$$

$$x_k'' = x_k \oplus r_k = x_k \oplus \prod_{i=0}^{i=k-1} \bar{x}_i \qquad r_{k+1} = \prod_{i=0}^{i=k} \bar{x}_i \tag{16.7}$$

$$\begin{cases} x_k'' = x_k \oplus \prod_{i=0}^{i=k-1} \bar{x}_i & (16.8) \\[2ex] r_{k+1} = \prod_{i=0}^{i=k} \bar{x}_i & (16.9) \end{cases}$$

The above Boolean relations describe the progression logic of an up-counter or down-counter in pure binary code. It will be noted that the passage from up-counting to down-counting is accomplished by the complementation of the letters appearing in the carry.

Complementation of the two members of the expression (16.8) gives:

$$\bar{x}_k'' = \bar{x}_k \oplus \prod_{i=0}^{i=k-1} \bar{x}_i \tag{16.10}$$

The expressions (16.9) and (16.10) are those for the counting with numbers corresponding to the complemented binary words. This is immediately evident. If the value of the number represented by A be denoted by $|A|$, then, effectively:

$$|X'| = |X| + 1$$
$$|X''| = |X| - 1$$
$$|\bar{X}'| = 2^n - 1 - |X'| = 2^n - 1 - |X| - 1 = |\bar{X}| - 1$$
$$|\bar{X}''| = 2^n - 1 - |X''| = 2^n - 1 - |X| + 1 = |\bar{X}| + 1$$
$$\tag{16.11}$$

16.2.1.3 Counting in pure Binary: Other Formulae

From the above formulae, where the input has been shown as E, the following relations can be deduced. The index n designates the clock-time ($n \geq 0$), the index k the bit position within the number ($k \geq 0$):

$$\begin{cases} x_k{}^{n+1} = x_k{}^n \oplus E^n \left(\prod_{i=0}^{i=k-1} x_i \right)_n & (16.12) \\ x_{k-1}^{n+1} \oplus x_{k-1}^n = E^n \left(\prod_{i=0}^{i=k-2} x_i \right)_n & (16.13) \end{cases}$$

whence:

$$x_k^{n+1} \oplus x_k^n = x_{k-1}^n \left(x_{k-1}^{n+1} \oplus x_{k-1}^n \right) \qquad (16.14)$$

Other formulae:

For a given sequence $X = (x_0, x_1, \ldots x_i \ldots)$, let us define two other sequences as follows:

$$\begin{cases} \overset{<}{X} = (-, \bar{x}_0 x_1, \bar{x}_1 x_2, \ldots, \bar{x}_{i-1} x_i, \ldots) & (16.15) \\ \overset{>}{X} = (-, x_0 \bar{x}_1, x_1 \bar{x}_2, \ldots, x_{i-1} \bar{x}_i, \ldots) & (16.16) \end{cases}$$

and, let

$$X' = (-, x_0 \oplus x_1, x_1 \oplus x_2, \ldots, x_{i-1} \oplus x_i, \ldots) \qquad (16.17)$$

The operators $<$ and $>$ must produce a 1 when X passes from 0 to 1 and from 1 to 0 respectively. They can be realised, in particular, with an RC derivation network with additional suppression of one of the two possible types of pulses, e.g. of the negative pulses. Assuming such an electrical implementation, the formulae for a binary counter can profitably be determined by use of the corresponding operators. In the text which follows the subscript will be assigned to the rank of the binary digit in question, the superscript to the clock time. Then:

Up-counting:

$$x_k^{n+1} = x_k{}^n \oplus \prod_{i=0}^{i=k-1} x_i{}^n \qquad (16.18)$$

whence:

$$\left(\overset{<}{x}_k \right)^{n+1} = \overline{(x_k{}^n)} \cdot (x_k{}^{n+1}) = \overline{(x_k{}^n)} \cdot \left(x_k{}^n \oplus \prod_{i=0}^{i=k-1} x_i{}^n \right)$$

$$= \overline{x_k{}^n} \cdot x_k{}^n \oplus \overline{x_k{}^n} \prod_{i=0}^{i=k-1} x_i{}^n = \overline{x_k{}^n} \prod_{i=0}^{i=k-1} x_i{}^n \qquad (16.19)$$

16.2 Pure Binary Counters

Also:

$$\left(\overset{>}{x}_k\right)^{n+1} = (x_k^n)\overline{\left(x_k^n \oplus \prod_{i=0}^{i=k-1} x_i^n\right)} = (x_k^n)\left(\overline{x_k^n} \oplus \prod_{i=0}^{i=k-1} x_i^n\right)$$

$$= x_k^n \cdot \overline{x_k^n} \oplus x_k^n \prod_{i=0}^{i=k-1} x_i^n = \prod_{i=0}^{i=k} x_i^n \qquad (16.20)$$

And similarly:

$$\left(\overset{>}{x}_{k-1}\right)^{n+1} = \prod_{i=0}^{i=k-1} x_i^n \qquad (16.21)$$

Whence:

$$\left(\overset{>}{x}_k\right)^{n+1} = x_k^n \left(\overset{>}{x}_{k-1}\right)^{n+1} \qquad (16.22)$$

(x_k passes from 1 to 0 if x_k equals 1 and if x_{k-1} itself passes from 1 to 0).
Reverting to the relation (16.19) it can be said that:

$$\left(\overset{<}{x}_k\right)^{n+1} = \overline{x_k^n} \prod_{i=0}^{i=k-1} x_i^n = \overline{x_k^n} \left(\overset{>}{x}_{k-1}\right)^{n+1} \qquad (16.23)$$

(x_k passes from 0 to 1 if $x_k = 0$ and if x_{k-1} passes from 1 to 0).
From the formulae (16.22) and (16.23) can be extracted:

$$\left(\overset{<}{x}_k + \overset{>}{x}_k\right)^{n+1} = (x_k')^{n+1} = (\overline{x}_k + x_k)^{n+1}\left(\overset{>}{x}_{k-1}\right)^{n+1} \qquad (16.24)$$

$$(x_k')^{n+1} = \left(\overset{>}{x}_{k-1}\right)^{n+1} \qquad (n = 0, 1, \ldots) \qquad (16.25)$$

(the figure x_k changes only if the figure x_{k-1} passes from 1 to 0).

Down-counting

In a similar manner:

$$\left(\overset{<}{x}_k\right)^{n+1} = \overline{(x_k^n)}(x_k^{n+1}) = \overline{(x_k^n)}\left(x_k^n \oplus \prod_{i=0}^{i=k-1} \overline{x}_i\right) = \overline{x_k^n} \prod_{i=0}^{i=k-1} \overline{x_i^n} = \prod_{i=0}^{i=k} \overline{x_i^n}$$

$$(16.26)$$

and

$$\left(\overset{>}{x}_k\right)^{n+1} = (x_k^n)\overline{(x_k^{n+1})} = (x_k^n)\overline{\left(x_k^n \oplus \prod_{i=0}^{i=k-1} \overline{x}_i^n\right)} = x_k^n \prod_{i=0}^{i=k-1} \overline{x_i^n} \qquad (16.27)$$

whence:

$$\left(\overset{<}{x}_k\right)^{n+1} = \overline{x_k^n}\left(\overset{<}{x}_{k-1}\right)^{n+1} \qquad (16.28)$$

$$\left(\overset{>}{x}_k\right)^{n+1} = x_k^n \left(\overset{<}{x}_{k-1}\right)^{n+1} \qquad (16.29)$$

$$(x_k')^{n+1} = (\overline{x_k^n} + x_k^n)\left(\overset{<}{x}_{k-1}\right)^{n+1} \qquad (16.30)$$

$$(x_k')^{n+1} = \left(\overset{<}{x}_{k-1}\right)^{n+1} \text{ for every } n\,(n = 0, 1, \ldots) \quad (16.31)$$

(x_k changes when x_{k-1} passes from 0 to 1 and only in this case).

488   16. Counters

Thus a counter can be constructed with the aid of flip-flops and of capacitive coupling, such that only the changes in one direction are transmitted from one stage to another: thus bringing out a classical property of binary counting: the flip-flop of rank k switches twice less frequently than that of rank $k - 1$, etc.

Examples

Figures 16.2a, b represent pure binary counters having 16 internal states. (The time base is not shown). The generalisation to the case of m binary stages (2^m internal states) is immediate. It will, however, be

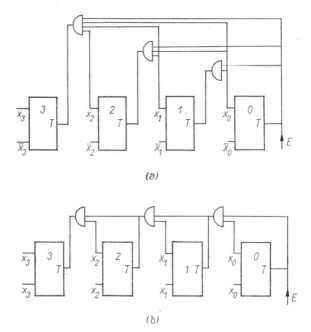

Fig. 16.2a, b. Pure binary counters

noted that for high m it is not possible to activate the flip-flops by means of the single-level logic as shown in Fig. 16.2a since, for rank $k - 1$, a gate with k inputs would be necessary. This leads to the realisation of the logical products $\prod x_i$ in recurrent form (e.g. in the manner shown in Fig. 16.2b). In this last configuration represented, it then suffices to have 2-input gates but, on the other hand, for rank k there is an input logic with k levels and, for the actuation of the flip-flops, the time needed is k times more than that required for a counter of type 16.2a. The situation here is very similar to that for the parallel adder, for which it

16.3 Decimal Counters

Examples of Decimal Counters ("Decades")

It is required to produce a counter which will accept input pulses, will count them and which, every time that the total becomes 10, will return to its initial state and transmit by way of its output line an 'overflow' impulse. Such a counter of capacity 10 is known as a decade. By interconnecting m decimal counting elements of this type, where the output of an element $j-1$ serves as an input to the element j, it is possible to construct a counter of capacity 10^m. Such counters are used in specialised computers working in decimal or in counting systems with visual display.

At least four binary digits are necessary to code 10 internal states. It was shown (Chapter 3) that when the corresponding decimal figure must be converted to analog, it can be of great help if the code in question be weighted. In the case of a conversion voltage for example, it suffices to perform the weighted sum of the currents corresponding with the 4 binary digits.

If n binary figures are used for coding 10 states, there are $2^n - 10$ unused combinations. The progression of code words, assigned to the 10 internal states, differs from that for counting in pure binary by a certain number of "jumps" which are usually shown by an arrow (e.g. Table 16.3). The total magnitude of these jumps is $2^n - 10$, which corresponds to 6 in a 4-position code.

Table 16.3

	I	II	III	IV
Q_0	0 0 0 0	0 0 1 1	0 0 0 0	0 0 0 0
Q_1	0 0 0 1	0 1 0 0	0 0 0 1	0 0 0 1
Q_2	0 0 1 0	0 1 0 1	0 0 1 0	0 0 1 0
Q_3	0 0 1 1	0 1 1 0	0 0 1 1	0 0 1 1
Q_4	0 1 0 0	0 1 1 1	0 1 0 0	0 1 1 1
Q_5	0 1 0 1	1 0 0 0	0 1 0 1	1 0 0 0
Q_6	0 1 1 0	1 0 0 1	0 1 1 0	1 0 0 1
Q_7	0 1 1 1	1 0 1 0	0 1 1 1	1 0 1 0
Q_8	1 0 0 0	1 0 1 1	1 1 1 0	1 0 1 1
Q_9	1 0 0 1	1 1 0 0	1 1 1 1	1 1 1 1
	(8 4 2 1)	(excess 3)	(2 4 2 1)	(5 1 2 1)

The code I is the code 8421.
The code II is the excess 3 code.
The code III is obtained by deliberately altering the transition $Q_7 - Q_8$ which is normally of one binary unit, by the addition of 6.
The code IV is obtained in a similar manner by twice increasing by 3 the value which would result from normal counting in pure binary (transitions $Q_3 - Q_4$ and $Q_8 - Q_9$).

Examples of Decades

a) *Code* I (8421)

The combination 1001 occurs only at Q_9. When the counter is in this state, it is necessary that the applied pulse return B_3, B_2, B_1 and B_0 to zero. If this were the case of a normal binary counter, the flip-flop B_0 would be returned to zero in any case. B_2, however, would remain at the state 0. It suffices, therefore, to prevent the switching of B_1 and to return B_3 to zero. Fig. 16.4a indicates a means of obtaining this result in the case of the *SRT* flip-flops.

b) *Code* II (excess 3)

In a similar manner, Fig. 16.4b shows an example of a decade for the code excess 3.

c) *Codes* III and IV

Figs 16.4c and 16.4d show examples of the way this can be done for each of these codes.

16.4 Reflected Binary Code Counters

16.4.1 Recurrence Relations in Gray Code Counting [12.61]

a) It is desired to determine the relations which link the two code-words Y_i and Y_{i+1} which, in reflected code, re-represent two consecutive numbers X_i and X_{i+1}, i.e. such that $X_{i+1} - X_i = 1$. It has been seen that Y_i and Y_{i+1} must be adjacent.

There are two cases:

— First case: X_i "even" then, in pure binary:

$$X_i = x_n x_{n-1} \cdots x_k \cdots x_1 0$$

and consequently:

$$X_{i+1} = x_n x_{n-1} \cdots x_k \cdots x_1 1$$

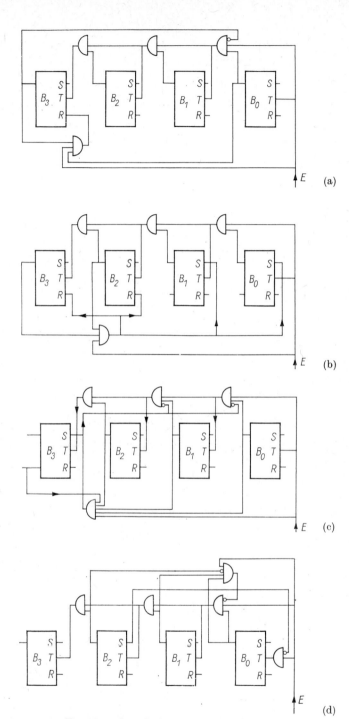

Fig. 16.4a, b, c, d. Some examples of decades

With these numbers correspond the following representations in the Gray code:

$$\begin{cases} Y_i = x_n(x_{n-1} \oplus x_n)(x_{n-2} \oplus x_{n-1}) \cdots (x_k \oplus x_{k+1}) \cdots (x_1 \oplus x_2)(0 \oplus x_1) \\ Y_{i+1} = x_n(x_{n-1} \oplus x_n)(x_{n-2} \oplus x_{n-1}) \cdots (x_k \oplus x_{k+1}) \cdots (x_1 \oplus x_2)(1 \oplus x_1) \end{cases}$$
(16.32)

Thus giving:
$$Y_i \oplus Y_{i+1} = 0\,0\,0 \cdots 0 \cdots 0\,1$$

Therefore:

If, in the Gray code, Y_i represents an 'even' integer, Y_{i+1} is obtained by complementation of the extreme right-hand digit.

— Second case: X_i 'odd'.

X_i commences with 1 on the right and can be expressed:

$$\begin{cases} X_i = x_n x_{n-1} \cdots x_{k+1}\,0\,1\,1\cdots 1\,1 \\ X_{i+1} = x_n x_{n-1} \cdots x_{k+1}\,1\,0\,0 \cdots 0\,0 \end{cases}$$
(16.33)

whence:

$$\begin{cases} Y_i = x_n(x_{n-1} \oplus x_n) \cdots (x_{k+1} \oplus x_{k+2})\,x_{k+1}\,1\,0\cdots 0\,0 \\ Y_{i+1} = x_n(x_{n-1} \oplus x_n) \cdots (x_{k+1} \oplus x_{k+2})\,\bar{x}_{k+1}\,1\,0\cdots 0\,0 \end{cases}$$
(16.34)

and
$$T_i = Y_i \oplus Y_{i+1} = 0\,0\cdots 0 \quad 1 \cdots 0\,0 \qquad (16.35)$$

whence:

If Y_i represents an odd number, the representation of the following number is obtained by complementation of the first figure on the left of the first "1" digit encountered when starting from the right.

b) We now denote the state of a counter in reflected binary by $R = (R_m \cdots R_k \cdots R_0)$.

From a knowledge of the parity of the number R^n, the logic equations defining the counter progression can be deduced.

But the parity of the number R^n in question is given by the digit B_0 of the pure binary representation.

$$B_0 = \bigoplus_{k=0}^{k=n} R_k \begin{pmatrix} B_0 = 0: R^n \text{ is even} \\ B_0 = 1: R^n \text{ is odd} \end{pmatrix} \qquad (16.36)$$

which thus provides the following relations

$$r_k^n = R_k^n \oplus R_k^{n+1} = R_{k-1}^n\,\bar{R}_{k-2}^n\,\bar{R}_{k-3}^n \cdots \bar{R}_1^n\,\bar{R}_0^n\,B_0^n\,E^n \quad (1 \le k \le m)$$
(16.37)

and
$$r_0 = R_0^n \oplus R_0^{n+1} = \bar{B}_0^n E^n \quad (k=0)$$

where E represents the input variable as above.

Putting:
$$R_{-1}^n = \bar{B}_0^n \begin{pmatrix} R_{-1}^n = 0, & R^n \text{ is odd} \\ R_{-1}^n = 1, & R^n \text{ is even} \end{pmatrix}$$

this gives for all ranks:

$$r_k^n = R_k^n \oplus R_k^{n+1} = R_{k-1}^n \bar{R}_{k-2}^n \bar{R}_{k-3}^n \cdots \bar{R}_1^n \bar{R}_0^n \bar{R}_{-1}^n E^n \quad (-1 \leq k \leq m) \tag{16.38}$$

with:
$$\bar{R}_{-1}^n = \oplus \sum_{k=0}^{k=m} R_k^n \tag{16.39}$$

Thus, in particular, relation 16.38 expresses the progression of the register with $m+1$ flip-flops $R_m, \ldots, R_0$.

The expression is, however, fairly complex on account of the term R_{-1}. The progression logic can, however, be simplified by the introduction of certain redundancies in the binary representation of the counter states. Consider a supplementary flip-flop to be known here as the parity flip-flop for which the output takes the value R_{-1}: this unit will switch every time that 1 is applied at its input. The performance of this counter with $(m+1)+1 = m+2$ flip-flops is always exactly defined by the equation (16.38) but the progression logic is considerably simplified by this addition of the redundant flip-flop R_{-1}.

It will be noted that the above formula for counting is then closely akin to that relating to counting in pure binary. It does, however, differ by the sign (⁻) on all the R_i except that of rank $k-1$ and by the variable stored in the supplementary parity flip-flop.

16.4.2 Down-Counting

It is easily seen that the progression relations for down-counting are obtained by replacing R_{-1} by the complement of its value for up-counting ($R_{-1} = 0$ for X_i 'even').

This circumstance makes the realisation of up/down counters very easy: to pass from up-counting to down-counting or vice versa it suffices to cause the parity flip-flop R_{-1} to switch to the opposite state, the remainder of the progression logic being unaffected.

16.4.3 Examples of Counters

Figures 16.5 represent 2 counters produced by the application of the formula 13.38, or by a recurrent decomposition of the latter in a similar manner to that adopted earlier for the pure binary (16.5b).

In the case of Fig. 16.5a where the formula 16.38 has been used directly, k (or $k+1$) inputs are necessary for the gate of rank $k-1$ driving R_{k-1}. In addition, from the electrical point of view, the loads

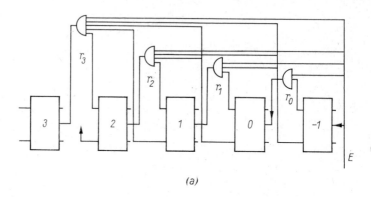

(a)

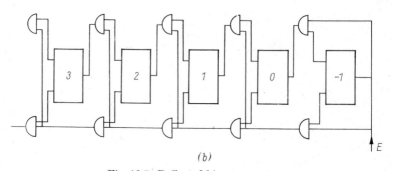

(b)

Fig. 16.5. Reflected binary counters

imposed on the flip-flops are unequal. This method is, therefore, only convenient for counters of very low capacity. It should be noted, finally, that provided there is a sufficiently high 'fan-in' for the gates, this device will need one gate for every stage.

In the case of Fig. 16.5b, use is made of the following formula derived from 16.38:

$$r_k{}^n = \left[R_{k-1}^n \left(\overline{R_{k-2}^n} \left(\overline{R_{k-3}^n} \left(\cdots \left(\overline{R_2{}^n} \left(\overline{R_1{}^n} \left(\overline{R_0{}^n} \left(\overline{R_{-1}^n} \right) \right) \right) \right) \right) \right) \right] E^n \quad (16.40)$$

16.4 Reflected Binary Code Counters

This layout thus requires 2 AND gates with 2 inputs per stage.

Other versions of the counter (Gitis) can be obtained by transforming the relation (16.38) by use of the De Morgan theorems. Giving

$$r_k^n = \left(\overline{R_{k-1}^n + \sum_{i=-1}^{i=k-2} R_i^n} \right) E^n \qquad (16.41)$$

an equation which, in its turn, can be represented in are recurrent form; e.g.:

$$r_k^n = \left[\overline{R_{k-1}^n + (R_{k-2}^n + (R_{k-3}^n + (\cdots (R_1^n + (R_0^n + R_{-1}^n)) \cdots)))} \right] E^n \qquad (16.42)$$

or

$$r_k^n = \overline{[R_{k-2}^n + (R_{k-3}^n + (\cdots (R_1^n + (R_0^n + R_{-1}^n)) \cdots))]} \, R_{k-1}^n E^n \qquad (16.43)$$

$$\left(r_k = R_{k-1} \overline{R_{k-2}} \cdots \overline{R_0 R_{-1}} E \right)$$

Other types of counter

As has been seen, thanks to the introduction of a redundant flip-flop, it has been possible to simplify the progression logic for the counter. But, the object of this logic is the detection of the rank k at which the transition occurs. This (unique) rank k is that of the zero following the block of 1's starting from the right (least significant digit) in the pure binary representation. A new type of counter with $(m + 1) + (m + 2)$ flip-flops $B_m, B_{m-1}, \ldots, B_0$ and $R_m, R_{m-1}, \ldots R_0, R_{-1}$, can be produced, where the flip-flops B_0 to B_m are those of a pure binary counter and the flip-flops R_{-1} to R_m those of a reflected counter. In this lay-out, proposed

$R_k' = \overset{<}{B_k}$

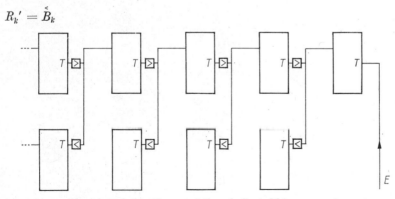

Fig. 16.6. Composite pure binary/reflected binary counter

by several authors (*Zavolokin, Bergelson, Beinchoker*), the pure binary counter assumes the function of a counter and of a detector of the unique transition $0 \to 1$; it controls the reflected binary counter whose task is to furnish a non-ambiguous representation. The liaison between the stages B_k and R_k is such that only the positive pulses which correspond to the passage of B_k from 0 to 1 are allowed to pass.

Figure 16.6 represents the general layout for such a counter, for which a particular design has been proposed by Zavolokin [61].

In a general way the progression logic is greatly reduced (at the expense of $m+1$ redundant flip-flops). Still other counters in the reflected binary code are possible, for the determination of the transition rank k (in reflected code) with the aid of an auxiliary counting in pure binary for each rank [61].

Bibliography

1. *Ajzerman, M. A., Gusev, L. A., Rosonoer, L. I., Smirnova, I. M., Tal', A. A.,* Logika, avtomaty, algoritmy, Gosudarstvennoe izdatel'stvo fiziko-matematičeskoj literatury. Moscow 1963.
2. *Arden, D. N.:* Delayed Logic and Finite-State Machines, Proc. AIEE Second Annual Symp. on Switching Theory and Logical Design, Detroit 1961, pp. 133—151.
3. *Arnold, B. H.:* Logic and Boolean Algebra, Englewood Cliffs: Prentice-Hall 1962.
4. *Arnold, R. F., Harrisson, M. A.:* Algebraic Properties of Symmetric and Partially Symmetric Boolean Functions, IEEE Transactions on Electronic Computers, Vol. EC-12, 244—251 (1963).
5. *Ashenhurst, R. L.:* The Decomposition of Switching Functions, Proceedings of an International Symposium on the Theory of Switching, April 2—5, Vol. 29 of Annals of Computation Laboratory of Harvard University, 74—116 (1959).
6. *Aufenkamp, D. D., Hohn, F. E.:* Analysis of Sequential Machines, (I) IRE Transactions on Electronic Computers, Vol. EC-6, 276—295 (1957).
7. *Aufenkamp, D. D.:* Analysis of Sequential Machines (II), IRE Transactions on Electronic Computers, Vol. EC-7, 299—306 (1958).
8. *Bartee, T. C.:* Digital Computer Fundamentals, New York: McGraw-Hill (1960).
9. *Bartee, T. C.:* Computer Design of Multiple-output Logical Networks, IRE Transactions on Electronic Computers, Vol. EC-10, 21—30 (1961).
10. *Bartee, T. C., Lebow, I. L., Reed, I. S.:* Theory and Design of Digital Machines. New York: McGraw-Hill 1962.
11. *Bazilevskii, Ju., Ja.:* The Theory of Mathematical Machines. New York: Pergamon Press 1963.
12. *Bergel'son, M. N., Vlasov, F. S.:* Postroenie reversivnykh sčetčikov dvoičnorefleksnogo koda, Vyčislitelnaja tekhnika, Sbornik No. 4, 74—81, Izdatel'stvo, Mašinostroenie, Moscow (1964).
13. *Birkhoff, G., McLane, S.:* A Survey of Modern Algebra. New York: MacMillan 1961.
14. *Biswas, N. N.:* The Logic and Input Equations of Flip-Flops, Electronic Engineering, 38, 107—111 (1966).
15. *Bodnarčuk, V. G.:* Sistemy uravnenij v algebre sobytij, Ž. Vyč. Mat. i Mat. Fiz, 3, No. 6, 1077—1088 (1963).
16. *Boole, G.:* An Investigation of the Laws of Thought, London 1854 (Reprinted Dover).
17. *Boucher, H.:* Organisation et fonctionnement des machines arithmétiques. Paris: Masson 1960.
18. *Brunin, J.:* Logique binaire et commutation. Paris: Dunod 1965.
19. *Brzozowski, J. A.:* Canonical Regular Expressions and Minimal State Graphs for Definite Events, Proceedings of the Symposium on the Mathematical Theory of Automata, Polytechnic Press of the Polytechnic Institute of Brooklyn, pp. 529—561 (1962).

20 *Brzozowski, J. A.:* A Survey of Regular Expressions and their Applications, IRE Transactions on Electronic Computers, Vol. EC-11, 324—335 (1962).
21 *Brzozowski, J. A., McCluskey, E. J.:* Signal Flow-graph Techniques and Sequential Circuit State Diagrams. IEEE Transactions on Electronic Computers, Vol. EC-12, 67—76 (1963).
22 *Brzozowski, J. A., Poage, J. F.:* On the Construction of Sequential Machines from Regular Expressions, IEEE Transactions on Electronic Computers, Vol. EC-12, 402—403 (1963).
23 *Brzozowski, J. A.:* Derivatives of Regular Expressions, J. Association for Computing Machinery, 11, 481—494 (1964).
24 *Buchi, J. R.:* Regular Canonical Systems. Arch. f, math. Logik u. Grundlagenforsch. H. 6/3—4, 91—111.
25 *Burks, A. W., Wright, J. B.:* Theory of Logical Nets. Proc. IRE, Vol. 41, pp. 1357—1365 (1953).
26 *Caldwell, S. H.:* Switching Circuits and Logical Design. New York: J. Wiley 1958.
27 *Carnap, R.:* Introduction to Symbolic Logic and its Applications. New York: Dover Publications 1958.
28 *Carvallo, M.:* Monographie des treillis et algèbre de Boole. Paris: Gauthier-Villars 1962.
29 *Carvallo, M.:* Principes et applications de l'analyse booléenne. Paris: Gauthier-Villars 1965.
30 *Castello, P.:* Clé des schémas électriques. Étude logique des circuits et automatismes. Paris: Dunod 1965.
31 *Chauvineau, J.:* La logique moderne, Presses Universitaires de France, 1957.
32 *Christiansen, B., Lichtenberg, J., Pedersen, J.:* Almene begreber fra logik, moengdeloere og algebra. Copenhagen: Munksgaard 1964.
33 *Chu, Y.:* Digital Computer Design Fundamentals. New York: McGraw-Hill 1962.
34 *Church, A.,* Introduction to Mathematical Logic, Vol. 1, Princeton: Princeton University Press 1956.
35 *Cohen, D.:* Truth Table Logic Speeds BCD-binary Conversions. Control Eng., 78—81 (1964).
36 *Constantinescu, P.:* Asupra reducerii numărului de contacte prin introducerea circuitelor în punte. Analele Univ. C. I. Parhon, Nr. 11-Seria Ştiinţelor Naturii, 45—69 (1956).
37 *Constantinescu, P.:* Ob analize i sinteze relejno-kontaktnykh mnogopoljusnikov c ventil'nymi elementami. Bull. Math. Soc. Sci. Math. Phys. de RPR, 3, (51), 21—64 (1959), 3 (51), 137—163 (1959).
38 *Constantinescu, P.:* Asupra sintezei schemelor cu funcţionare in mai multi timpi pulsatorii. Analele Universităţii Bucureşti, Seria Ştiinţele Naturii, Matematică-Fizică, pp. 129—136 (1961).
39 *Copi, I., Elgot, C., Wright, J.:* Realization of Events by Logical Nets. J. Association for Computing Machinery 5, 181—196 (1958).
40 *Croy, J. E.:* Rapid Technique of Manual or Machine Binary-to-Decimal Integer Conversion Using Decimal Radix Arithmetic. IRE Transactions on Electronic Computers, Vol. EC-10, 777 (1961).
41 *Curtis, H. A.:* A New Approach to the Design of Switching Circuits. Princeton: D. Van Nostrand 1962.
42 *Dean, K. J.:* An Introduction to Counting Techniques and Transistor Circuit Logic. London: Chapman and Hall 1964.

43 *Denis Papin, M., Faure, R., Kaufmann, A.:* Cours de calcul booléien appliqué Paris: Albin Michel 1963.
44 *Dubisch, R.:* Lattices to Logic. New York: Blaisdell 1964.
45 *Dugue, D.:* Ensembles mesurables et probabilisables. Paris: Dunod 1958.
46 *Eichelberger, E. B.:* Hazard Detection in Combinational and Sequential Switching Circuits. IBM J. of Research and Development, 9, 90—99 (1965).
47 *Elgot, C. C.:* Decision Problems of Finite Automata Design and Related Arithmetics. Trans. Am. Math. Soc. 98, 21—51 (1961).
48 *Fenstad, J. E.:* Algoritmer i matematikken: en inforing in den rekursive matematikk og dens anvendelse. I, II, Nordisk matematisk tidskrift, 12, 17—35, 99—111 (1964).
49 *Flegg, H. G.:* Boolean Algebra and its Applications. London: Blackie 1964.
50 *Flores, I.:* Computer Logic, Englewood Cliffs: Prentice Hall 1960.
51 *Flores, I.:* The Logic of Computer Arithmetic. Englewood Cliffs: Prentice Hall 1963.
52 *Francois, E.:* Analyse et synthèse de structures logiques selon le critère de fiabilité. Une méthode de synthèse particulière aux logiques séquentielles. Annales de Radioélectricité, tome **XX**, 310—323 (1965).
53 *Garner, H. L.:* The Residue Number System, IRE Transactions on Electronic Computers, Vol. EC-8, 140—147 (1959).
54 *Gazale, M. J.:* Les structures de commutation à m valeurs et les calculatrices numériques. Paris: Gauthier-Villars 1959.
55 *Ghiron, H.:* Rules to Manipulate Regular Expressions of Finite Automata, IRE Transactions on Electronic Computers, Vol. EC-**11**, 574—575 (1962).
56 *Gill, A.:* Introduction to the Theory of Finite-state Machines. New York: McGraw-Hill 1962.
57 *Ginsburg, S.:* A Synthesis Technique for Minimal-state Sequential Machines. IRE Transactions on Electronic Computers, Vol. EC-8, 13—24 (1959).
58 *Ginsburg, S.:* A Technique for the Reduction of a Given Machine to a Minimal State Machine, IRE Transactions on Electronic Computers, Vol. EC-8, 346—356 (1959).
59 *Ginsburg, S.:* Synthesis of Minimal State Machines, IRE Transactions on Electronic Computers, Vol. EC-8, 441—449 (1959).
60 *Ginsburg, S.:* An Introduction to Mathematical Machine Theory. Reading, Massachusetts: Addison-Wesley 1962.
61 *Gitis, E. I.:* Sčetčiki v dvoičnom cikličeskom kode, Avtomatika i Telemekhanika, Tom **XXV**, 1104—1113 (1964).
62 *Gluskov, V. M.:* Nekotorye problemy sinteza cifrovykh avtomatov, Vyčisl. matem. i matem. fizika, No. 3, 371—411 (1961).
63 *Gluškov, V. M.:* Abstraktnaja teorija avtomatov, Uspekhi mat. nauk. T. XVI, vyp. 5(101) 3—62 (1961).
64 *Gluškov, V. M.:* Sintez cifrovykh avtomatov, Fizmatgiz, Moscow 1962.
65 *Gluškov, V. M.:* Vvedenie v kibernetiku. Izdatelstvo akademii nauk ukrainskoj SSR. Kiev 1964.
66 *Goering, O.:* Radix Conversion in a Hexadecimal Machine, IEEE Transactions on Electronic Computers. Vol. EC-**13**, 154—155 (1964).
67 *Gradštejn, I. S.:* Prjamaja i obratnaja teoremy. Elementy algebry logiki, Izdatelstvo nauka, Glavnaja redakcija fiziko-matematičeskoj literatury. Moscou 1965.
68 *Gray, H. J.:* Digital Computer Engineering. Englewood Cliffs: Prentice Hall 1963.

69 *Haas, G.:* Design and Operation of Digital Computers. Indianapolis-New York: Howard W. Sams and Co., The Bobbs-Merrill Company 1963.
70 *Harrison, M. A.:* Introduction to Switching and Automata Teory. New York: McGraw-Hill 1965.
71 *Hennie, F. C.:* Iterative Arrays of Logical Circuits. New York: MIT Press and J. Wiley 1961.
72 *Hill, C. F.:* Nor-Nand Logic, Control Engineering, **11**, 81—83 (1964).
73 *Hohn, F. E.:* Applied Boolean Algebra, an Elementary Introduction, New York: Macmillan 1960.
74 *Huffman, D. A.:* The Synthesis of Sequential Switching Circuits, Journal of the Franklin Institute, **257**, 161—190 (1954), 275—303 (1954).
75 *Huffman, D. A.:* The Design and Use of Hazard-free Networks, Journal of the Association for Computing Machinery, 4, 47—62 (1957).
76 *Humprey, W. S.:* Switching Circuits with Computer Applications. New York: McGraw 1958.
77 *Hurley, R. B.:* Transistor Logic Circuits. New York: John Wiley 1961.
78 *Ioanin, G.:* Über die Mindestzuständezahl von Mehrtaktschaltungen, Note. Institute of Mathematics in Bucharest, 1963.
79 *Jauquet, C.:* Méthodes de synthèse de circuits logiques permettant l'emploi d'éléments logiques statiques, Revue A Tijdschrift, No. 6 (1959).
80 *Kalužnin, L. A.:* Što takoe matematičeskaja logika, Izdatel'stvo nauka, Glavnaja redakcija fiziko-matematičeskoj literatury, Moscow 1964.
81 *Karnaugh, M.:* The Map Method of Synthesis of Combinational Logic Circuits, Communications and Electronics, No. 9, 593—599 (1953).
82 *Khambata, A. J.:* Introduction to Integrated Semi-Conductor Circuits. New York: John Wiley 1963.
83 *Kintner, P. M.:* A Simple Method of Designing NOR Logic, Control Engineering, No. 2, 77—81 (1963).
84 *Kitov, A. I., Krinicki, N. A.:* Elektronnye cifrovye mašiny i programmirovanie, Gosudarstvennoe izdatel'stvo fiziko-matematiceskoj literatury. Moscow 1961.
85 *Kleene, S. C.:* Representation of Events in Nerve and Finite Automata, Automata Studies, Annals of Mathematics Studies, No. 34, 3—41 (1956).
86 *Kobrinskij, N. E., Trakhtenbrot, B. A.:* Vvedenie v teoriju konečnykh avtomatov, Gosudarstvennoe izdatel'stvo fiziko-matematičeskoj literatury, Moscow 1962.
87 *Kuntzman, J.:* Algèbre de Boole. Paris: Dunod 1965.
88 *Kuznecov, O. P.:* Predstavlenie reguljarnykh sobytij b asinkhronnykh avtomatakh, Avtomatika i telemekhanika, Tom **XXVI** (1965).
89 *Lantschoot, E. van:* L'algèbre logique et son application à la synthèse des circuits combinatoires à relais, Revue A Tijdschrift, No. 6 (1959).
90 *Lee, C. Y.:* Switching Functions on the N-dimensional Cube, Transactions of the AIEE, Vol. **73**, Part I, 289—291 (1954).
91 *Lespinard, V., Pernet, R.:* Arithmétique. Lyon: Devigne 1962.
92 *Lindaman, R., Cohn, M.:* Axiomatic Majority Decision Logic, IRE Transaction on Electronic Computers, Vol. EC-**10**, No. 1 (1961).
93 *Livovschi, L.:* Automate finite a elemente logice penumatice ši hidraulice, Monografii asupra teoriei algebrice a mecanismelor automate, Editura Academiei Republicii Populare Romîne, 1963.
94 *Lytel, A.:* ABC's of Boolean Algebra. Indianapolis-New York: Howard W. Sams and Co., The Bobbs-Merrill Company 1963.
95 *McCluskey, Jr., E. J.:* Minimization of Boolean Functions, Bell System Technical Journal, **35**, 1417—1444 (1956).

96 *McCluskey, Jr., E. J., Bartee, T. C.:* A Survey of Switching Circuit Theory. New York: McGraw-Hill 1962.
97 *McCluskey, Jr., E. J.:* Transients in Combinational Logic Circuits, Redundancy Techniques for Computing Systems, pp. 9—46, Spartan Books Co. (1962).
98 *McCluskey, Jr., E. J. Schorr, H.:* Essential Multiple Output Prime Implicants, Proceedings of the Symposium on the Mathematical Theory of Automata, Polytechnic Press of the Polytechnic Institute of Brooklyn, pp. 437—457, New York (1962).
99 *McCluskey, Jr., E. J.:* Introduction to the Theory of Switching. New York: McGraw-Hill 1965.
100 *McNaughton, R., Yamada, H.:* Regular Expressions and State Graphs for Automata, IRE Transactions on Electronic Computers, Vol. EC-**9**, 39—47 (1960).
101 *McNaughton, R.:* The Theory of Automata, a Survey, Advances in Computers, Vol. 2. New York: Academic Press 1961.
102 *Marcus, M. P.:* Switching Circuits for Engineers. Englewood Cliffs: Prentice Hall 1962.
103 *Marcus, M. P.:* Relay Essential Hazards, IEEE Transactions on Electronic Computers, EC-**12**, 405—407 (1963).
104 *Marcus, M. P.:* Derivation of Maximal Compatibles Using Boolean Algebra, IBM Journal, pp. 537—538 (1964).
105 *Mealy, G. H.:* A Method for Synthesizing Sequential Circuits, Bell System Technical Journal, **34**, 1045—1079 (1955).
106 *Miller, R. E.:* Switching Theory, Vol. 1: Combinational Circuits, Vol. 2: Sequential Circuits and Machines. New York: John Wiley 1965.
107 *Moisil, Gr. C.:* Teoria algebrică a mecanismelor automate. Bucharest: 1959. Russian: Algebraičeskaja teorija diskretnykh avtomatičeskikh ustrojstv. Moskow: Izdatelstvo inostrannoj literatury 1963.
108 *Moisil, Gr. C.:* Functionarea in mai multi timpi a schemelor cu relee ideale, Editura academiei republicii populare romîne, 1960.
109 *Moisil, Gr. C.:* Lectii asupra teoriei algebrice a mecanismelor automate, Editia III-a, Bucharest 1961.
110 *Moisil, Gr. C.:* Teoria algebrică a circuitelor cu transistori, Curs de automatizare industriala, Tip. B., Ediţia II-2, V, Probleme speciale, Bucharest 1961.
111 *Moisil, Gr. C.:* Circuite cu transistori, I, II Monografii asupra teoriei algebrice a mecanismelor automate, Editura academici republicii populare romîne I, 1961, II 1962.
112 *Moore, E. F.:* Gedanken-experiments on Sequential Machines, Automata studies, Annals of Mathematical Studies, No. 34, 129—153, Princeton (1955).
113 *Moore, E. F.:* Sequential Machines, Selected papers, Reading, Massachusetts: Addison-Wesley 1964.
114 *Naslin, P.:* Les aléas de continuité dans les circuits de commutation à séquences, Automatisme, No. 6, 220—224 (1959).
115 *Naslin, P.:* Circuits logiques et automatismes à séquences. Paris: Dunod 1965.
116 *Netherwood, D. B.:* Minimal Sequential Machines, IRE Transactions on Electronic Computers, Vol. EC-8, 367—380 (1959).
117 *Nichols, III, A. J.:* Multiple Shift Register Realizations of Sequential Machines, AD 60 81 55 (1964).
118 *Pelegrin, M.:* Machines à calculer électroniques, Application aux automatismes. Paris: Dunod 1964.
119 *Pfeiffer, P. E.:* Sets, Events and Switching. New York: McGraw-Hill 1964.
120 *Phister, M.:* Logical Design of Digital Computers. New York: John Wiley 1958.

121 *Pichat, E.:* Décompositions simples disjointes de fonctions booléennes données par leurs composants premiers, R. F. T. I. — Chiffres, Vol. 8, 63—66 (1965).
122 *Polgar, C.:* Technique de l'emploi des relais dans les machines automatiques. Paris: Eyrolles 1964.
123 *Pressman, A. I.:* Design of Transistorized Circuits for Digital Computers. New York: John Francis Rider 1959. Circuits à transistors pour calculateurs numériques. Paris: Dunod 1963.
124 *Quine, W. V.:* The Problem of Simplifying Truth Functions, Am. Math. Monthly, **59**, 521—531 (1952).
125 *Quine, W. V.:* A Way to Simplify Truth Functions, Am. Math. Monthly, **62**, 627—631 (1955).
126 *Quine, W. V.:* On Cores and Prime Implicants of Truth Functions, Am. Math. Monthly, **66**, 755—760 (1959).
127 *Rabin, M. O., Scott, D.:* Finite Automata and Their Decision Problems, IBM Journal of Research and Development, **3**, 114—125 (1959).
128 *Reza, F. M.:* An Introduction to Information Theory. New York: McGraw-Hill 1961.
129 *Richalet, J.:* Calcul opérationnel booléen, L'onde électrique, t. **XLIII**, 1003 — 1021 (1963).
130 *Richards, R. K.:* Arithmetic Operations in Digital Computers. Princeton: D. Van Nostrand 1955.
131 *Roginskij, W. N.:* Grundlagen der Struktursynthese von Relaisschaltungen. Munich: R. Oldenbourg 1962 (Tranlated from Elementy strukturnogo sinteza relejnykh skhem upravlenija, Moskow, 1959). Also in English: The Synthesis of Relay Switching Circuits. New York: Van Nostrand 1963.
132 *Rosenbloom, P. C.:* The Elements of Mathematical Logic. New York: Dover Publications 1950.
133 *Roth, J. P.:* Minimization Over Boolean Trees, IBM Journal of Research and Development, **4**, 543—558 (1960).
134 *Rudeanu, S.:* Axiomele laticilor și ale algebrelor booleene, Monografii asupra teoriei algebrice a mecanismelor automate. Bucharest: Editura academiei republicii populare romîne 1963.
135 *Russell, A.:* Some Binary Codes and a Novel Five-Channel Code, Control, pp 399—404 (1964).
136 *Scott, N. R.:* Analog and Digital Computer Technology. New York: McGraw-Hill 1960.
137 *Shannon, C. E.:* The Synthesis of Two-Terminal Switching Circuits, Bell System Technical Journal, **28**, 59—98 (1949).
138 *Shannon, C. E.:* A symbolic Analysis of Relay and Switching Circuits, Transactions of the AIEE, **57**, 713—723 (1938).
139 *Siwinski, J.:* Uklady przekaznikowe w automatyce, Warsaw, Wydawnictwa Naukowo-Techniczne 1964.
140 *Soubies-Camy, H.:* Les techniques binaires et le traitement de l'information. Paris: Dunod 1961.
141 *Spivak, M. A.:* Algoritmy abstraktnogo sinteza avtomatov dlja rasširennogo jazyka reguljarnykh vyraženij, Izvestija akademii nauk SSSR, Tekhničeskaja kibernetika **1**, 51—57 (1965).
142 *Spivak, M. A.:* Traktovka teorii avtomatov metodami teorii otnošenij, Problemy kibernetiki, **12**, 68—97, Moskow (1964).
143 *Susskind, A. K., Haring, D. R., Liu, C. L., Menger, K. S.:* Synthesis of Sequential Switching Networks, AD 60 88 81, 1964.

144 *Šrejder, Ju. A.:* Što takoe rasstojanie, Gosudarstvennoe izdatel'stvo fiziko-matematičeskoj literatury, Moskow 1963.
145 *Stibitz, G. R., Larrivee, J. A.:* Mathematics and Computers. New York: McGraw-Hill 1957.
146 *Stoll, R. R.:* Sets, Logic and Axiomatic Theories. San Francisco: W. H. Freeman and Company 1961.
147 *Talantsev, A. D.:* Ob analize i sinteze nekotorykh električeskikh skhem pri pomošči special'nykh logičeskikh operatorov, Avtomatika i telemekhanika, **XX**, 898—907 (1959).
148 *Tarski, A.:* Introduction à la logique. Paris Nauwelaerts Louvain: Gauthier-Villars 1960.
149 *Tassinari, A. F.:* A Bi-directional Decade Counter with Anticoincidence Network and Digital Display, Electronic Engineering, pp. 173—177 (1964).
150 *Torng, H. G.:* Introduction to the Logical Design of Switching Systems. Reading Massachusetts: Addison-Wesley 1964.
151 *Touchais, M.:* Les applications techniques de la logique. Paris: Dunod 1956.
152 *Traczyk, W.:* Elementy i uklady automatyki przekaznikowej, Wydawnictwa Naukowo-Techniczne, Warsaw 1964.
153 *Trakhtenbrot, B. A.:* Algoritmy i mašinnoe rešenie zadač. Moscou: Gostekhizdat 1957. Also in english: Algorithms and Automatic Computing Machines. Boston: D. C. Heath and Company, 1963 and in french: Algorithmes et machines à calculer. Paris: Dunod 1963.
154 *Unger, S. H., Paull, M. C.:* Minimizing the Number of States in Incompletely Specified Sequential Switching Functions, IRE Transactions on Electronic Computers, Vol. EC-8, 356—366 (1956).
155 *Urbano, R. H., Muller, R. K.:* A Topological Method for the Determination of the Minimal Forms of a Boolean Function, IRE Transactions on Electronic Computers, Vol. EC-5, 126—132 (1956).
156 *Veitch, E. W.:* A Chart Method for Simplifying Truth Functions. Proc. ACM, 127—133, 2—3 (1952).
157 *Vorob'ev, N. N.:* Priznaki delimosti. Moscou: Gosudarstvennoe izdatel'stvo fiziko-matematičeskoj literatury 1963.
158 *Walker, B. S.:* A Control Engineer's Way into Logic, Control, **9**, 384—386 (1965). (Part I), **9**, 446—448 (1965). (Part II).
159 *Weeg, G. P.:* Uniqueness of Weighted Code Representation, IRE Transactions on Electronic Computers, Vol. EC-**9**, 487—490 (1960).
160 *Weyh, U.:* Elemente der Schaltungsalgebra. Munich: R. Oldenbourg 1961.
161 *Whitesitt, J. E.:* Boolean Algebra and Its Applications. Reading: Addison-Wesley 1961.
162 *Wilkes, M. V.:* Automatic Digital Computers. New York: John Wiley 1956.
163 *Wood, P. E.:* Hazards in Pulse Sequential Circuits, IEEE Transactions on Electronic Computers, Vol. EC-**13**, 151—153 (1964).
164 *Young, J. F.:* Counters and Shift Registers Using Static Switching Circuits, Electronic Engineering, **37**, 161—163 (1965).

Index

Addition (binary), 25
Adder, 224
  serial, 227, 330, 402
  parallel, 227
Adjacent, adjacence, 187
Algebra, 2, 18
  of Boole, Boolean, 2, 116
  of classes, 87
  of logic, 87
  o regular expressions, 456
  Sheffer, 161
Alphabet, 341
Analogy, analog, 1
AND (funtion-), 157
  AND-INV, 158, 175
  NAND, 158, 175
Applicable, 416
Arden, 399
Ashenhurst, 271, 281, 300
Associativity, 87, 88
Asynchronous, 320, 335, 352, 371, 374
Aufenkamp, 348
Automaton, 323

Base, 22, 456
Binary
  (numeration), 23, 24
  (operation), 6
  pure, 24
  reflected, 50
Boolean, 114, 115, 138

Calculus:
  propositional, 105
  relations, 106
Canonical (form), 131
Carry, 26, 227
Cartesian, 8
Characteristic (vector), 128, 140
Checks, 56
  parity, 231

Class:
  equivalence, 14
  residual, 17
Code, 44
  BCD, 44, 45, 46
  bi-quinary, 49
  excess, 3
  excess E, 49
  pure binary, 44
  reflected, 50
  two-in-five, 48
  unit distance, 54
Coding, 2, 47
  digital, 2
  binary, 44
  of the internal states, 407, 478
Combinational, 213, 219, 376
Commutativity, 88, 89
Compatible, 418
Complement, 5, 70
  (-ation), 6, 120
  restricted (to 1), 228
  true (to 2), 229
Complete (functionally-set), 174
Computer:
  analog, 1
  digital, 1
Concatenation, 380
Congruence, 14, 15
Constant, 18
Contact, 67
Continuous, 1
Contradiction, 101
Conversion, 24, 29, 30, 43
  binary decimal, 34
  pure binary-reflected binary, 52, 333
  reflected binary-pure binary, 51, 332
Cost, 246, 406
Counter, 61, 313, 330, 483
  decimal, 489
  pure binary, 484
  reflected binary, 490

Index

Counter-domain, 17
Counting, 10
Cover, 7
  closed, 420
  minimal, 253
Cube
  n-dimensional, 185
Curtis, 271
Cycles, 353, 354
Cyclis, 50

Decimal, 23
Decomposition:
  disjoint, 272
  non-disjoint, 300
Definite (-event), 389
Derived, 389
Digit, 25
Dipole, 72
Discontinuous, 2
Discrete, 1, 2, 319
Disjunctive (canonical form), 134, 151
Distributivity, 89
Division, 28
Domain, 17
Dual, duality, dualization, 115
  (function), 156
Duodecimal, 23

Element (of a set), 4
Encoder, 54
  V-Scan, 235, 333
Excluded middle, 103

Fan-in, fan-out, 221
Field, 19
Flip-flop, 309, 329,
Function (cf. mapping too)
  Boolean, 138, 185
  Characteristic, 184
  incompletely specified, 183
  logic, 97
  majority, 173, 174
  output, 341
  Peirce, 175
  propositional, 106
  quorum, 173
  Sheffer, 175
  symmetric, 174
  threshold, 172
  transition, 341
  transmission, 72

Gates, 212
Graph, 17, 345
Gray, 50
Group, 19
  commutative, 19

Hammer, 161
Hamming, 50, 187, 190
Hazards, 353, 355
Hohn, 348
Huffman, 322, 366

Idempotence, 90, 119
Implication, 99
  reciprocal, 97
Inclusion (relation) 5
Incompatibility, 100
Input, 1, 321
Intersection, 6
Involution, 120
Ioanin, 424
Iteration, 381, 382
Ivanescu, 161

Karnaugh, 195
  k-cube (maximal-), 253
  (essential), 254
Kleene, 386

Logic:
  (-circuit), 213
  (-al product), 92
  (-al sum), 94
  (Diode-), 216
  DCTL, 217
  DTL, 218
  functional, 100
  propositional, 105
Lukasiewicz, 3

Mapping, 17
  bi-univocal, 18
  of A into B, 17
  of A onto B, 18
Marcus, 426
Matrix, 12
  connection-, 348
  partition-, 275
Maximal k-cube, 253
Maxterm, 151
McCluskey, 251
Mealy, 322, 345, 400, 465

Member of a set, 3
Minterm, 147
Modulo, 14
  (addition-), 167
Moisil, 3
Moore, 322, 347, 399, 464
Morgan (de), 121
Multiplication, 28
Mutually exclusive (-set), 7

n-dimensional cube, 185
Negation, 95
  (principle of double-), 102
Non-equivalence, 99, 167
NOR (function-), 101, 158, 175
Numeration, 21
  (decimal-), 20
  (positional-), 21

Octal, 23
Operation, 19
  AND 92
  OR-inclusive, 94
OR:
  (function-), 158
  -INV, 158, 175
  NOR, 158, 175
  exclusive, 97, 167
Ordered pairs, 8
Output, 1, 321

Part, 5
Partitions, 7
Paull, 375, 418, 423
Peirce, 101, 158,
Phase, 306, 307, 319, 338, 343, 349
Pichat, 292
Points, 3
Povarov, 271
Product, 380
  cartesian, 8
  fundamental, 134, 147
Proposition, 91
  compound, 95
Pseudo-associativity, 165

Quantifier, 107
Quantization, 1
Quine, 251
Quotient, 16

Race, 353, 357, 363
Rank, 15
Reflexivity, 13

Reflected, 50
Relays, 67
Relation, 11
  binary, 11
  equivalence, 13
  inclusion, 11
  unary, 11
  univocal, 17
Remainder, 16
Residual modulo m, 17, 57

Separable (function), 272
Sequence, 341
Sequential machine, 323
Sequential network, 303
Set, 3
  characteristic, 184
  empty, 7
  universal, 7
Sexagesimal, 23
Shannon, 131
Sheffer, 100, 159, 161
Spivak, 458
Structure, 18
Subset, 5
Subtraction, 26
Successor, 342
Sum:
  fundamental, 134, 151
  logical, 94
Switching, 213
Symmetrical, 12
Symmetry, 13
Synchronous, 319, 323, 352, 370, 374
Systems:
  digital, 1
  numeration, 21

Tautology, 101
Transitivity, 13
Truth table, 94, 99, 138

Unary operation, 6
Unger, 375, 418, 423
Union, 6
Universe, 7

Vector, 126
Venn (-diagrams), 8, 192

Weight, 24, 46
Word, 341
  Binary, 5, 44
  code, 44